Humans as Components of Ecosystems

Springer
New York
Berlin
Heidelberg
Barcelona
Budapest
Hong Kong
London
Milan
Paris
Santa Clara
Singapore
Tokyo

Mark J. McDonnell
Steward T.A. Pickett
Editors

Humans as Components of Ecosystems

The Ecology of Subtle Human Effects
and Populated Areas

Preface by Gene E. Likens
Foreword by William J. Cronon

With 59 Illustrations

Springer

Mark J. McDonnell
Bartlett Arboretum
151 Brookdale Road
Stamford, CT 06903-4199
USA

Steward T.A. Pickett
Institute of Ecosystem Studies
Mary Flagler Cary Arboretum
Millbrook, NY 12545-0129
USA

Text preparator: Marjory Spoerri
Acquiring editor: Robert Garber

Library of Congress Cataloging-in-Publication Data
Humans as components of ecosystems : the ecology of subtle human
  effects and populated areas / Mark J. McDonnell and Steward T. A.
  Pickett, editors : preface by Gene E. Likens ; foreword by William
  J. Cronon.
      p.   cm.
  Includes bibliographical references and index.
  ISBN 0-387-94062-6 (hardcover) — ISBN 0-387-98243-4 (softcover)
  1. Human ecology — Congresses.   I. McDonnell, Mark J.
  II. Pickett, Steward T., 1950–
  GF3.H86   1993
  304.2—dc20   93-10444

Printed on acid-free paper.

First softcover printing, 1997.

Production managed by Bill Imbornoni; manufacturing supervised by Jacqui Ashri.
Camera-ready copy supplied by the editors.
Printed and bound by Edwards Brothers, Inc., Ann Arbor, MI.
Printed in the United States of America.

9 8 7 6 5 4 3 2 1

ISBN 0-387-94062-6 Springer-Verlag New York Berlin Heidelberg   (hardcover)
ISBN 0-387-98243-4 Springer-Verlag New York Berlin Heidelberg SPIN 10576833 (softcover)

We dedicate this book to the memory of

Julie C. Morgan,

friend and humanist.

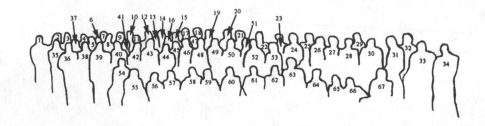

Cary Conference, 1991. 1. James E. Ellis; 2. Stephen B. Baines; 3. Joseph S. Warner; 4. Hélène Cyr; 5. Paul A. Bukaveckas; 6. Gary M. Lovett; 7. Martin Christ; 8. David R. Foster; 9. Peter R. Jutro; 10. Peter J. Richerson; 11. David L. Wigston; 12. Richard J. Borden; 13. Alan R. Berkowitz; 14. Ray J. Winchcombe; 15. John Aber; 16. Robert G. Lee; 17. Robert H. Gardner; 18. William R. Jordan III; 19. Lindsay R. Boring; 20. Stuart E. G. Findlay; 21. Edward A. Ames; 22. Charles D. Canham; 23. Dan Binkley; 24. Fredric H. Wagner; 25. Michael L. Pace; 26. Kimberly E. Medley; 27. James F. Kitchell; 28. James T. Callahan; 29. Orie L. Loucks; 30. Robert Howarth; 31. Stephen W. Pacala; 32. Gene E. Likens; 33. Forest Stearns; 34. Nels E. Barrett; 35. Jane V. Hall; 36. Juan Carlos Castilla; 37. Emily W. Russell; 38. Michael Williams; 39. Wolf Dieter-Grossmann; 40. David L. Strayer; 41. David R. Foster; 42. Kathleen C. Weathers; 43. Mark J. McDonnell; 44. Kathryn Lajtha; 45. Bertrand Boeken; 46. Robert J. Naiman; 47. Stephen Boyden; 48. Frank B. Golley; 49. Steward T. A. Pickett; 50. Christine R. Padoch; 51. Andrew P. Vayda; 52. Eduardo H. Rapoport; 53. Dolors Vaqué; 54. Bill L. Turner II; 55. Mark Nelson; 56. Margaret M. Carreiro; 57. Richard S. Ostfeld; 58. John E. Hobbie; 59. John H. Lawton; 60. D. Alexander Wait; 61. Edward P. Bass; 62. Linda Leigh; 63. Frank N. Egerton; 64. Jonathan J. Cole; 65. Nina M. Caraco; 66. Clive G. Jones; 67. Richard V. Pouyat. Absent from photo William Robertson IV.

Foreword: The Turn Toward History

The past two decades have seen an exciting convergence among the many intellectual disciplines that study relationships between human beings and the natural systems they inhabit. Scholars and scientists who in the past would have had little to do with one another have come to share a growing awareness of the inadequacies of their own particular fields, and have sought to rectify those inadequacies by reaching out across disciplinary boundaries to explore perspectives other than their own. One reason for this convergence of interests has of course been the complexity and dramatic scale of contemporary environmental problems, which are more than ready to defeat any simple-minded theories or methodologies that offer too facile an explanation of their causes. But another reason is less obvious, and perhaps even more interesting—the growing sense that to understand the role of human beings in promoting ecological change one must turn to the past and develop a much more historical sense of how such change occurs.

This turn toward history has not come easily to all the fields that have participated in it. In ecology, a long tradition of viewing historical change as abnormal, as "disturbance," dates back to the first generation of American ecologists, when Frederic Clements postulated that the tendency of natural systems was to progress through a series of stages toward a stable state called the "climax" which would persist indefinitely unless something intervened to disrupt it. Although Clementsian ecology did have a time dimension—the succession of pre-climax community types that would gradually prepare the way for the stable state to come—the teleology of the system (like the Hegelian dialectic which was one of its intellectual roots) was toward stasis, what we might call "the end of history." Clements' climax and the definitions of ecological "community" on which it rested soon underwent a withering critical assault from scientists like Gleason and

vii

Tansley, but the ahistorical legacy of his approach long outlived his own theories. For over half a century, the bias of American ecology was to try to describe and analyze natural systems in their "pristine state," as they would have existed in the hypothetical absence of human activity and long-term historical change. Even after World War II, when Clementsian approaches were no longer driving the major research agendas of the field, attempts to create a more dynamic ecology were usually conducted on such constrained geographical and historical scales—the interactions of a few species on a few acres over a few seasons or at most a few years—that the problem of long-term historical change remained on the back burner. Ecologists continued to view the human modification of ecological systems mainly as a source of disturbance, a factor to be controlled and eliminated from scientific analysis if they had to acknowledge it at all.

The many problems with this ahistorial bias of ecology began to become apparent by the 1960s. One source of change came from the allied field of palynology, whose researchers had been carefully assembling a body of data since at least the 1930s that irrefutably showed ecological stability to be an illusion, at least on extended time scales. In the long run, even the apparent continuity of the assemblages that an earlier generation had confidently labeled as "climaxes" proved indefensible. Another new perspective came from the earliest long-term ecological research sites, the most influential of which, Hubbard Brook, took as its starting point the dramatic and inten-tional human manipulation of an entire watershed. The result of Bormann and Likens' clearcutting of a small New Hampshire valley was to demon-strate that the aggregate behavior of a "disturbed" system was far more dynamic and complex than had been previously appreciated. Only empirical observation over an extended time could even describe such behavior, let alone explain it. No theory that lacked a serious time dimension could do justice to an event like the cutting of the Hubbard Brook forest, for the answers to its riddles were inextricably bound to its history, and could only be explored by adopting historical modes of analysis. Hubbard Brook also mirrored on a very small scale the vast human manipulation of earthly ecosystems which in the wake of Rachel Carson's *Silent Spring* in 1962 and the first Earth Day in 1970 became the obsession and battle-cry of the modern environmental movement. Despite the discomfiture that many ecologists felt about the misuse of their science by the new popular movement, a growing number of them nonetheless responded by trying more seriously to include anthropogenic change as a legitimate part of their research. All of these tendencies have encouraged ecologists to recognize that their field, like geology and evolutionary biology, is a genuinely *historical* science, with analytical problems and approaches that are unavoid-ably different from those of less time-bounded fields like physics or chemistry.

This turn toward history has occurred in other fields as well. The first generation of anthropologists shared an analogous bias to the first generation of Clementsian ecologists, seeking to describe "primitive" tribes in their "pristine" state, inhabitants of an eternal "ethnographic present" in which historical change began to occur only after tribal societies had been "disturbed" by intruding Europeans. The validity of such approaches came under increasing attack as anthropologists began to acknowledge the dynamics of non-western cultures, and collapsed altogether as colonialism faded in the years following World War II. Although some bias toward static theory persists in anthropology as well as ecology, the field has become far more historical over the past two decades, and practitioners of ecological anthropology have made real contributions to our understanding of the ways human and natural systems interact in long-term cultural and ecological change.

Geography is another field with much to offer our study of long-term environmental change as it relates to human beings. Unlike ecology, it has never been without a historical component, although historical geographers have often felt beleaguered in their efforts to defend time to their colleagues as an analytical category co-equal with space in their analyses. Like ecology, geography also underwent a long period during the 1940s, 1950s, and 1960s when the dominant nomothetic approaches and abstract models showed little interest in taking the time dimension seriously. Only as the limits of such models and approaches began to be apparent, and as geographers responded to the same external pressures that were influencing ecologists, did the importance of a more historical perspective seem compelling.

Examples of this turn toward history could be gathered for virtually all of the disciplines that are now making serious efforts to study long-term environmental change as it relates to human cultures and behaviors. Even the field of history itself, which can hardly be accused of having ignored the importance of time in its arguments and narratives, went through several decades in the post-war period when the human relationship to nature moved to the extreme margins of its analyses. The reasons for this indifference to natural context were complicated. In part, it reflected a powerful reaction against what many scholars saw as the excessively materialist explanations of human behavior—Marxist and non-Marxist alike—that had held sway with an earlier generation of scholars. In the 1950s, historians tried to assess the causes of social, economic, and political change by appealing mainly to human ideas and human culture, not to the changing human relation to nature. But the subsequent growth of environmentalism and a new effort to try to integrate materialist and idealist modes of explanation reversed this tendency as well. By the 1970s a new subfield of "environmental history" had emerged and was starting to offer powerful arguments for why one could not hope to understand the human past if one wrote out of one's story the *non*-human past.

And so the time now seems ripe for all these different disciplines—ecology, history, geography, anthropology, and others—to acknowledge that their intellectual journeys have been carrying them toward a common path. Clearly, no single one of these perspectives is adequate by itself to the task at hand. But the fact that our work seems increasingly to point toward common questions suggests how much we have to learn from one another if only we can scale the walls that separate us so as to start working together on the problems we all find fascinating. For this reason, I'm immensely heartened by the publication of *Humans as Components of Ecosystems*. Not only does it make a compelling case for the need to study anthropogenic ecological change, but it implicitly demonstrates that a historical approach can serve as a bridge that can link our various disciplines. All those disciplines will benefit if this book helps erode the boundaries between them. With luck, future historians will eventually recognize this volume as an important early step toward creating a genuinely historical ecology and a genuinely ecological history.

William J. Cronon
Frederick Jackson Turner Professor of History,
Geography, and Environmental Studies
University of Wisconsin-Madison

January 1993

Preface

This book, a product of the fourth Cary Conference, amply demonstrates the achievement of a major goal of all Cary Conferences. That is, Cary Conferences were conceived to provide a forum for comprehensive discussion of major ecological issues from more philosophical and comprehensive perspectives. The Institute of Ecosystem Studies is proud to make its staff and facilities available on a biennial basis for the purpose of initiating and fostering these important discussions.

On the one hand, the influence of humans on ecosystems may be quite obvious, such as when a forest is cut. On the other, humans live within and among ecosystems, but ecologists largely have avoided making detailed and rigorous analyses of the more subtle effects that human activities generate. Indeed, most ecologists have sought out rather pristine or remote areas for study, rather than those which include humans. This book brings together a stimulating mix of disciplines to highlight and evaluate these problems, as well as the opportunities of including humans as integral components in studies of "natural" ecosystems. An exciting dialogue has ensued. Hopefully this book will convey some of the excitement and challenges that were generated at this Cary Conference.

Gene E. Likens
Director
Institute of Ecosystem Studies
Millbrook, New York

January 1993

Acknowledgments

Because this book owes so much to the interaction and stimulation of the fourth Cary Conference, we owe a deep debt of gratitude to the staff of the Institute of Ecosystem Studies (IES) who contributed so much to the success of that meeting. In particular, Jan Mittan, Conference Coordinator, Owen Vose, Henry Behrens, Carol Boice, Dick Livellara, Bob Meyers, Reb Powell of the IES Operations Department, Dave Bulkeley, Greenhouse Manager, Joseph Warner, IES Administrator, Kathleen Weathers, Laboratory Manager, and Jill Cadwallader were especially helpful. Stephen Baines, Nels Barrett, Martin Christ, Hélène Cyr, Richard Pouyat, and Alexander Wait helped with logistics and transportation. We are also grateful to the Research Assistants and graduate students who suffered dislocations and delays in accommodating the Conference.

For intellectual leadership and encouragement, we thank the members of the Steering Committee, Jon Cole, Bill Cronon, Clive Jones, Gene Likens, Gary Lovett, Christine Padoch, and Gilbert White.

We thank Marj Spoerri, Assistant to the Scientific Staff, for her competence and good cheer in the herculean task of preparing the text for publication. We are grateful to Annette Frank, IES Librarian, for her assistance in tracking down bibliographic materials. Sharon Okada prepared the cover art from a logo design subtly affected by Clive Jones.

Financial support for the Conference was provided by the Mary Flagler Cary Charitable Trust and the National Science Foundation. M.J.M. was supported by the Lila Wallace Readers' Digest Fund during the preparation of this book.

Contents

Contributors

John D. Aber Institute for the Study of Earth, Oceans, and Space, University of New Hampshire, Durham, NH 03824, USA

Edward P. Bass Philecology Trust, 201 Main Street, Fort Worth, TX 76102, USA

Richard J. Borden College of the Atlantic, Bar Harbor, ME 04609, USA

Stephen Boyden Centre for Resource and Environmental Studies, Australian National University, GPO Box 4, Canberra, ACT, Australia 2601

Nina F. Caraco Institute of Ecosystem Studies, Box AB, Millbrook, NY 12545, USA

Stephen R. Carpenter Center for Limnology, University of Wisconsin-Madison, Madison, WI 53706, USA

Juan C. Castilla Departamento de Ecología, Pontificia Universidad Católica de Chile, Casilla 114-D, Santiago, Chile

Jonathan J. Cole Institute of Ecosystem Studies, Box AB, Millbrook, NY 12545, USA

William J. Cronon Department of History, University of Wisconsin-Madison, Madison, WI 53706, USA

Frank N. Egerton Department of History, University of Wisconsin-Parkside, Kenosha, WI 53141, USA

David R. Foster Harvard Forest, Harvard University, Petersham, MA 01366, USA

xix

Robert H. Gardner Environmental Sciences Division, Oak Ridge National Laboratory, Oak Ridge, TN 37831-6036, USA

Wolf-Dieter Grossmann Environmental Research Center, Permoserstrasse 15, O-7050 Leipzig, Germany

Jane V. Hall Department of Economics, California State University, Fullerton, CA 92634, USA

William R. Jordan III University of Wisconsin-Madison Arboretum, 1206 Seminole Highway, Madison, WI 53711, USA

Peter R. Jutro Terrestrial Effects Research, United States Environmental Protection Agency, RD-682, Washington, DC 20460, USA

Charles E. Kay Ecology Center and Department of Fisheries and Wildlife, Utah State University, Logan, UT 84321, USA

James F. Kitchell Center for Limnology, University of Wisconsin-Madison, Madison, WI 53706, USA

Linda Leigh Space Biospheres Ventures, P.O. Box 689, Oracle, AZ 05623, USA

Gene E. Likens Institute of Ecosystem Studies, Box AB, Millbrook, NY 12545, *USA*

Mark J. McDonnell Institute of Ecosystem Studies, Box AB, Millbrook, NY 12545, USA

William B. Meyer George Perkins Marsh Institute, Clark University, Worcester, MA 01610-1477, USA

Mark Nelson Space Biospheres Ventures, P.O. Box 689, Oracle, AZ 05623, USA

Robert V. O'Neill Environmental Sciences Division, Oak Ridge National Laboratory, Oak Ridge, TN 37831-6036, USA

Stephen W. Pacala Department of Ecology and Evolutionary Biology, Princeton University, Princeton, NJ 08540, USA

Michael L. Pace Institute of Ecosystem Studies, Box AB, Millbrook, NY 12545, USA

Christine R. Padoch Institute of Economic Botany, The New York Botanical Garden, Bronx, NY 10458-5126, USA

Benjamin L. Peierls Curriculum in Marine Sciences, University of North Carolina, Chapel Hill, NC 27599-3300, USA

Steward T.A. Pickett Institute of Ecosystem Studies, Box AB, Millbrook, NY 12545, USA

Richard V. Pouyat USDA Forest Service, Northeast Forest Experiment Station, c/o Institute of Ecosystem Studies, Mary Flagler Cary Arboretum, Box AB, Millbrook, NY 12545, USA

Eduardo H. Rapoport Departamento de Ecología, Universidad Nacional del Comahue, C.C. 1336, Bariloche 8400, Argentina

Peter J. Richerson Division of Environmental Studies, University of California-Davis, Davis, CA 95616, USA

Emily W.B. Russell Department of Geological Sciences, Rutgers University, Newark, NJ 07102, USA

B.L. Turner II Graduate School of Geography, George Perkins Marsh Institute, Clark University, Worcester, MA 01610, USA

Monica G. Turner Environmental Sciences Division, Oak Ridge National Laboratory, Oak Ridge, TN 37831-6036, USA

Andrew P. Vayda Department of Human Ecology, Cook College, Rutgers University, P.O. Box 231, New Brunswick, NJ 08903, USA

Frederic H. Wagner Ecology Center and College of Natural Resources, Utah State University, Logan, UT 84322-5205, USA

Michael Williams School of Geography, University of Oxford, Mansfield Road, Oxford OX1 3TB, England

1
Introduction: Scope and Need for an Ecology of Subtle Human Effects and Populated Areas

Mark J. McDonnell and Steward T.A. Pickett

Focus on Basic Ecology

The science of ecology is the study of the factors that influence abundance and distribution of organisms, the interactions between these organisms, and relationships between organisms and the fluxes of energy and materials in the environment. Basic ecology is an extremely broad science that can encompass any system on Earth. Yet, in pursuing their basic science, North American ecologists have ignored many of the ecological manifestations of humans.

The large, obvious, and direct influences of humans in ecosystems have been recognized, of course. Ecological studies of the causes, patterns, and effects of air and water pollution, habitat destruction, accelerated species extinction, the introduction of exotic species and genetic material, and the like, are well known. Indeed, many of these major effects have extended to the global scale and have earned a place in the headlines. Recognition of such large and conspicuous effects of humans on natural systems has motivated a multitude of ecological studies. Entire journals are devoted to reporting the results of applied studies of the severe and conspicuous ecological effects of humankind.

But there are other kinds of human interactions in ecosystems, and these have been largely neglected in *basic* ecological research, especially in North America. To understand the difference between the kinds of environmental problems that have been included in ecological studies, and those that have been relatively neglected, a simple classification can be used. With apologies to Sergio Leone and the genre of "Spaghetti Westerns," the ecological influences of humans can be divided among "the good, the bad, and the subtle." Research in all of these realms can be turned to both basic and applied questions.

1

The Traditional Perception of Human Influences

"The bad" refers to those obvious, negative influences of human activities that traditionally have been studied by ecologists. The belching smokestack, the toxic waste spill in a lake, the damming of a wild river, and the introduction of an exotic pest are just a few examples. Ecologists have worked hard and creatively to quantify the spread and impacts of such events, have addressed their basic scientific implications, and have tried to make their conclusions available to policy-makers and managers. In fact, ecologists have been instrumental in bringing grave problems and their solution to the attention of the public. There is little to quibble with in this important aspect of ecology.

The Ecology of Populated Ecosystems and Landscapes

"The good" refers to environmental influences that are associated with areas populated by humans. We use a positive term not because all results of human habitation are ecologically good, but because on the face of it, the concentration of people, from villages to metropolitan areas, is apparently a socially desirable end. However, the ecological data base concerning such environments is neither extensive, comprehensive, nor long-term. Despite the attention given to certain conspicuous, negative effects of human population centers, there is a host of other ecological questions that have scarcely been asked about populated areas. For instance, how do natural areas embedded in cities function? Are such natural areas sustainable? Are the ecological effects of the neighborhood or the metropolitan region preeminent in determining the structure and function of natural areas? More radically, urban and suburban areas, as well as less densely populated areas, can be treated as ecosystems or biotic communities of which people are a part. Populated landscapes and metropolitan areas can serve as massive surrogate experiments for addressing basic ecological questions. The well-documented history of populated areas is an advantage for their use in ecological research. It would be extremely difficult for ecologists on their own to arrange for such experiments! Such unplanned experiments can be used to test whether general ecological principles survive application to novel environments such as those represented by population centers. Indeed, environmental variables that are usually confounded may be decoupled in populated systems and, therefore, expose the limiting factors in fundamental but complex ecological processes. The questions above are only illustrative; there are undoubtedly many important and interesting ecological questions appropriate to populated areas that have yet to be imagined. Ecological research needs to be focused more directly on populated areas to yield basic scientific understanding as well as to generate expertise for policy-makers.

Subtle Human Effects

The third kind of human influence can be labeled "the subtle." Subtlety covers a wide variety of often inconspicuous or unexpected interactions of humans with ecosystems. Indirect effects, historical effects, echoes of the past, biological legacies, lagged effects, and unexpected action at a distance are important kinds of subtle causes and effects (Russell, Chapter 8, this volume). These will be defined more fully and exemplified in this volume. However, to give shape to the scope of this volume, we offer hypothetical examples below.

Indirect effects are those where the focus is on two ecological entities of interest, but the outcome of interaction between them is mediated by a third party. Examples are the mediation of competition among plants by microbes or the anthropogenic introduction of an organism that ultimately changes the trophic structure of an ecosystem.

Historical effects are the alteration of a contemporary interaction as the result of a prior state of a system. For instance, the current structures of many forests are actually the result, at least in part, of land use a century or more ago; or the undulating microtopography of a forest floor may be a persistent effect of a prior natural catastrophe. Biological legacies and echoes of the past are subsets of historical effects.

Biological legacies are persistent alterations in the biotic component of the environment. Such phenomena include the presence of large, decay-resistant woody debris in a forest or stream, or the deposition of an organic layer in soil by a partial vegetation harvest.

An *echo of the past* is a historical effect that is apparent only sporadically. Such echoes would include the destruction of a species refuge used only during drought, or the removal of a particular sapling cohort by human land use.

Lagged effects are those that are triggered at some time before they appear. The lag phase of logistic population growth is a well known example. An anthropogenic example is the gradual destruction of stratospheric ozone by chlorofluorocarbons (CFCs).

Unexpected action at a distance. Action at a distance, of course, can include blatant effects such as pollution released at a distance from its locus of impact. Of greater interest here are distant subtle effects. Oddly, many current regional or global environmental problems are actions at a distance which were unexpected. Indirect effects, since they involve a third party, can be caused at a distance. Pest eradication programs often have subtle, distant effects, as when migratory or widely ranging animals are involved. The accumulating generalizations about the importance of indirect effects can be tested by extension to cases in which humans are involved.

Purpose and Scope of the Book

Subtle influences of humans have been mostly ignored by ecology. The environments built by people and the indirect effects of such environments have not been a concerted focus of basic research in ecology, especially in North America. This book grew out of a Cary Conference designed to map this largely unexplored frontier of ecology, and to begin to equip ecologists to explore the territory effectively. Both important applications and the refinement and testing of basic ecological principles may result from this exploration.

It will undoubtedly take a long time to develop effectively an ecology of the subtle and of human-populated areas. However, we believe the discipline of basic ecology is now ready to incorporate humans as components of ecosystems. The recent discussions concerning a paradigm shift in ecology indicate this readiness (Botkin 1990; Pickett et al. 1992). What might be labelled a "non-equilibrium" paradigm emphasizes the openness and historical contingency of ecological systems. Such a viewpoint is much more accommodating to the role of humans than the older view embodied in the "balance of nature" idea.

The specific objective of the book was to educate ecologists in general about the importance of incorporating humans and their effects in ecological concepts, models, and studies. We will especially highlight the neglected or non-obvious roles of humans. Hence, the book will focus on subtle human effects and the ecology of populated areas.

Although the title is cast in terms of ecosystems, the insights and exhortations are appropriate to many other levels of ecological organization and scales of study. So, subtle human effects must be sought when studying individuals, populations, communities, ecosystems, and landscapes. Furthermore, humans will function as important parts of ecological communities, ecosystems, and landscapes. Thus, the domain of our focus is large.

The book is divided into four sections. Part I will summarize key points from the social sciences and humanities concerning the nature and magnitude of human effects on the environment. In order to encourage ecologists to collaborate with specialists who focus on humans, several chapters in Part I explicitly examine the assumptions and approaches of social scientists, historians, geographers, economists, and anthropologists. Several of these chapters sound the recurring theme of the book: that virtually no ecological system, regardless of how apparently remote or pristine, lacks a human impact or component.

Part II will present case studies that illustrate the discovery and nature of human effects, including subtle ones, and those usually neglected because they are associated with the populated areas American ecologists usually avoid. The concept of subtlety is refined in Part II.

Part III presents case studies that are especially rich in management and policy implications, and that point out the potential to learn much about humans as components of ecosystems by restoring or constructing ecosystems. Indeed, several of the studies in Part III were motivated by policy needs.

Although the division of the core of the book into parts highlights important concepts, methodological steps, and policy applications, the contributions illustrate the highly integrated and multifaceted nature of studies that account for the ecology of humans. Therefore, in reality, there is much overlap among parts in the insights and issues raised. The integration also reflects the communication among authors that resulted from the gestation of the book during the fourth biennial Cary Conference.

Part IV caps the book with two kinds of overviews. Chapter 22 is comprised of the commentaries by four participants in the Cary Conference from which this book grew. We have tried to preserve a sense of the excitement and interaction that characterized the Cary Conference, and expose different personal perspectives on the issues and opportunities that can guide future research on humans as components of ecosystems. The essays represent anthropology, human ecology, marine ecology, and terrestrial ecology. The second overview is our own encapsulation of the themes and insights that emerge from the remainder of this volume.

This book differs from other efforts to assess the role of humans in the environment. The classic by Thomas (1956), and the contemporary milestone by Turner et al. (1990a) illustrate the duration, richness, and depth of concern with humans as ecological, geographical, and geomorphological agents. However, we have tried to compile a book that has an explicitly ecological viewpoint, and which will be of particular value in exposing all ecologists to the human frontier of their discipline. We hope our effort provides an explicitly ecological complement to the existing literature on the knowledge of how humans interact with their environment.

Section I   The Human Factor: Perception and Processes

The social sciences and humanities have humans as their professed focus of study. Ecologists should profit from the experience, approaches, and conclusions of fields that focus on humans as individuals, societies, and environmental agents. Hence, the first goal of Section I is to illustrate the insights into humans, as components of ecosystems, that emerge from the social sciences and humanities.

In order to facilitate the interaction of ecologists with researchers in the social sciences and humanities, the methods and approaches of these other disciplines must be made familiar to ecologists. Therefore, the second goal of Section I is to expose ecologists to some of the questions and modes of study used by specialists on humans.

Of course, history, geography, and anthropology, among other fields, have documented many kinds of intentional and unintentional environmental effects of humans that ecologists must come to account for in their studies of natural, modified, and managed environments. The chapters in this part of the book thus have a third goal—documenting subtle human causes and effects, and highlighting the ubiquity of human population, both now and in the past. This task will continue throughout the book.

Finally, in order to increase the understanding of humans as components of ecosystems, it may be helpful to understand why mainstream ecology has ignored them as components of all manner of ecological systems. So, the fourth goal of Section I is to examine some important biases and constraints that have led ecologists to neglect populated areas and to miss subtle human causes and effects, and that also lead to sometimes problematic interactions between ecology and society. Egerton (Chapter 2) begins Section I with this goal in mind.

Ecology is not alone in its ambivalence concerning the place of humans relative to nature. Indeed, the founder of modern human geography, George

Perkins Marsh, wrestled with the persistent and problematic "great question," of humankind's relationship to nature (Williams, Chapter 3). Nor is the struggle of ecology with the role of humans new, as indicated by the quotation of Tansley (1935) in the title of Chapter 3. Furthermore, the great question, of whether humankind is part of or apart from nature, suggests the need to determine the correct proportion of anthropocentric and nature-centric explanations in any given situation (Turner and Meyer, Chapter 4). Even economic models take a fundamental but hardly recognized stance about the relationship of people to nature (Hall, Chapter 5). Such basic assumptions will affect how ecologists can use economic approaches, and even affect policy as well (Hall, Chapter 5). Vayda (Chapter 6) proposes a process-oriented approach to the study of human agency, while Boyden (Chapter 7) illustrates a systems-oriented approach.

2
The History and Present Entanglements of Some General Ecological Perspectives

Frank N. Egerton

Since the 1970s the word "ecology" has been applied not only to a science which was named in 1866 and formally organized in the 1890s and early 1900s, but also to the environmental movement (which would be better called either "the environmental movement" or "environmentalism"). This terminological confusion cannot be easily resolved, however, because many of the environmental movements have taken a serious interest in the science of ecology and want to see it develop to meet the needs of civilization—as they define them.

The civil rights movement and the Vietnam War served to politicize many Americans and others who were disturbed by aspects of western civilization. During the 1960s and 1970s many concerned citizens throughout the world came to believe that the world's most important problems do not stand in isolation and usually cannot be solved in isolation. If the atomic bombs dropped on Japan in 1945 had demonstrated that physics was no longer just a science belonging to physicists, the technologies used in the 1960s and 1970s to destroy not just southeast Asians, but other rebellious peoples of the third world and their environments, and parts of our environments, generalized that lesson: all the sciences must belong to all the people, because the sciences help us to control our future, for better or worse.

Since public concern about the development of science is legitimate, how, beyond providing funds and facilities for education and research, should that concern be expressed? This interesting question is explored here by examining some of the efforts to relate ecology to civilization.

The most basic issue concerns the objectivity and verifiability of science. Several decades ago these issues seemed well understood and practically beyond controversy, but then Thomas Kuhn (1962, 1970) argued in *The*

Structure of Scientific Revolutions that scientists construct paradigms which define the rules and problems for investigating a science and that later paradigms can define new rules and problems. There seemed to be more than one way to skin a cat. In ecology, this lesson seems especially apt; one could choose to do community ecology, systems ecology, ecosystems ecology, population ecology, or various other ecologies, and it is impossible to "prove" that any one of them is "right" and the others "wrong." This situation would be difficult enough if all ecologists were locked away in an ivory tower and had to confront only nature and each other. However, even the most reclusive of ecologists also have to deal with civilization in some fashion, and many ecologists choose to confront questions about the relationships between science and society. When society includes environmentalists who have learned the lessons of Kuhn and of the Vietnam era, the confrontation can seem bewildering. The easy response for ecologists is to throw up their hands and walk away, knowing that ecologists can handle their own differences and leave it to science educators and popularizers to handle society's questions and concerns. But those who create knowledge should also assume some responsibility for putting it into clear perspective. This is particularly urgent for ecologists studying humans as components of ecosystems or the ecology of populated areas, because public cooperation is important, and that cooperation is most readily obtained when scientists and the public understand and respect each other.

There are three general issues in ecology of interest to both scientists and the concerned public which are the focus of the remainder of this chapter: the economy (or balance) of nature, evolution and ecology, and theoretical versus applied ecology. Most ecologists probably have similar perspectives on each of them which interconnect within "mainstream" ecology. Currently, there are also alternative perspectives on each issue which interconnect to produce an "alternative" ecology.

It will help to understand this situation by considering the distinction between mythopoeic and critical thinking. Mythopoeic thinking, which means myth-making, was the only kind of thought which existed before the rise of Greek natural philosophy (Frankfort et al. 1946). The world and our relationship to it were explained by anthropomorphic myths which satisfied curiosity and reduced anxiety about what might happen in the future. For example, the seasons were explained in a Greek myth which told of Persephone (= Kore), daughter of the goddess of fertility, Demeter, being kidnapped by Hades, god of the underworld. Demeter's sorrow at the loss of her daughter plunged the earth into perpetual winter, causing the people to pray to Zeus to intervene. He was able to negotiate a compromise with Hades, who allowed Persephone to return to earth from the underworld for part of the year, during which time Demeter's happiness caused the springtime and summer, but Persephone's annual return to the underworld caused a return of Demeter's unhappiness, resulting in autumn and winter (Guirand 1959).

A philosopher of science, Karl Popper (1962), has observed that all science begins in myth. The change from mythopoeic to critical thinking was one of the most remarkable developments in our intellectual history, and yet it is not widely known and appreciated. It occurred among the Greek city-states along the eastern shore of the Aegean Sea (Kirk et al. 1983; Lloyd 1970). It began with Thales of Miletus, who was active for several decades after 600 B.C. He and his followers turned their backs on the myths which claimed to explain everything, but in reality obscured the understanding of both nature and humanity. They began a long dialogue known as natural philosophy, which had two implicit rules: explanations must be based upon observable phenomena; and any plausible explanation is discarded when difficulties are found and a better one is developed. Yet scientific progress has not been a simple process of conjecture and refutation, as it began with the early natural philosophers. One of the great achievements of the early natural philosophers was the atomic theory of matter, but it was not widely accepted, because it explained all events in nature as the result of random motions. The offspring of plants and animals closely resemble their parents, and so it is implausible that heredity is a series of random events. Since the atomic theory did not explain these and other regularities, it was rejected in favor of a less sophisticated theory of change (van Melsen 1952).

Furthermore, myths can arise and linger within science alongside critical thinking. The balance of nature provides a good example. The early indications of it are found in the works of Herodotos of Halicarnassos, the father of Greek history, who wrote in the mid 400s B.C., more than a century after natural philosophy began. He may have been influenced by it in developing his own interest in natural phenomena, but he did not fully share its naturalistic perspective. He described Egyptian plovers picking leeches from the open mouths of crocodiles and noted that "timid" animals such as hares produce more offspring than do predatory species which eat them (Herodotos bks. II, III; Egerton 1973). He saw this as evidence of the wise planning of Divine Providence.

Almost a century later, while natural philosophy was still flourishing, Plato (428/427-348/347 B.C.) included two creation myths in his *Dialogues*. Since the creation of the world was not something anyone really knew anything about, it seemed appropriate to offer two possibilities, which were not mutually exclusive, and either one or both could be accepted or not as one pleased. In *Timaeus* (30c-d), God was said to have made the works into "one visible animal comprehending within itself all other animals of a kindred nature," and in *Protagoras* (320d-321b), the god Epimetheus was said to have made each species of animal equipped with some means of preservation, such as speed, weapons, and wings.

These brief comments by Herodotos and Plato provided the literary foundation for the balance-of-nature myth which continued to grow in Greco-Roman antiquity and later. I will not follow here the details of that story down through the ages as I have done so elsewhere (Egerton 1973). Let us skip down to the time of John Ray (1627-1705), who was not only an Anglican clergyman, but also a most industrious observer and great theoretician of natural history. His contemporary, Robert Hooke, had made a convincing argument that fossils represent the remains of once-living plants and animals. Since the fossil forms seldom matched the living forms of species, the question arose as to whether species became extinct. Ray remarked in *Three Physico-Theological Discourses* (1693:147) that if extinction occurred, it would contradict the wisdom of the ages (which we have seen goes back to Plato's *Dialogues*) that all species have a means of survival. He suggested instead that since the world was still incompletely explored, species known only as fossils might still live in unexplored regions.

Other naturalists of his time and afterwards followed his example of elaborating aspects of the balance of nature until Carl Linnaeus in 1749 used various examples from the literature and from personal observations to formally organize the subject as a special branch of natural history which he called the "economy of nature." Although it was a descriptive system rather than an ecological theory, it had its dynamic aspects. He could envision gradual changes both in the landscape and in vegetation of an area over time due to soil erosion and the decay of organic matter into soil (Linnaeus 1775). He also discussed the checks which populations use to prevent some species from becoming so numerous they exterminate other species (Linnaeus 1760; trans., 1781).

These writings, from Ray to Linnaeus, must be seen as a mixture of mythic elaborations and critical thinking, and that mix persisted. Lamarck was perhaps first to offer a new answer to the concern about how the balance could be preserved if species have become extinct. By 1801, when he published *Système des animaux sans vertèbres*, it was no longer easy to believe, as Ray had, that all species known as fossils still lived in unknown parts of the world. Lamarck (1801) suggested instead that fossils represent early forms of species which live on in altered forms. He may not have seen the significance of this idea for the balance of nature in 1801, because he discussed that concept in a different part of that book from his discussion of evolution (Lamarck 1801), but in *Philosophie zoologique* (1809), he explicitly related them.

By overestimating the potentials of organisms for change and the inheritance of such changes, Lamarck neglected the importance of competition in nature. The first naturalist to appreciate this importance was Augustin-Pyramus de Candolle. He explained the distributions of plant species as being partly due to competition between them (de Candolle 1809, 1820). There is a possibility that his appreciation of the importance of competition was influenced by Malthus (1798) who suggested briefly that biological species are subjected to the same kind of population pressures as humans. De Candolle had studied under Pierre Prevost, French translator of Malthus, and was also a friend of French economist Jean-Baptiste Say, who was interested in Malthus' theory.

The merits of de Candolle's 1820 essay were fully appreciated by Charles Lyell, who drew upon it in his own even more comprehensive discussion of the distribution of species. As a geologist, Lyell was interested in both temporal and spatial distribution, and the scope of his *Principles of Geology* enabled him to bring together the various aspects of the economy of nature. In rejecting Lamarck's theory of evolution, he also rejected an uncritical belief in the great capacity of individuals to change and to pass on those changes to a new generation. In exploring the possible causes of extinction, Lyell found Cuvier's idea of periodic catastrophes no more convincing than Lamarck's ideas, and he placed his emphasis upon competition as a leading cause.

Lyell did not stop with de Candolle's essay, however. He also read and was impressed by Linnaeus' essays. He absorbed Linnaeus' perspective on the economy of nature, but modified it somewhat to accommodate both extinction and competition. That Lyell failed to detect some of the implications of his own synthesis would be realized by Alfred Russel Wallace who wrote in his "species notebook," "Some species exclude all others in particular tracts. Where is the balance? When the locust devastates vast regions and causes the death of animals and man, what is the meaning of saying the balance is preserved?" (McKinney 1966). Unfortunately, he never published these comments made about 1855.

In his famous essay, "On the Tendency of Varieties to Depart Indefinitely From the Original Type," (1859) which he sent in 1858 to Darwin to publish, Wallace also discussed the balance of nature, but from a different perspective. Under the influence of Malthus, he assumed all species produce more offspring than can survive, and his interest was primarily in what the checks to their increase were and what were the consequences of their operation. In modern terms, his evolutionary concerns inhibited his ecological curiosity.

Charles Darwin's "Notebooks on Transmutation" (1837-1839) indicate that, despite Lyell, he was strongly influenced by naturalists who minimized competition in nature—until he read Malthus in October 1838. Darwin then gained a new appreciation of de Candolle and Lyell's perspective (Darwin 1987; Egerton 1967). His theory of natural selection implicitly corrected the exaggeration of niche specialization found in the writings of earlier naturalists. Darwin's evolutionary theory also provided a new way to explain the differential reproductive capacities of different species, in terms of their survival requirements rather than in terms of physiological necessity or "benevolent design."

There was a significant incompatibility between important elements of Darwin's theory and Linnaeus' concept of the economy of nature, but he never turned his attention to the differences. He appreciated the usefulness of Linnaeus' perspective on the interrelations of species and also the way in which Lyell had improved upon this perspective by interjecting both the extinction of species and an emphasis on competition to explain it. Lamarck was wrong to dismiss all extinctions, but Lyell was also wrong in dismissing all transformations of species. Darwin saw that he could continue down the road of modifications which Lyell had followed, and therefore, he never saw a need to reject the Linnaean concept altogether. To do so would have been to reject what we call an ecological perspective, and Darwin knew he needed that perspective.

When Ernst Haeckel read Darwin's works he also saw the need for that perspective and coined its modern name, "oecologie" (1866). However, Haeckel was already involved by then with the sciences of evolution, anatomy, and systematics; he did not become an ecologist. Bramwell (1989) argues that he was at least an environmentalist. Stephen A. Forbes, an American zoologist, did acquire this new perspective as his main outlook. He believed that, despite extinctions, nature existed in a stable equilibrium which was maintained by the evolutionary process, and he defended this idea in a well-known essay, "The Lake as a Microcosm" (Forbes 1887). For almost a century most ecologists were content with Forbes' theoretical perspective.

As Wallace asked in his note, but not in print, if organisms compete and species become extinct, where is the balance? Although mythopoeic science is based upon some data about nature, it is never constructed in a way in which it can be refuted. According to the understanding of Ray and Linnaeus, species were provided with safeguards to ensure their survival, but when it became clear that species do become extinct, it did not destroy the balance of nature myth. Nor was it destroyed by Darwin's theory of evolution, which rejects the notion that species were preadapted to a stable environment. A myth cannot be destroyed by facts. Myths can only be abandoned when people realize that, despite the claim of explaining everything, they can never be subjected to critical testing because the claims are too vague.

Abandonment of the balance of nature myth by ecologists does not seem to have occurred because of any direct challenge, such as Wallace's question, but because of the indirect influence of four other aspects of ecology. One of these was the gradual loss of confidence in the community concept. This concept also has its origin in Linnaeus' economy of nature essay of 1749, but it did not achieve much prominence until Karl Möbius's study on the oyster biocoenosis in 1877. As is well known, Henry A. Gleason expressed skepticism of the utility of the community concept in 1917 and later, but it is difficult to discard one concept until an adequate alternative is developed (Gleason 1917). By the 1960s, the continuum concept of vegetation was well enough developed to challenge the biotic community concept (McIntosh 1967, 1975, 1985).

Another important specialization was population ecology, which has been concerned throughout this century with developing mathematical models to describe population changes (Kingsland 1985; McIntosh 1985). An early hope was to describe both the causes and patterns of population cycles in nature. Although progress was made in both respects, the "cycles" turned out to be tendencies rather than regular patterns.

A third development tending to undermine the balance of nature myth was evolutionary ecology. Both ecology and genetics began to emerge as scientific disciplines in the 1890s, and not until the early 1940s was there a synthesis achieved between genetics and evolution (Mayr 1982). Since ecology is far more diverse a subject than genetics, it is not surprising that an effective synthesis of ecology and evolution was not achieved until the 1970s. The struggle for this achievement is much less well known than that of the rejection of the community concept and the development of population ecology, but a conference was held on this subject in 1985, and the resulting set of papers which appeared in the Summer 1986 issue of the *Journal of the History of Biology* provide an excellent survey on the synthesis of ecology and evolution.

Fourth, there has been a sustained dialogue among ecologists throughout this century concerning scientific methods. This dialogue has been stimulated by the diversity of ecology, and it has led to a widespread rejection of hypotheses which lack potential falsifiability. This criterion is deadly to any mythology still lurking within science. I am unaware of any history of ecological methodology, but there is a start toward one in chapter 7, "Theoretical Approaches to Ecology," of Robert P. McIntosh, *The Background of Ecology: Concept and Theory* (1985).

Before discussing further the balance of nature and ecology and evolution, let us turn to our third general perspective—theoretical versus applied ecology. Thus far we have examined aspects of the history of theoretical ecology. Applied ecology arose from agricultural concerns. Many of the early authors who explained why species do not become extinct also wrote about how to suppress agricultural pests (Egerton 1973).

The application of abstract knowledge to practical problems may once have seemed so natural that it required little comment. But in the aftermath of atomic bombs, chemical pollution, and environmental transformations effected by artificially-powered machines, we now feel the need for a philosophical perspective to guide our thinking. Both those who trust and mistrust applied science begin their quest for perspective with the Scientific Revolution. In the mid 1500s, different approaches to science were in competition. A century later, there were two methodologies which predominated, an experimental method advocated by Francis Bacon and Galileo, and a mathematical-reductionist method advocated by Galileo and Descartes. According to Hall (1983), Newton, in his *Philosophiae naturalis principia mathematica* of 1687, showed how these two methodologies could be combined into what is often considered "the" methods of modern science.

However, the "mechanical philosophy," as it was often called, led to oversimplifications in animal physiology (Hall 1969; Rothschuh 1973), and it had little impact upon natural history. Nevertheless, in retrospect Bacon has been criticized for his desire to give humanity mastery over nature (Eiseley 1958, 1973), and the Scientific Revolution has been criticized for having eliminated the organic perspective of how nature works (Merchant 1980).

However, neither Bacon's vision nor the agricultural concerns of early naturalists led to a strong early development of applied ecology. The reason seems to be that agricultural sciences, forestry, and fish and wildlife management all got started before ecology did, and then after ecology was established, it took a good while before it had achieved either the sophistication or the prestige to exert much influence on these other disciplines.

The first step, therefore, in understanding the relationship between theoretical and applied ecology is to realize that scientists and managers who practice applied ecology include only a relatively few ecologists (Egerton 1985a; Magnuson 1991; Polunin 1986). Barton Worthington (1983) is one of those few and he has published the story of his career. He seemed happy to accept the priorities of his employers. He was practical-minded when working in Africa and preservation-minded when he returned to Britain. Sometimes resource managers, such as Aldo Leopold, have changed their priorities as a result of their professional experience (Meine 1988). The great majority of natural resource managers must have taken college courses in ecology, but this exposure would not suffice to instill in them the priorities of professional ecologists, even if most theoretical ecologists share a set of values.

Applied ecologists clearly do not all think alike or respond to the same influences. American managers of sport fishing and hunting are more likely to respond to political than economic pressures. Sometimes there have been struggles between the representatives of theoretical and applied science, but even so, a capitalistic economy has not always defined the issues. It is interesting to compare the history of the struggle between resource managers and theoretical ecologists in Russia from the 1890s into the 1930s (Weiner 1988) with the history of the struggle between federal pest controllers and mammalogists in America from 1890 to 1985 (Dunlap 1988): the similarities may be more impressive than the differences. There can also be different perspectives and priorities between the managers of different agencies of the same government. This can be seen in both the Russian example and in America between the managers in the U.S. Departments of Agriculture and Interior.

Even within the same agency perspectives may differ. A historian of the California fisheries found that federal managers favored generous commercial fish harvests whereas state managers wanted a more restricted annual catch (McEvoy 1986). On the Great Lakes, where the situations were not parallel, sometimes the federal managers were the ones favoring a restricted catch in opposition to state managers (Egerton 1985b, 1987, 1989; Egerton and Christy 1993).

Our three themes—balance of nature, ecology, and evolution, and theoretical versus applied ecology—come together in the historical interpretations of Donald Worster. His *Nature's Economy: The Roots of Ecology* (1977) was the first book-length history of ecology by a single author and may be the most widely read. His perspective in it and also his whole style of writing history seems heavily indebted to Eiseley.

Loren Eiseley (1907-1977) is remembered as a popularizer of science who had a marvelous writing style and capacity to help others to appreciate nature. His considerable talents as a writer have attracted the attention of two biographers (Angyal 1983; Carlisle 1983), and as far as they go, they describe the real Eiseley. But neither emphasizes Eiseley as a critic of science. He rightly insisted that science and scientists should not be placed on a pedestal beyond the reach of criticism. Good critics can provide a service to both science and society. But what I find in Eiseley's criticism is not the claim that modern evolutionists have distorted the picture of nature, though there are hints that this may be true. His real complaint is that these evolutionists destroyed the mythopoeic mental world which we found appealing and replaced it with a world view toward which we feel alienated. He implied that scientists like Darwin who defended the new perspective are insensitive and perhaps otherwise morally deficient (Eiseley 1958, 1979). In one book he contrasted Darwin and Henry David Thoreau as interpreters of nature (Eiseley 1964) with Thoreau receiving the more favorable evaluation. I do not see Eiseley as a popularizer of science, but as an alienator from science. Yet, he was too knowledgeable to pretend he could refute the world view he disliked. So what did he leave with his readers? Mainly, an admiration for guru Eiseley and nostalgia for a pre-Darwinian world view.

The 1960s and 1970s when Eiseley's influence was strongest was also a time of general discontent, for reasons discussed above. Paul Sears (1964) had called ecology subversive because it does not seem compatible with modern conceptions of material progress, whether capitalistic or communistic. His essay inspired Paul Shepard and Daniel McKinley to compile *The Subversive Science: Essays Toward an Ecology of Man* (1969). For these three authors "mainstream" ecology seemed to offer an alternative perspective to the artificiality and pollution of material progress.

Simultaneously, another world view emerged, the New Left (Long 1969). Unlike both capitalism and Marxism, the New Left saw more losses than gains from the Industrial Revolution, it disapproved of either capitalistic or Marxist warfare, it lacked confidence in "mainstream" science of either capitalism or Marxism, and the science it favored was holistic, not reductionistic. Theodore Roszak is an influential New Left cultural historian who thought science was a major cause of the modern world crisis (Wade 1972). Perhaps Shepard and McKinley encouraged him to think that ecology might not succumb to the same crass influences which he thought had corrupted other sciences, but he warned: "Ecology stands at a critical crossroads. Is it, too, to become another anthropocentric technique of efficient manipulation, a matter of enlightened self-interest and expert, long-ranging resource budgeting? Or will it meet the nature mystics on their own terms and so recognize that we are to embrace nature as if indeed it were a beloved person in whom, as in ourselves, something sacred dwells?" (Roszak 1972:403).

These comments about Eiseley and the New Left are background for understanding Worster's perspective on the history of ecology. He believed even Darwin was to be alternately pilloried for his theory of natural selection and praised for his love of nature. Furthermore, Worster could make far more out of the economic connection than Eiseley had. Eiseley had pretty much limited his jaundiced discussion of this subject to Darwin's use of Malthus, but Worster found the sins of the ecologists ran much deeper. Charles Elton's food chain could be viewed as an economic model; August Thienemann spoke of producers, consumers, and reducers; Edgar Transeau calculated solar energy input and sugar production in a corn field; and Raymond Lindeman and G. Evelyn Hutchinson calculated the flow of energy from one trophic level to another (Worster 1977). The chapter discussing these researches is preceded by one describing the U.S. Department of Agriculture's campaign against predatory mammals and rodents and is followed by another describing the decline of holistic studies in favor of analytic studies.

One gets the strong impression from reading all this that ecologists who fling quadrats or measure productivity are victimizing nature and have no capacity to love the unspoiled landscape in the way that Gilbert White and Thoreau did. Since writing *Nature's Economy*, Worster has read Ronald Tobey's *Saving the Prairies* (1981), Ronald Engel's *Sacred Sands* (1983), and has taken another look at Paul Sear's *Deserts on the March* (1935, 3rd ed. 1959). (To his list could now be added Runte [1990] on Joseph Grinnell.) Worster could then no longer sustain his argument that ecologists who measure are unfeeling about the unspoiled environment. Nevertheless, he was reluctant to admit error: "The conclusions in that book still strike me as being, on the whole, sensible and valid ... scientific analysis cannot take the place of moral reasoning ... ecology promotes, at least in some of its manifestations, a few of our darker ambitions toward nature and, therefore, itself needs to be morally examined and critiqued from time to time" (Worster 1990).

In 1972 an English medical chemist, James E. Lovelock, published his early thoughts on a new kind of economy of nature concept which he called "Gaia," after a Greek earth goddess. The earlier balance of nature concept had focused attention upon the life histories of plants and animals and the interactions among their populations. This new concept emphasized instead the chemical cycles which flow from the earth to the waters, atmosphere, and living organisms.

The potential for this biogeochemical approach had actually existed ever since Joseph Priestly (1772) reported that green plants produce a gas that keeps animals alive while using in turn a gas animals produce. Priestly's observations were carried further by another chemist, Jean-Baptiste-Andre Dumas (1841), whose article prompted the philosopher Herbert Spencer (1844) to publish some interesting speculation on the carbon cycle. As interesting as these three reports appear in retrospect, they failed to establish a continuous research tradition. Another aspect of the idea which did develop steadily throughout the 1800s was the study of agricultural chemistry—in soils and in plants and with some attention also to animal manure (Browne 1944; Rossiter 1975; Wines 1985). By the 1880s, marine ecologists (usually trained as zoologists or bacteriologists) began trying to discover whether what the chemists and plant physiologists had learned about the vegetation-land-air nitrogen cycle could also be documented in the sea (Mills 1989).

The founder of the science of biogeochemistry was a Russian, Vladimir I. Vernadsky, who had studied chemistry, mineralogy, and soil science. He published a textbook on geochemistry in 1924 and *Biosfera* in 1926 (Fedoseyev 1976; Grinevald 1988; Polunin and Grinevald 1988; Sagan 1990). Since *Biosfera* was translated into French (1929) and German (1930), it was accessible in the West and thus a precursor to, but not a direct intellectual ancestor to, the Gaia concept which Lovelock and biologist Lynn Margulis developed in the 1970s (Grinevald 1988; Mann 1991). Lovelock's first book-length treatment of the Gaia thesis appeared in 1979, and he updated it in 1988.

The studies of Lovelock and Margulis convinced them that the biogeochemical cycles in, on, and above the surface of the earth are not random, but exhibit homeostasis, just as some animals exhibit homeostasis in body heat and blood concentrations of various substances. That homeostasis must indicate that the Earth is alive. A parallel can be drawn between the thin living layer on the surface of the Earth and the cambium layer of a tree. The wood inside and the bark outside that layer are dead. Most of a mature tree is dead and yet that thin layer of cambium makes the tree very much alive and interacting with its environment.

Analogy once played an important role in scientific methodology, and it still does have a more limited role of serving as a guide to possibilities. However, it can no longer be taken as a proof of anything, and if it is, then we are back to the mythic concept of Plato's *Timaeus*, except that there are now data to provide some plausibility.

In 1988 the advocates of the Gaia hypothesis had won enough scientific attention for the National Science Foundation, National Aeronautics and Space Administration, and the American Geophysical Union to sponsor a conference of 150 scientists from around the world to evaluate the hypothesis. Margulis and Gregory Hinkle produced a precise enough explanation to give it some scientific respectability (Joseph 1990), but without providing any clear way to establish their claims. They believe that life, rather than inanimate forces, mainly controls the earth's environment, and that organisms maintain earthly homeostasis just as a mammalian brain maintains homeostasis within a mammal's body. We more or less understand how homeostasis works when it is controlled by a brain within an organism, but not how it could work in a world "system" which lacks a brain.

There is a diverse environmental movement in North America, Europe, Australia, and elsewhere (Young 1990) which supports a wide variety of organizations, ranging from the conservative National Wildlife Federation to the moderate National Audubon Society to the radical Greenpeace (just to mention three American examples; Gifford et al. 1990). Professional ecologists are inclined to work with such groups since they usually more or less share their goals. Unfortunately, there is an important environmental magazine, *The Ecologist*, founded and edited in England by Edward Goldsmith, and handled in America by MIT Press, which often portrays professional ecologists as part of the problem, not as part of the solution. Rather than seeking support from conventional ecologists, Goldsmith and his associates seek to build an alternative science based on the perspectives of Worster and Lovelock. Some characteristics of this alternative ecology are found in a symposium volume, *Gaia: The Thesis, the Mechanisms, and the Implications* (Bunyard and Goldsmith 1988).

What basis is there for professional ecologists to take this group of radical environmentalists seriously? In the first place, they have attracted the support of some well-positioned scholars (e.g., Grinevald 1988; Margulis 1988; Merchant 1980, 1989; Ravetz 1988; Sagan 1988). In the second place, scientists are disposed to "work within the system," partly because they obtain grants from governments and other "establishment" agencies and partly because scientists believe that the advancement of knowledge will influence policies of government (Gillispie 1968). Most ecologists can be viewed as part of the "establishment." In the third place, there is a convergence of evolutionary and ecological theory and techniques with those of economics which has attracted the notice not just of Marxist scholars like Pepper (1984) and Young (1985).

This convergence allows radicals to claim that ecologists are "tools" of capitalism or of "the establishment." Two young historians of ecology who seem inspired by Worster have carried out studies which offer a negative perspective on the careers of "establishment" ecologists Eugene P. Odum and Howard T. Odum (Taylor 1988; Hagen 1990). Both were "guilty" of accepting research grants from the Atomic Energy Commission. Neither historican claims that this made either Odum brother an uncritical booster of atomic power or weapons. Rather, their "crime" was that they added some respectability to the "atomic establishment" by associating with it. (One wonders what historians would say about ecologists if they had never bothered to study the impact of radiation on the environment.) A good treatment of radiation ecology history is Bocking (1991).

In addition, Howard T. Odum is guilty of "technocratic" reasoning. This sounds fairly sinister when discussed in the abstract. Is he inventing new ways to harness wild or domesticated nature to meet the greedy demands of wealthy elites? A way to test Taylor's vague apprehensions is to assess the fruits of Odum's work at a slightly later date than Taylor did. A good means for this test is to survey the chapters in a book entitled *Ecological Engineering* which is dedicated "to Howard T. Odum, our teacher and colleague, and a pioneer in ecological engineering" (Mitsch and Jorgensen 1989). One finds that the chapters discuss how to lessen the environmental harms of pollution and other human activities. What seems morally significant is not the methodology used but the goals toward which it is directed.

Not all complaints against "mainstream" ecology can be answered pragmatically. What about the convergence of evolutionary and ecological theory and techniques with those of economics? Does this not show that these sciences have been "captured" by the capitalistic ethos?

Critics of capitalism have complained that it is too competitive, and that this competitiveness creates disparities of wealth and bankruptcies which wreak havoc on society. Marxist critics expected capitalism to collapse by now, and they explain its persistence as due to the existence of some welfare programs and of police and armies to protect the property of the wealthy. I think these claims are correct but do not tell the whole story. Both the success of capitalism and the convergences between economics and these two sciences (i.e., evolutionary biology and ecology) can be explained by an important characteristic which critics and defenders of capitalism tend to overlook.

Both sociologists and followers of the Gaia thesis criticize ecologists and evolutionists for emphasizing competition in nature and society; instead, they emphasize the importance of cooperation. However, in nature and in capitalistic society, competition and cooperation are both important. Lichens are made up of cooperating algae and fungi, but two lichens can compete with each other for their needs. Capitalism has not collapsed, not only because of welfare programs and coercive powers, but also because it involves cooperation as well as competition. If capitalism involved only competition, its critics would be right—it would have collapsed by now. However, there is enough cooperation between workers and management, between business and government, and between business and customers to keep the economy going. Since competition and cooperation exist in both nature and capitalism, it is not surprising that there are theoretical convergencies between capitalism and science.

However, just because capitalism is in some sense "natural" does not mean that our capitalistic economy is good for either society or the environment. Cancer is also in some sense "natural," but we are eager to be done with it. Practically all historians, myself included, who study the environmental history of capitalistic society view it with dismay. The only environmental historian I can think of who defends capitalism is Anna Bramwell (1989), who believes socialism will not work and that capitalism can be reformed. The great majority of environmental historians believe that the virtues of capitalism are outweighed by its vices. It encourages overconsumption, pollution, great disparity in wealth, and indefinite population growth.

Thus, the inclination of scientists to "work within the system" means that ecologists run a risk of being co-opted by the capitalistic establishment, just as radical critics claim. My counterclaim is merely that these critics have not yet demonstrated that ecologists and scientific ecology have been corrupted by the system. It seems fair to ask the radical environmentalists, who imagine the practitioners of this science have been corrupted, that they produce more than innuendos and guilt by association to substantiate their claims.

The ecological sciences, to whose history I have devoted 30 years of study, may not be civilization's greatest achievement, but I think they are among the greatest. Ecology is the most diverse of the sciences, and that diversity has resulted in enormous growing pains. Most ecologists are working to achieve a unified science. Despite much skepticism both within and without, I see the *self*-correcting process of the ancient natural philosophers alive and well within ecology.

When ecologists study either humans as components of ecosystems or the ecology of populated areas, they are apt to find that they must deal not only with some members of society who have little concern for ecology, but also with others who have an alternative vision of what ecology should be. A dialogue needs to be maintained with both groups. This chapter attempts to explain the perspective of many who urge an alternative to "mainstream" ecology. With understanding, effective communication becomes possible.

Acknowledgments

I thank two friends: Jacques Grinevald, a participant in the 1987 Gaia conference, for bringing its proceedings and other Gaia references to my attention and for extensive comments on an early draft of this paper; and Jean-Marc Drouin for information on the indirect connections between Malthus and de Candolle and on the 1809 de Candolle statement on competition.

Recommended Readings

Egerton, F.N. (1973). Changing concepts of the balance of nature. *Quart. Rev. Biol.* 48:322-350.

Egerton, F.N. (1983, 1985a). The history of ecology: achievements and opportunities, part one, *J. H. Biol.* 16:259-311; part two, *J. Hist. Biol.* 18:103-143.

McIntosh, R.P. (1985). *The Background of Ecology: Concept and Theory*. Cambridge University Press, New York.

Worthington, E.B. (1983). *The Ecological Century: A Personal Appraisal*. Clarendon Press, Oxford.

3
An Exceptionally Powerful Biotic Factor

Michael Williams

The Human Component

The questions raised by the relationship between humans and their physical environment are as old as western thought. As Glacken (1967) has pointed out, people throughout the ages have persistently asked three broad questions:

1) Is the Earth a purposely made creation, or, put another way, is it a divinely designed Earth for humans?
2) Have climate, relief, and configuration had an influence on the physical, moral, and social nature of human beings, or again, put another way, does the environment have an influence on human behavior? and
3) In their long tenure of Earth, have humans changed it from its hypothetical pristine condition, or are humans agents or components of ecological and geographical change?

Of these, the first two questions have been explored frequently from Greek civilization onward. The first assumed an Earth designed (by God) for humans alone, and they, moreover, were at the apex of the hierarchy of life. The first approach delved into questions of mythology, theology, and philosophy. The second was a determinist view that environments (especially temperature) affected individual and cultural characteristics. It delved into medicine, pharmaceutical law, and weather observations. But the third question had a subject matter that was far less specific and yet, was more all-embracing. It simply encompassed the ideas, activities, and technologies of everyday life.

While the first two questions have receded in the intellectual consciousness of western society from the latter part of the nineteenth century onward, the third question about humans as agents of environmental and ecological change has loomed larger and larger, so that at present it is a major global preoccupation. In an age of scientific thinking where society and nature have been separated as two different fields of study, the role of humans as components has not always been appreciated. Therefore, it might be profitable to trace briefly how humans have gradually been included in ecological studies.

There are many possible starting points, but certainly, Darwin turned a few millennia of assumptions about the habitable world upside down. Darwin believed nature to be "a web of complex relations" and that no individual organism or species could live independently of that web. Although he did not explicitly exclude humans from that ecological interdependence, he did include them implicitly when he said, "We are all netted together" (Darwin 1859). Darwin seemed as though he was groping towards the concept of the ecosystem when he wrote about "the struggle for existence," but never quite formulated the concept. Nonetheless, this biocentric view of a single shared Earth was new, for humans had always been regarded as superior to nature.

The term ecology was first introduced by Ernst Haeckel in 1866 to describe that branch of biology that dealt with interrelationships. It was a vague term that was as much philosophical as scientific, and it was loosely organized around the idea of the "community." But the big advance in work on communities came during the later years of the nineteenth century with the work of Henry C. Cowles of Chicago and Fredric E. Clements of Nebraska. Cowles developed ideas of succession and climax vegetation after studying the patterns of vegetation growing along the successive dunes and swales on the south shore of Lake Michigan. Patterns of ecological succession in space seemed to parallel the development of vegetation through time (Cowles 1899). Clements's work in Nebraska (1904) stressed the dynamics of ecological succession in the plant communities that would reach stable climax communities in given climatics unless radically disturbed either by humans or fire.

In stressing association, climax, and succession, both Clements and Cowles provided a conceptual framework for the description and understanding of vegetation. However, the framework suggested a predictability, almost an inevitability, about the stages of development of grassland and forest which tended, in time, to become a rigid orthodoxy that excluded the appreciation of such complex influences as humans, animals, wind, fire, insects, and disease. Taken to its logical conclusion, it implied that if either the forest or the prairie were left alone, they would progress to a stable equilibrium and self-perpetuating state, and that conversely, there must have once existed a wonderful primeval forest or grassland until European settlers devastated it, which simply was not true.

Clements' view of nature as a pure state unsullied by humans led him to regard modern humans as "an alien in a natural world, a bull thrashing around in a china shop" (Worster 1977). In fairness, of course, Clements had grown up in a world that had once had some semblance of that glorious primeval climax until it was broken by the sod-busters. Nonetheless, the point is that his concepts met a ready acceptance, as the contrast between nature and civilization ran deep in American literature and culture, and, one can add, science; it was "etched on the American consciousness" (Worster 1977). The Indian, by definition in that earlier view, was not civilized, therefore did not count.

The distinction between nature and civilization had no such relevance in the long-settled countries of Europe, honed to their present biological state and visual appearance by centuries, if not millennia, of human effort. Any climax vegetation had probably been extinguished hundreds of years ago.

It is an over-simplification to suggest that Arthur G. Tansley was the first person to put humans back into nature, but he certainly placed humans fairly and squarely into ecological studies when he came up with his new, all-inclusive concept of the ecosystem in 1935. Throughout the previous 10 years, Tansley had delivered a number of cogent counter-arguments to the succession-climax school and its notion of mono-climax. He claimed that humans made an anthropogenic climax that could be just as stable and balanced as the ideal Clementsian primeval, climatically-induced climax. In his celebrated paper "The Use and Abuse of the Vegetational Concepts and Terms" (1935), he elaborated his views:

"It is obvious that modern civilized man upsets the 'natural' ecosystems or 'biotic communities' on a very large scale. But it would be difficult, if not impossible, to draw a natural line between the activities of human tribes which presumably fitted into and formed parts of 'biotic communities' and the destructive human agencies of the modern world. Is man part of 'nature' or not? Can his existence be harmonized with the concept of the 'complex organism'? Regarded as an exceptionally powerful biotic factor which increasingly upsets the equilibrium of pre-existing ecosystems and eventually destroys them, at the same time forming new ones of very different nature, human activity finds its proper place in ecology."

Thus the ecosystem developed as a more scientific term to replace the anthropomorphic community. In the ecosystem, the biological and the non-biological were interdependent, and it is clear that Tansley regarded humans as an integral part of that concept and major agent of change.

In more recent times, the role of humans in nature has been taken further. Anderson (1956) has commented on the existence of whole landscapes of humanly-dominated and even humanly-created floras, while Van der Maarel (1975) has gone further and suggested that humans are not only a part of nature but have created nature. Thus, it is possible to take the view that God may have created the world, but humans have made modern natural ecosystems.

Cultural Landscapes

The role of humans in the creation of ecosystems has been more readily accepted in the longer-settled (and better documented) lands of Europe than in the more recently settled areas of America. A recent interdisciplinary symposium held in Norway on the occasion of the hundredth anniversary of the founding in 1886 of the Botanical Institute of Bergen addressed the theme of human impacts. The meeting and proceedings were entitled, *The Cultural Landscape: Past, Present, and Future* (Birks et al. 1988).

The papers revolved around the idea that in Scandinavia, in the words of the Director of the Institute, "a virgin landscape is a fiction" (Birks et al. 1988). Simply, there is no wilderness. With some small and doubtful exceptions in remote areas where axe or human-caused fire might not have penetrated, the whole landscape had to be considered as "one great cultivation effect." Consequently, he suggested, there was a need to move away from the usual dichotomies of uncultivated/cultivated, nature/human-made, or wilderness/settlement so prevalent in North America to the idea of "a gradient of human impact." In his view, the impacts were so persuasive and subtle that what were considered "natural landscapes" in previous generations had to be understood for what they were—"relics of earlier types of land-use." These relics had been maintained by massive imputs of human-power, and, therefore, at the present time had become uneconomical to maintain. Once traditional methods of farming and landscape maintenance were abandoned the vegetation regenerated. The landscape was dynamic. As a consequence, the record of the composition and structure of past vegetation was in danger of being lost, as were the plant successions that occurred once the old management techniques were abandoned.

The emphasis of the Bergen symposium, therefore, was on the origins and working of human practices, such as forest management (pollarding), meadow management, and fodder production that created and maintained these distinctive landscapes from at least 5000 B.P. to the near present. In order to understand and possibly help to preserve these particular aspects of the traditional, valued landscape, the contributors to the symposium used many techniques and approaches at a variety of spatial and temporal scales. For example, papers employed pollen analysis, LANDSAT imagery, air photographs, descriptive phyto-sociological approaches, energy and nutrient budgets, and multivariate techniques such as canonical correspondence. All these entered into the quest to understand these particular resultant cultural landscapes.

Perhaps the most striking thing of all the implications of the Bergen symposium was the use of the term "Cultural Landscape" to describe the ecological outcomes. It was apt, but to a geographer's ear, strange. The participants said that they did not know when or where the term originated. Yet, the concept of "cultural landscape" is part of the heritage of every human geographer, especially in North America. It is associated with the writings of the American scholar Carl Sauer and the whole Berkeley school of human-historical-cultural geography that he spawned from 1925 onwards. Simply, in his own words: "The cultural landscape is fashioned out of a natural landscape by a cultural group. Culture is the agent, the natural area is the medium, the cultural landscape the result" (Sauer 1925).

The aim was to trace the development of the "natural landscape" into a "cultural landscape," that is to say, the landscape made up of cultures (Mikesell 1968). A characteristic of that school was the close interdisciplinary links those geographers had with anthropologists, historians, and botanists, but the ethos was always intensely humanist in the broadest sense (Williams 1983).

The Bergen gathering was composed predominantly of ecologists and paleobotanists, with only a smattering of historians and historical geographers. Yet they used the term "Cultural Landscape" with a new meaning. They were unaware of the origin and rich literature of the term they were using. They devised a new methodology of a cultural landscape, which, it is suggested, is not without its significance in what follows in this book.

The American Forest

The Bergen symposium threw up many parallels, contrasts, and points of reference that could be applied to the American scene. To illustrate these, examples are drawn from *The Americans and Their Forests* (Williams 1989).

The aim of that volume was to understand the geographical, historical, aesthetic, cultural, and economic relationships of Americans with what was probably the most prominent visual biotic feature of the continent—its forests—and its removal. Broadly, therefore, it was a study in the creation of a cultural landscape in Sauer's sense, although such an approach was not consciously being followed.

Nor was ecological change the primary concern, but ecological factors could not be ignored because forests are a dynamic vegetational complex. In that dynamic, the dominant Cowles/Clements succession and climax model of American ecology had its limitations. There is abundant evidence of human interaction, regulation and alteration of the forest. Far from being the epitome of wilderness, of unspoiled beauteous nature, it was, and is (except in its remotest parts) a cultural phenomenon.

That this should be so is not surprising. Humans have found an ecological niche in every part of the Earth. Their increasing numbers, intellectual and technical manipulation of the environment, and the extra-somatic energy which they can call upon have enabled humans to dominate everywhere (Goudie 1986; Simmons 1989; Mannion 1991). Moreover, the massive turnover of biomass needed in order to feed, clothe, shelter, and warm humans through so many millennia must be conducive to fundamental ecosystemic change. Ecosystems do not exist in isolation from their surroundings, neither do humans.

In looking for change, therefore, one must abandon the old stereotypes of a primitive landscape transformed to a domestic and carefully humanized one. As in Scandinavia, perhaps we, too, should be looking for "gradients of human impact." The changes are not necessarily good or bad, but they are possibly subtle. It is a debatable point whether North American subtlety is more brutal in the more uniform landscapes of the continent compared to the more minutely varied and culturally older landscapes of Europe.

Of course, some ecologists have already been well aware of the human impact, and one can cite, for example, the work of Ahlgren and Ahlgren (1983), Brush and Davis (1984), Hazel Delcourt (1987), Chapman, et al. (1982), and Forman and Godron (1986), to mention but a few, who have variously been aware of the effects of perturbations such as fire, storm, windthrow, ice storm, disease, and human change, so that, in the words of Loucks (1979), "The features we study are truly ephemeral in the light of the history of forests; what persists is little more than a mosaic of restless change."

The Indian Impact

One of the greatest myths about North America is that it was a lightly populated region on the coming of the Europeans. Kroeber's long-accepted estimate of a total of just over 1 million Amerinds is now discredited and the population estimates have been boosted to between 9.8 and 12.25 million by Dobyns (1966) and the Berkeley scholars Cook (1973), Simpson and Borah (see Cook and Borah 1972). The majority of the Amerinds were resident in the eastern forests and not the plains where they were driven later. In 1492, parts of the eastern seaboard had population densities in excess of those of the closely settled parts of Western Europe. Old world pathogens, particularly smallpox after 1518, destroyed them (Crosby 1986), but in its heyday the whole of the western hemisphere may have had a population about as great as that of Europe.

When assembled, the contemporary evidence of fields and clearings is overwhelming. Two comments must suffice, one a description, one an illustration. Said one settler in 1622 (quoted in Maxwell 1910): "...the objection that the country is overgrown by woods and consequently not in many years to be penetrated for the plow, carried a great feebleness about it for there is an immense quantity of Indian fields cleared ready to hand by the natives, which, till we are grown over-populous, may be very abundantly sufficient."

Figure 3.1. The town of Secota, Virginia, as depicted in John Smith's *Narrative of The First Plantation in Virginia* [1620], reproduced by Thomas Hariot [1893], courtesy of State Historical Society of Wisconsin. The sketch depicts various aspects of life in an actual Indian village.

The illustration of the village in Secota, Virginia, from John Smith's *Narrative of the First Plantation in Virginia* (Fig. 3.1), shows a complex but orderly settlement, the nature of which is well substantiated by dozens of contemporary descriptions. The illustration also suggests a hierarchy of land uses which can be transformed into a schematic representation of forest use (Fig. 3.2).

On the assumptions made by anthropologists and others that cropland and an extended fallowing cycle needed an average of 2.3 acres of land per person (Kroeber 1939; Heidenreich 1977) then anything between roughly 22.5 and 28 million acres of forest would have been affected at any one

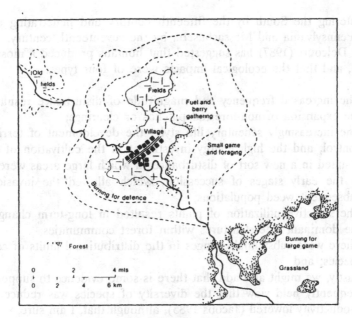

Figure 3.2. Schematic representation of Indian forest use.

time. This is nearly 20% of the current total of 270 million acres of land in crops in the eastern-most 31 states today. Against this sort of speculation one must place the current lack of palynological evidence of crop pollens and weeds, except in a few locations.

The impact of Amerinds on grasslands has not been considered, but one school of thought, based on actual eye witness accounts and subsequent botanical investigation and theorizing, suggests that most of the prairie openings and even large anomalous areas like the Kentucky Blue Grass country are a consequence of Indian burning (Day 1953; Maxwell 1910; Pyne 1982; Rostlund 1957; Russell 1983).

The very prairies themselves may be a consequence of repeated conflagration. There is substantial evidence that from Wisconsin in the north to Texas in the south, repeated firing by the nomadic hunting cultures of the Plains Indians sustained the grassland vegetation against forest encroachment on its eastern edge and, indeed, caused a gradual extension of the grassland. The Indians' object was to extend the range of game by continually burning the forest edge and also by burning abandoned fields to prevent forest regeneration. This occurred from post-glacial times onwards. Thus, one can view Indian incendiarism as either an attempt to maintain the grassland and the game on it as the forest advanced with the improvement of climate after the retreat of glaciers, or as an attempt to remove the forest and extend the grasslands. Either way, fire was critical.

It is thought that the buffalo, for example, spread throughout the continent as the forests burned, crossing the Mississippi in about A.D.

1000, entering the South by the fifteenth century and penetrating as far east as Pennsylvania and Massachusetts by the seventeenth century.

Hazel Delcourt (1987) has suggested that humans produced a mosaic of land use, and that the ecological impacts were of four types:

1)   The increased frequency and magnitude of disturbance resulted in the expansion of non-forested patches or clearings;

2)   The increasingly sedentary life style, the development of territorial control, and the high energy investment in the cultivation of crops resulted in a new sort of disturbance in which large areas were kept in the early stages of succession, which allowed the invasion of subsequent weed populations;

3)   The selective utilization of plants resulted in long-term changes in the dominant tree structures within forest communities;

4)   There were substantial changes in the distributional limits of certain species; and

5)   Lastly, we might also add that there is some evidence to support the frequently held view that the diversity of species was reduced and productivity lowered (Jacobs 1985), although that, I am sure, was not always true.

Perhaps the grasslands together with the changed extent and composition of the forest over large areas may be regarded as the greatest and most enduring cultural contribution of the Indians to the continent as a whole. In every way, the impact of early humans on vegetation was greater than suspected, and greater than many would care to admit.

European Agricultural Clearing

It goes without saying that clearing eliminates one vegetation cover and replaces it by another. My calculations suggest that in the 250-odd years of European colonization before 1850, some 113.74×10^6 acres were "improved," which invariably meant cleared of timber, and that in the succeeding 60 years to 1910, a further 190.5×10^6 acres were affected.

After about 1910, abandonment and regrowth began to outpace new clearing, particularly in the South, although it had started much earlier during the late eighteenth century in the marginal areas of upland New England. During the twentieth century, the *net* loss of farmland in the 31 eastern-most states between 1910 and 1958 was 43.8×10^6 acres, and this can be translated almost wholly into woodland gain (Hart 1968). A further 16.9×10^6 acres of farmland went out of production between 1959 and 1979, much of that now being included in the suburban expansion of the urban areas of the eastern seaboard, although massive clearing (about 6.5×10^6 acres) in the bottomland hardwood forests of the Mississippi and tributaries between 1960 and 1973 in order to create land for growing soybean and other crops has partially compensated for this reversion (Williams 1991).

The power of the temperate forest to regenerate has always been underestimated. It is impossible to believe that a forest returned with exactly *the same* floristic composition. The question cannot be answered definitively, but there is plenty of secondary evidence of changing composition in the Adirondacks and throughout the South, and the point has been proved by the modeling of stand history according to different assumptions and management practices (Dale and Doyle 1987).

Whatever the composition, the forest is not a "natural" climax. The redwood forests around Monterey Bay north to San Francisco are second and sometimes even third growth. The forest of the Harvard Forest at Petersham, Massachusetts, which looks for all the world like primeval forest to the untutored eye, grows out of abandoned fields that still have their surrounding stone walls and which can be dated precisely from the 1760s onwards (Raup 1966).

Logging and Selective Cutting

In most cases, logging was the prelude to the creation of farmland, except in the cutover land of the extreme north of the lake states, the Pacific Northwest and parts of the South. However, in its earlier pre-industrial days, logging was highly selective and usually led to a new disturbance regime. The outstanding example of this was the prejudice in favor of white pine (*Pinus strobus*) in the northern forests from New England to Minnesota, and later for red pine (*P. resinosa*), and later still for hemlock (*Tsuga canadensis*). In pre-Revolutionary days white pine was valued as a source for masts because of its straightness, strength, and height. Subsequently, it was favored over all other timbers for construction. Its discovery in Michigan was a potent reason for the westward migration of the lumber industry. Similarly, white cedar (*Chamaecyparis thyoides*) was used for shingles, red and white oak (*Quercus borealis* and *Q. alba*) for barrel staves for the Caribbean molasses trade, and hickory (*Carya spp.*) for firewood.

Surrounding trees were often destroyed indiscriminately in the drive to get the desirable trees. Timothy Dwight reported in 1821 that many Europeans who visited New England during the late eighteenth century were surprised at the small girth of the trees, something which many of them attributed to "the sterility of the soil." In fact, Dwight said, "The real cause was the age of the trees, almost all of which are young." Simply, the older trees were eliminated by the late eighteenth century and what was left was second or third growth (Dwight 1821).

Similar examples of the positive qualities of particular timbers and of prejudice against others can be cited for the selection of hardwoods such

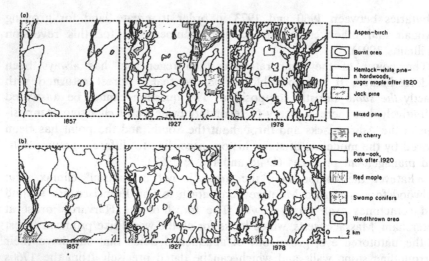

Figure 3.3. Changes in areal extent of forest cover types of (a) Township 28N, Range 4W (originally a hemlock-white pine—northern hardwoods area) and (b) T28N R2W (originally a representative mixed pine area), Crawford County, Michigan, USA. Unshaded areas are largely grasslands and open shrublands. Information based on Government Land Office field survey notes and 1857 township plat maps, 1927 Michigan Land Economic Survey map for Crawford County, and 1978 MIRIS township maps for Crawford County (Whitney 1987).

as white oak, chestnut oak (*Q. prinus*), and black locust (*Robina pseudoacacia*) for railroad cross-ties. Their greater resistance to rotting ensured their selection. Not until deep penetrating creosoting was perfected were lesser quality timbers used. By 1910, just over 600,000 acres of these hardwoods were being culled annually out of the Appalachian and eastern forests for railroad cross-ties (Olson 1971).

The post-disturbance vegetations have been clearly documented for the forest in at least two parts of the country. In the Allegheny Plateau, the dominant beech (*Fagus grandifolia*) and hemlock were cut intensively from 1880 to 1930 followed by severe overbrowsing by deer and the dominant trees have been replaced by rapidly growing, shade-intolerant species such as black cherry (*Prunus serotina*) and red maple (*Acer rubrum*) (Whitney 1990). In north-central lower Michigan, the cutting out of the white pine, and later of the hemlock and the better hardwoods, led to their replacement by sugar maple (*Acer saccharum*) and a subsequent fire-sequence of aspen (*Populus tremuloides*), oak, and jack pine (*P. banksiana*) that have become the basis of a new pulp-oriented lumber industry (Whitney 1987). The distribution of forest cover types at three time intervals, 1857, 1927, and 1978 in two townships in Crawford County, Michigan, reflect these different harvesting strategies (Fig. 3.3).

Fire

As both instigators of fire and suppressors of fire, humans have had an enormous effect on forest vegetation. Fire has been a major anthropogenic factor in ecology since earliest times (Pyne 1982; Russell 1983; Stewart 1956). Understandably, the suppression of fire, particularly after the devastating and life-taking fires of Michigan and Peshtigo, Wisconsin in 1871, Michigan again in 1881, Hinckley, Minnesota in 1891, and Baudette, Minnesota and Idaho in 1910, was seen as the obvious way to impose management and control on the nation's forests and increase timber supply at a time when it was thought that the country was entering a period of "timber famine" at the turn of the century.

But as Ashley Schiff has pointed out in *Fire and Water: Scientific Heresy in the Forest Service* (1962), fire suppression became an orthodoxy or dogma on the part of the Forest Service that took no account of the pyrophytic nature of so many of America's trees. These pyrophytic species include ponderosa pine (*P. ponderosa*), Douglas-fir (*Pseudotsuga menziesii*), western white pine (*P. monticola*), longleaf pine (*P. palustris*), and sequoia (*Sequoia- dendron*) (Kozlowski and Ahlgren 1974; Brown and Davis 1973; Kilgore 1976).

At the other end of the scale, of course, the extensive areas of aspen and jack pine varieties throughout the states bordering the Great Lakes and parts of the Rockies are not natural types. These species are aggressive, post-fire reproducers and have been successful only because of human-made fire disasters in the past (Fig. 3.3). Similarly, in the South there was the observation of Mrs. Ellen Long as early as 1889 that wildfire was "...the prime cause and preserver of the *Pinus palustris* [longleaf] to be found there; but for the effects of these burnings...the maritime pine belt would soon disappear and give place to a jungle of hardwood and deciduous trees." Indeed, some 30 x 10^6 acres may have been so invaded by post-fire hardwoods. The composition of the nations's forests has changed permanently with the incidence of fire, and in so doing has deteriorated significantly.

Charcoal Making

One final example of human impacts on forest ecology is taken. The pre-coke iron smelting industry (that is to say roughly before 1855, though it was by no means dead as late as 1945) was highly localized by factors of production, such as supplies of ore, water for motive power and cooling, and of course, wood for charcoal. Wood for fuel was relatively abundant and plentiful in most eastern states; nevertheless, the demand for charcoal in the immediate vicinity of a furnace was enormous. Few would have disagreed with N.W. Lord in 1884 in his assessment of what was happening in the Hanging Rock district of southeastern Ohio, a major center of the nineteenth century iron industry when he said: "The disappearance of the forests under the demands of the furnaces, which is now so apparent throughout the region, increases every day the difficulty of obtaining the necessary fuel, and marks very plainly the fate of the iron industry" (Lord 1884). The effect of the furnaces on the immediate surrounding forests was enormous. A detailed study of 837 square miles of Jackson and Vinton counties in the Hanging Rock district, Ohio, shows that 60% of the forest was clear-cut between 1850 and 1860 down to four-inch diameter trees, and that forest regenerated sufficiently for recutting to be carried out during the early part of this century (Beatley 1953).

On the modest assumption that 150 acres of well-stocked woodland is sufficient to smelt 1000 tons of ore, then 4800 square miles of forest must have been felled between 1855 and 1910, which, if rotational practices incorporating regrowth are taken into account, falls to about 3000 square miles. Impressive as it is, however, it is a mere pin-prick in forest change compared to agricultural clearing. It is only 1.3% of the forest cleared for agriculture, and a mere 0.8% if the regrowth factor is taken into account. Locally, however, it was noticeable and assumed an importance as a destroyer of forest out of all proportion to the amount of forest affected. Agricultural clearing, on the other hand, was considered "natural" and part of "progress," and therefore, was rarely commented on.

One could go on taking examples of human impacts on forest size, density, and composition brought about by, for example, the grazing of stock in the forest, domestic fuelwood cutting (probably the biggest user of wood), and forest cutting and management techniques, the latter having been proved conclusively to affect composition stand and diversity (Franklin and Forman 1987). But probably enough has been said to suggest that the human factor must of necessity be included in any consideration of the continent's ecosystems.

"Another Landscape"

If Americans want to draw on a tradition of looking into the ways humans have entered into nature, they do not have to go many miles from the location of the Institute of Ecosystem Studies at Millbrook in upstate New York, the site of the Cary Conference that stimulated this book.

Perhaps the first real appreciation in the western world of the role of humans in the environment came with the writings of that versatile Vermonter, George Perkins Marsh. His book, *Man and Nature: or Physical Geography as Modified by Human Action*, which was published in 1864, was originally entitled, *Man, the Disturber of Nature's Harmonies*. The title worried his publisher in Scribners, who adhered to the age-old concept that humans were a part of a divine design, and anything they did was predetermined because they were partners with God in improving an earth created by Him for them. "Is it true?" he objected. "Does not man act in harmony with nature? And with her laws? Is he not part of nature?

"No," Marsh replied, "...nothing is further from my belief, that man is 'part of nature' or that his action is controlled by the laws of nature; in fact a leading spirit of the book is to enforce the opposite opinion, to illustrate the fact that man, so far from being...a soul-less, will-less automaton, is a free moral agent working independently of nature." (Quoted from Lowenthal's reprint edition of Marsh in 1965).

There are two ideas implicit here. First that nature left alone is largely in harmony and static, and second, that the actions of humans are not necessarily benign and part of some divine plan. Human actions could be deleterious. Way back in 1847 Marsh had seen the evils of excessive and "injudicious" clearing and burning of the forests, the damming of streams and even changes in precipitation in his native Vermont. He said then, "Every middle-aged man who revisits his birth place after a few years of absence, looks upon another landscape than that which formed the theatre of his youthful toils and pleasures" (Marsh 1848).

So Marsh ended his book with what he called "The Great Question"—"Whether man is of nature or above her," which was, incidentally, almost the identical phrase used by Tansley in 1935 when he coined the term ecosystem: "Is man part of nature or not?" Simply, the answer to that "Great Question" is the subject of this book.

Conclusion

In conclusion, there are a number of questions about the role of humans in ecosystems that should be addressed:

1) Was the Bergen symposium (Birks et al. 1988) correct in labeling the focus of their ecological studies a "cultural landscape"?

2) If so, has ecology in America, because of its origins and ethos, been guilty of ignoring the human factor?

3) If that is accepted, then should ecologists be collaborating with environmental historians, historical geographers, and the like, and be looking for "gradients of human impact" in nature? In other words, are there human effects in nature?

4) In those human effects, are there gradients of impact and are there some key quantifiable processes, such as removal, replacement, species alteration, species simplification, and the curtailing of regeneration? Also, are there some key methodologies? In addition, are the impacts indirect, lagged, or historical?

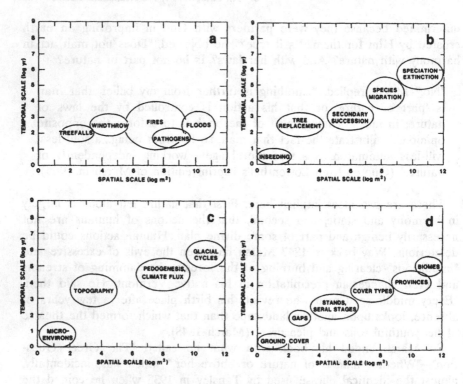

Figure 3.4. (a) Disturbance regimes; (b) forest processes; (c) environmental constraints; and (d) vegetation patterns, viewed in the context of space-time domains. From Urban et al. (1987).

5)   Finally, are these impacts sequential, complementary, or on some scale or gradient of increasing effect that parallels the intensification of human disturbance? Could they be plotted in something like the way Urban et al. (1987) used to order disturbance regimes, forest processes, environmental constraints, and vegetation patterns by scale and by time (Fig. 3.4)?

It is all too easy to ask such rhetorical questions, and difficult to answer them successfully, but certainly, the vast canvas of the American forest through time and space provides abundant material for the partial if not complete resolution of these questions and brings us face to face with humans as components of ecosystems.

Recommended Readings

Anderson, E. (1956). Man as a maker of new plants and new plant communities. In: W.L. Thomas, ed. *Man's Role in Changing the Face of the Earth*, pp. 763-777. University of Chicago Press, Chicago.

Delcourt, H.R. (1987). The impact of prehistoric agriculture and land occupation on natural vegetation. *Ecology* 34:341-46.

Marsh, G.P. (1864). *Man and Nature: Or Physical Geography as Modified by Human Action*. Scribner, New York. Reprinted 1965 with introduction by D. Lowenthal, Harvard University Press (Belknap Press), Cambridge, Massachusetts.

Williams, M. (1989). *Americans and Their Forests: A Historical Geography*. Cambridge University Press, New York.

Worster, D. (1977). *Nature's Economy: A History of Ecological Ideas*. Cambridge University Press, New York.

4
Environmental Change: The Human Factor

B.L. Turner II and William B. Meyer

A substantial literature from a variety of subfields that study the interaction of nature and society documents the human influence on the physical environments of the Earth. The evolving character of this influence can be traced through a series of major syntheses from *Man and Nature* (Marsh 1864) to *La biosphère* (Vernadsky 1929) to *Man's Role in Changing the Face of the Earth* (Thomas 1956) to the most recent effort, *The Earth as Transformed by Human Action* (Turner et al. 1990a). Virtually every corner of the biosphere, these works show, has been touched in some ways by human action. The long history of human occupance has left few, if any, ecosystems in a pristine or fully natural condition. So large has our overall impact been that we may speak of an Earth transformed.

In stressing this point, of course, one runs the risk of exaggerating the human impact. If ecology has tended to neglect subtle human influences, so have some conservationists tended to overstate them in a manner equally requiring correction. George Perkins Marsh, writing in the mid-nineteenth century, ascribed to the natural world a stability "which, without the action of man, would remain, with little fluctuation, for countless ages" (Marsh 1864). As a result, he interpreted most changes in the natural environment occurring on less than a geological time-scale as the result of human interference. Of course, such a sharp temporal division errs; natural and human forces operate in tandem and what or how much can be attributed to either is often difficult to determine. For example, some collapses of marine fish stock once blamed on overfishing are now reinterpreted as natural fluctuations (Hilborn 1990). Claims formerly taken for granted that deforestation has aggravated downstream flooding in the Ganges-Brahmaputra and Amazon river systems have recently been reexamined and disputed (Ives and Messerli 1989; Sternberg 1987). The imbalance in ecology away from the recognition and study of subtle human influence should be corrected, not reproduced in the opposite direction.

Humankind, as this volume emphasizes, leaves both obvious and subtle impacts on ecosystems. It does in places distant from, as well as close by, large human populations. In the totality of nature-society relations, the difference between the obvious and subtle effects is often an important one. Subtle impacts may escape attention until they reach a threshold that makes them obvious, by which time they may be far more difficult to manage (see Russell, Chapter 8, this volume). Effects that are subtle are often difficult to trace to their causes, and without demonstration of cause, repair or mitigation is unlikely. Human-induced global climatic change, for example, is or would be a subtle impact because it is so difficult to identify against a background of natural variation. With conclusive proof of human impact lacking, many are unwilling to address policies to counteract it. (On the other hand, not all human-induced environmental change need be viewed as undesirable or requiring mitigation, a point discussed later in the chapter (see also McDonnell and Pickett, Chapter 1, this volume.)

In the interest of clarity, researchers studying the human dimensions of environmental change separate—while linking—the questions of human impact, human perception of that impact, and human response to it. Our three realms overlap, needless to say, but they do mark out broadly distinct concerns. The focus here is on the first—the human activities that generate change (whether subtle or obvious). For this narrower purpose, no rigid distinction between subtle and non-subtle effects is needed.

In this chapter, the human influence on the environment and ecosystems is explored from several angles. The discussion is derived from studies of long-term nature-society relationships and of contemporary global environmental change. This exploration exposes several lessons that are important in understanding human impacts on ecosystems.

Historical Assessments

The ability of humankind to effect significant environmental change, especially at the ecosystem level, is ancient. For the most part, premodern human-induced changes were localized or regional in spatial extent (Thomas 1956), with several possible exceptions. There is some evidence that early hunter-gatherers eradicated the megafauna of the Western Hemisphere (Martin and Klein 1984), and burning and other flora-altering activities of paleolithic societies affected the continental ranges of biota (Sauer 1956, 1961). It has become apparent that early societies substantially altered the floristic composition of many biomes through subtle manipulations of land cover. Dense populations modified large areas of the Amazonian rainforest before the Columbian Encounter (Balée 1989; Roosevelt 1989). It is highly probable that many forests throughout the world were similarly affected to varying degrees by paleolithic societies.

An escalation of environmental change in the global aggregate began with the Columbian Encounter. The invasion by Europeans of the Western Hemisphere prefigured the subsequent, less dramatic incorporation of Africa, Australia, Oceania, and Asia into a world system dominated by the nations of the North Atlantic realm. The immediate ecosystemic effects of this explosion of transcontinental interactions were major biotic changes produced by both deliberate and inadvertent introduction of species (Crosby 1986), including enormous declines in human populations from "virgin soil" diseases (Denevan 1976; Whitmore 1991). The impacts wrought on biota and land cover by European expansion are well known. In the Americas, besides the deliberate transfers and transformations, they include: the invasion of new and sometimes continental-scale areas by Old World weeds; the ecosystemic impacts of the introduction of livestock that accelerated erosion on slopes, promoted the extensive use of arid and semi-arid areas, and ultimately led to the replacement of forest and savanna with African grasses suitable for these domesticates (Parsons 1972); the lessened use and even abandonment of large areas depopulated by the collapse in Amerindian numbers, resulting, for example, in the afforestation of *tierra caliente* throughout much of tropical America (Turner and Butzer 1992); and the preferences of the invading European population for particular environments and particular ways of using them.

The processes of globalization that assumed their modern shape and scale in the Columbian Encounter led to impacts in Europe as well. Food production, for example, the introduction of the potato (*Solanum tuberosum*) and maize (*Zea mays*), in particular altered land uses and supported increased populations. The wealth that Europe accumulated, moreover, was a contributing factor—one of many—to the process of industrialization. Fossil-fuel based industry has forever altered the scale and kind of environmental change that humankind can effect, leading to substantial impacts on the energy and material flows that sustain the biosphere (Kates et al. 1990).

A study of modern impacts covering the past 300 years tends to emphasize the Industrial Revolution as the principal watershed of change in the global nature-society relationship; one covering the past five centuries, on the other hand, underlines the landscape-type changes typified in the Columbian Encounter (Adams 1990). Whichever of the two one chooses to stress—and it is clear that they were closely interrelated—their results today are evident in the rise of mass-consuming societies whose reach in affecting the global environment in general and specific ecosystems in particular is enormous. Actual and threatened species extinction has gone from being largely an island phenomenon to one now affecting continental biota (Peters and Lovejoy 1990; Prance 1990). Forests in remote places are felled to provide wood and wood products for distant consumers, and monolithic plantations displace diverse patterns of land cover. Of the net human clearance of forested area since postglacial times, half occurred by the middle of the nineteenth century and half has occurred since (Kates et al.

1990)—a fact that suggests the long-term character of human impact on Earth's environments, but also testifies to the accelerated pace of modern human-induced change.

These examples suggest that environmental change has escalated as human numbers have grown and other aspects of human society have changed. They may hint at more of the same as global numbers continue to climb. The doomsday implications of modern population growth, however, are not necessarily supported by studies of population-environmental relationships. What happens if the examination is conducted at a smaller spatial scale: if place is held constant and the long-term relationships of society and nature are traced throughout its history? (see Richerson, Chapter 11, this volume). One such assessment compared the Tigris-Euphrates lowlands, the Egyptian Nile Valley, the Basin of Mexico, and the central Maya lowlands of Mexico and Guatemala at the millennial time scale (Whitmore et al. 1990). Some interesting insights emerge from this and related studies:

1) Dense settlement and long-term high rates of population growth are not exclusively recent phenomena, at least at the regional level, and the human transformation of regional environments is of equal antiquity;

2) For the most part, premodern changes involved the landscape elements basic to the agricultural resource base (soil, vegetation, surface water, and so forth), and they often substantially altered these elements;

3) Nevertheless, over millennia of intensive use of the regions, sustained occupation did not necessarily lead to persistent degradation of the environment to the point of land abandonment, and the sustained use and/or recovery of land long cultivated and highly altered was not uncommon;

4) Human populations have experienced periods of sharp decline as well as sustained growth in all of the regions. In the extreme case of the central Maya lowlands, the highest regional population was attained by A.D. 800, a figure several dozen times that of today;

5) Recovery of the environment following population loss rarely involved a return to the system that had existed prior to human use. For example, though the deciduous tropical forests of the central Maya lowlands returned upon abandonment after years of clearance and intensive land use, the relative abundance of species (e.g., *Brosimum alicastrum*) apparently changed; and

6) While population increase through time, regardless of the technological base and social organization, invariably led to major ecosystemic changes, it is not at all clear that the transformation and biotic simplification of the environment translated into socioeconomic deterioration.

Such lessons from the long term are important to consider in assessing the human impacts on ecosystems, subtle or not, in populated areas. Again, though, an important caveat must be entered. Contemporary environmental change at a global level and in most regions and locales of the world is novel in pace, magnitude, and kind. The study of global change has enlarged the scope of environmental impacts from a set that can be seen as a cumulative kind—localized change that attains a global magnitude through widespread occurrences—to include a complementary set of a systemic kind: impacts on realms of the biosphere, such as the atmosphere, climate, and oceans, that operate as fluid, connected global systems (Turner et al. 1990b). Owing to the worldwide reach and implications of these kinds of change, they have focused attention on the global scale. Nevertheless, they also have both direct and indirect impacts at the ecosystem scale, and, therefore, should be considered in the study of human-induced changes in ecosystems.

The Human Causes of Environmental Change

The sources of human impact on the environment can be traced along a chain from proximate sources through goals to driving forces (Turner 1989; see also Clark 1988). Proximate sources are the human actions that directly affect the physical environment, such as forest cutting and burning, release of pollutants, and transfer of species. The goals of change are the immediate purposes for which those actions are taken. Forest is cleared for such varied purposes as cultivation, fuelwood and timber collection, settlement, and hydroelectric power generation; it may also be affected by proximate sources of air pollution or fire that are generated by yet other goals. The driving forces of change, finally, are those underlying aspects of human society from which the goals are derived and by which they are defined as possible and desirable.

It is at the level of driving forces that contention begins to build among social scientists as to the prime cause of change or the conceptual ordering of interacting factors (Stern et al. 1992; Turner and Meyer 1991). The reasons for disagreement are several. Conflicting perspectives within the social sciences diverge over the very question of what constitutes evidence, demonstration, or proof (Sack 1990; Turner 1991). Even where agreement on this point exists, moreover, little empirical work has been done on a global aggregate or comparative level to support the universal primacy or significance of any one candidate force; indeed, the evidence, such as it is, suggests that most cause-impact relationships vary by context (see Vayda, Chapter 6, this volume). Strong theoretical arguments have been advanced for the worldwide dominance or, alternatively, insignificance of a number of candidate forces. The search for pan-global cause-impact relationships will undoubtedly be stimulated by the current interest in global environmental change. Different factors, however, seem to differ in importance according to the social and environmental settings in which they operate, and recent global-regional assessments indicate that the global patterns of impact are often poorly mirrored at the regional scale (Kates et al. 1990).

These qualifications notwithstanding, various efforts are underway to determine how the major human driving forces of (global) environmental change—singly or in combination—contribute to the goals and proximate sources of environmental change. The candidate forces are controversial. Some investigators have narrowed the categories of driving forces to three: population, production, and consumption, a scheme resembling the "$I = PAT$" equation of Commoner (1972) and Ehrlich and Ehrlich (1990), in which human impact (I) is a function of population (P), affluence (A), and the technology (T). Others view these categories, with the exception of population, as simply too diffuse to be meaningful; the relation of changes in population, production, and consumption (however specified) to environmental change is a truism. The important issue is that of the forces driving these changes themselves: of *why* levels of resource demand and waste emissions rise or fall. Shifts in production and consumption are less the explanations of environmental change than they are themselves among the things to be explained. The causes of these shifts, in this view, are to be sought in the categories detailed below: population growth; state of economic development; technological change; social organization; and beliefs and attitudes. Population change is included for the reasons noted.

Population growth, expounded most recently and thoroughly by Ehrlich and Ehrlich (1990), refers to increases in the human numbers dependent on the resources of the Earth's surface or portions thereof. Such increases both raise the pressure on resources to satisfy basic biological needs and magnify the effect of increases in per capita consumption. Population growth is undeniably important in driving environmental change, but it is not the only or necessarily even the principal factor. As a rule of thumb, increasing numbers, *ceteris paribus*, lead to increasing environmental impacts. It must be borne in mind, however, that population pressures are mediated by other broad socioeconomic and cultural conditions, such that a given level of population does not always result in the same level or kind of environmental impacts. Population pressures interact with and are modified by a range of other social factors that affect the degree of pressure exerted on resources.

The population change-forest conversion relationship is a case in point. It is commonly held that population growth is a, or the, key factor in forest clearance, especially in the tropical Third World. Aggregate statistical studies lend some support to the contention (Allen and Barnes 1985; Rudel 1989). Closer examination, however, indicates that the matter is complex. Deforestation or reforestation may occur in areas experiencing either growth or decline in population, for example, and within any regional economy, population-forest relationships vary further depending on the forest ecosystems involved (Young et al. 1990). Studies of Southeast Asia demonstrate that logging by international operations, not clearance by poor farmers, is the proximate source of much of the rapid deforestation that is going on in the region (Brookfield and Byron 1990). Such is apparently the case even in the Philippines where rural population growth is high; there the timber industry deforests regardless of local population pressures, though this logging may then make farming a more attractive option for the landless who colonize the newly cleared land rather than move to the cities (Kummer 1992). Of course, without the population pressures, or with a higher level of economic development, the denuded land might be left alone and reforestation take place. International market demand from distant populations, too, contributes to clearance, but to ascribe all of the clearance to market forces is simplistic, for government activities such as concessions and subsidies in Southeast Asia have, as in many countries, distorted market incentives and promoted more clearance than would be efficient or socially desirable (Boado 1988; Gillis 1988a, 1988b).

Population pressures, in short, occur within a social context shaped by other factors that can act—interactively or independently—as driving forces of environmental change. Differences in the level and the distribution within society of wealth and per capita resource consumption characterize different states of *economic development*. Arguments that invoke these variables (see, e.g., Bienen and Leonard 1985; Hecht and Cockburn 1989; Sack 1990; WCED 1987) trace the association of development and environmental impact in various directions. Affluent, mass-consuming societies may generate high resource demands, but may also be able to afford measures for the mitigation of environmental damage. Impoverished nations may have relatively low per capita demands, but may be forced into unsustainable patterns of resource use for survival.

Technological change refers to the capacity to extract, process, and deliver resources, and to deal with the wastes of consumption. Change in technological capacity leads to both obvious effects (the ability to produce goods at lower costs or exploit resources once uneconomical) and subtle ones (incidental effects). Technological change in itself may be more important in accounting for net global change than in explaining differences in impact between different regions of the Earth. To a great extent, innovations are not fully independent variables in accounting for environmental change; the innovation is necessary but not sufficient. Its differential adoption across the Earth's surface reflects differences in who can afford to adopt it. It is perhaps the side-effects of technologies—the differing environmental consequences of fossil fuel vs. nuclear power wastes, for example—that are of the greatest significance here. For discussions of technological change as a driving and mitigating force, see Ruttan (1971); Commoner (1972); and Ausubel and Sladovich (1989).

Social organization is perhaps the broadest category of candidates for driving forces. Its most prominent themes deal with the political-economic structures of ownership and distribution (see Hall, Chapter 5, this volume). Orthodox economic theory identifies structures that distort the free-workings of market allocations as significant driving forces. When institutions fail to allocate costs and benefits in an efficient manner, resources can be wasted and pollutants emitted to excess, causing environmental degradation.

Some criticisms of this approach identify the operation, rather than the failure, of free markets—through private ownership or the commodification of resources—as the prime culprit, encouraging degradation through rapid depletion and unsustainable use geared to immediate profits; others claim that market failure is so pervasive in the envionmental realm as to make orthodox economic approaches unrealistic and misleading. For discussions of these issues in the context of biodiversity, see Hanemann (1988) and Randall (1988).

More concretely, social organization as a cause of environmental and ecosystem change is illustrated in the following examples:

1) Degradation of marginal lands inadequately managed by the displaced poor who have lost their former holdings through government policy or changes in land tenure rules;

2) Clear cutting of species-rich forests, driven by international market demands and assisted by local government permits and taxation of forest land;

3) Forest damage from acid rain, where the spatial partitioning of authority prevents the "receiving" state from enforcing controls in the "producing" state; and

4) Contamination of coastal estuaries and wetlands by waste dumped into drainage systems because of free access to non-privately owned and poorly regulated disposal areas.

Finally, we adopt a distinction between *beliefs*, correct or mistaken, about the environment and how it operates, and *attitudes* or cultural valuations, either directly regarding the environment or of importance in shaping environmental behavior (R.C. Mitchell, pers. comm.). Both are sometimes cited as driving forces of change. Erroneous beliefs regarding the richness of tropical rainforest soils under permanent cultivation, it is sometimes claimed, help to drive excessive clearing (e.g., Guppy 1984). Attitudes that accord cattle ranching high prestige in Central American societies have likewise been ascribed part of the responsibility for a degree of forest conversion to pasture exceeding what the market would dictate (Myers and Tucker 1987); other attitudes cause grass to be maintained on lawns, cemeteries, and other such locales in water short areas (e.g., the American Southwest).

The impacts of these driving forces depend not only on social but on environmental context. Certain environments—mountains, islands, tropical rain forests, and the high latitudes are often cited as examples—are more fragile than others, and they may be more vulnerable than would different zones to given combinations of driving forces. The impacts of driving forces may also differ by type of environmental change; the principal driving forces of deforestation may not be those of trace-metal pollution.

The investigation of driving force-environmental change relationships is further complicated by the fact that changes are less confined than ever to the immediate vicinity of the agents—whether proximate sources or driving forces—that are ultimately responsible for them. Two kinds of transfer of change occur. One is transfer of impacts through a physical system: acid rain falling hundreds of kilometers from industrial regions, or water pollution transferred downstream through river networks. Globally systemic environmental changes represent the extreme form of spatial transfer, where localized changes alter fluid, globally functioning systems (e.g., the Antarctic ozone hole generated by CFCs principally emitted in the Northern Hemisphere; globally uniform increases in tropospheric CO_2 due to developed-world industry and to tropical deforestation; eustatic sea-level rise brought on by global climate change).

The second, more complex form is transfer of impact through social processes. Here the changes occur as a result of decisions or demands from elsewhere; for example, Japan successfully reforests much of its area while Japanese companies clear forests in Southeast Asia to satisfy domestic demand. Such examples are less straightforward cases of spatial transfer than those occurring through purely physical media. External demand is necessary, but it may not be sufficient; the consent to the transaction of agents in both source and receiving regions may be required. Clearance by Japanese companies degrades or devalues forest environments in Southeast Asia only because the host governments permit it. Conditions in both places thus function as driving forces, further complicating the issue of causation.

A final point of importance is the distinction between environmental change and environmental degradation wrought by the driving forces. "Degradation" as an evaluative term is necessarily one of human definition, even if human actors define the term to refer to any change from the natural state of an ecosystem. If the term is restricted to environmental changes that (over some time scale) cause a decline in the wealth and well-being of the human population, then much environmental change, past and present, may not constitute degradation. From this point of view, the impacts of population growth (holding other factors equal) or of increase in any of the other driving forces may be positive ones. In some cases, inputs generated by population growth increase the output of the system, though altering it still further from its original condition, and population decline can lead to the abandonment of productive and conservationist, if highly "artificial," systems that required large inputs of labor. Of course, much and perhaps most contemporary environmental change may represent degradation even by an anthropocentric standard. That conclusion, however, must be reached by investigation rather than assumed as a given. Human responses to environmental change, in any case, will be strongly affected by the perception of the extent to which changes are damaging or innocuous to society.

From Human Dimensions to Ecosystem Impacts: Lessons

This brief discussion has several simple messages for understanding human impacts on the environment and on particular ecosystems. Most environments and environmental conditions have been altered, many even transformed, by human action. The landscape elements of the environment—vegetation cover, fauna, soil, water—have a particularly long history of alteration. That history is so ancient and subtle in many instances that it may not be recognized unless the scientist deliberately looks for it—though it may also be falsely imagined if the investigator looks too hard for it. Despite the antiquity and increasing ubiquity of human use, the capacity of the environment to recover—albeit altered—or to sustain human occupance, even in larger numbers at higher standards of living, has been the norm. To make this generalization is not necessarily to uphold a "cornucopian" view of modern nature-society relationships, however, because not only is environmental change more intense today than in the past, but it also has expanded to affect the basic biogeochemical flows of the earth.

Varied human causes drive environmental and ecosystem change. These causes generally interact in a systemic way and rarely operate independently of one another. The impacts of these interactions vary by socioeconomic and environmental context, and it is increasingly common for the source and the impact of change to be distant from one another. Following from this last point, a methodological lesson for both natural and social scientists relates to the scale of inquiry. The ecosystem approach, valuable in stressing the importance of local complexity and interconnections, tends by the same token, toward the micro-scale in perspective and toward emphasis on the individuality of the unit and of the processes under examination. Such a focus may obscure the importance of meso- and macro-scale influences just as one on individual social systems—be they firms, communities, or even nation-states—may hide broader processes in the area of human driving forces. Unresolved spatial-scale issues rank among the principal challenges to the study of environmental and ecosystem change.

Recommended Readings

Clark, W.C., coordinator. (1988). The human dimensions of global change. In: Committee on Global Change. *Toward an Understanding of Global Change*, pp. 134-200. National Academy Press, Washington, D.C.

Stern, P.C., O.R. Young, and D. Druckman, eds. (1992). *Global Environmental Change: Understanding the Human Dimensions*. National Academy of Sciences Press, Washington, D.C.

Thomas, W.L., Jr., ed. (1956). *Man's Role in Changing the Face of the Earth*. University of Chicago Press, Chicago.

Turner, B.L., II, W.C. Clark, R.W. Kates, J.F. Richards, J.T. Mathews, and W.B. Meyer, eds. (1990a). *The Earth as Transformed by Human Action*. Cambridge University Press, Cambridge and New York.

Turner, B.L., II and W.B. Meyer. (1991). Land use and land cover in global environmental change: considerations for study. *Int. Soc. Sci. J.* 130:669-679.

5
The Iceberg and the Titanic: Human Economic Behavior in Ecological Models

Jane V. Hall

Introduction

In the long run the planet has the upper hand, in the short run humans act as if they do and as if this will continue to be the case. This aspect of human behavior is manifested in economic institutions and policies, and in individual responses to economic signals. The implications for how human economic behavior can be incorporated in ecosystem research are founded in the nature of economic models and how they are used to formulate policy.

The title of this chapter reflects the author's perception that the body politic often founds decisions on the seemingly-precise prognostications of economic models, finding comfort in estimates carried out to the third (or greater) decimal. Reliance on guidance from these models in turn generates policies (and consequently affects human behavior and choice) based on the assumptions of the models that produce the estimates. The assumptions, therefore, become embedded in the behavior they are used to predict—in short, a classic self-fulfilling prophecy. This sort of collective social choice suggests a nation that regards itself as being as impregnable as was the iceberg, blithely assuming that any large clouds on the horizon are really only on the scale of the Titanic and that, in any event, ordinary economic decisions will correct the steering before any solid object is encountered.

Examples of this behavior abound. From the recent cover of *Time* asking whether we can do without nuclear power to the on-going debate over the National Acid Precipitation Assessment Program (NAPAP), we see a willing view of the growth economy as an iceberg, when it ultimately is not even on the scale of the Titanic, but more like a lifeboat, fragile and adrift among the complex biological and geochemical systems that typical econo n-ic assumptions and models treat so cavalierly.

51

Given this not altogether optimistic view of how economic understanding drives some basic policies which *in turn* introduce the human factor in ecosystems in new ways (for example, construction of a new highway opens up undeveloped land to housing in ways and on a scale that the choice of a rail system would not), how can an understanding of economics enable scientists to better incorporate the human factor in basic ecosystem studies?

The answer rests on the most basic concepts and assumptions of general economics, observations from the fields of natural resource and environmental economics, and the degree to which these merely reflect dominant social values.

Basic Tenets of Economics

The majority of economists argue that humans react in specific and predictable ways to economic signals, in a mechanistic scheme often referred to as the price system. In our economy we do rely very heavily on prices to convey tremendously complex information in a very distilled form. The most fundamental tenets are the following:

Optimization. Each individual pursues his own self-interest in a way that maximizes his welfare. By this atomistic process, the system is also optimized, in the sense that no one can be made better off without making someone else worse off. When all producers and consumers follow this rule, scarce resources are allocated in the most efficient way. Two important assumptions that are embedded in most economic models fall out of this. One is that the pursuit of individual self-interest leads naturally and automatically to the best possible outcome for the system, given resource constraints. This then becomes a matter of definition; if individuals are free to pursue their self-interest and do so, the outcome is optimal. (This is also known as Reaganomics.) The other is that maximizing is equivalent to optimizing. Consumers maximize satisfaction, producers maximize profits which means minimizing costs. There is a corollary—the impact of an individual's choices (good or ill) falls on the individual—all costs and benefits of a choice are internal to the decision-maker.

Value. What something is worth is measured by how highly it is valued by humans, and generally only those values that can be monetized are really recognized. Hence the scramble to put a price tag on the ecosystems of Prince William Sound in the aftermath of the Exxon Valdez oil spill. This anthropocentric approach to determine value has been called "economic imperialism" by Herman Daly, but it is the basis for our price system of exchange and, therefore, for production and consumption. Beyond its anthropocentrism, this approach also places greatest weight on the values held by those who own or manage assets in the current time period.   By comparing everything in present value terms, the future is discounted to obscurity.

Rationality. We assume all players in this game know their own self-interests, now and in the future, and the best way to successfully pursue them, i.e., to maximize given prices and their private resource constraints. A corollary is that information must be complete, accurate, and fully understood so that all implications of a choice are clear to the individuals.

Marginalism. All of the infinite number of decisions that individuals must make in this rational optimizing process takes place at the margin. The consequences of previous choices are history, and the next decision is forward looking. In short, what's past is past, not prologue. Lest this look like a completely absurd view of the world, economics also assumes that past decisions do enter into future ones in the sense that relative prices change in response to collective market forces. Past decisions to invest, produce, or consume will, therefore, influence current and future prices. The decision to fill your gas tank today is based only on the prospective utility to you of overcoming the inertia of your vehicle, but the collective impact of all of our past decisions about consuming and producing fuels is reflected in current prices and is, therefore, not completely out of the picture. If past choices lead to a change in relative prices, your current (but still marginal) choice will be affected, otherwise it will not. The aggregate impact of all of our marginal choices and the cumulative impact over time can be enormous, but the margin is what counts. This situation sends this message: don't worry about the aggregate in any physical sense. If it is important in terms of human welfare, as measured by value embedded in prices, prices will change and influence the marginal choice appropriately.

Impersonal transactions. All of this works through a complex scheme of production, consumption, specialization, and exchange. Most transactions are at a very distant arm's-length. When I buy an orange, I am nowhere near the grove where the orange was produced and hundreds of miles from the river that was diverted to irrigate the grove. I won't buy the orange directly from the grower or pick it directly from the tree. Our economic system is based on billions of such daily transactions. The implications of this, ecologically, are perhaps reflected best in the recent Goldman Awards, including one to the man who filmed dolphins drowning in tuna nets. That visual report very quickly reduced the arm's-length relationship between the fishery and the folks buying canned tuna.

Continuing and unlimited growth. Most economists, and our political system, at virtually all levels of government, assume that economic growth, and both the associated increased rate of resource mobilization and increased pressure on natural sinks and other species, is inevitable and unlimited (Costanza 1989). The process of substituting away from economically scarcer resources to less scarce ones is assumed to be unlimited. New deposits will be discovered, better technologies will be invented, an altogether different material will be found or developed to do the same thing, etc.

This is politically seductive because it implies that even a larger population can (will) always be materially better off than its antecedents were as long as growth is sustained. It also appeals to economists who would otherwise have to face the messy issues of equity, distribution, and anthropocentric values. It is not a fluke that many winners of the Nobel Prize for economics were recognized for their work in championing the virtues and possibility of perpetual growth or demonstrating the mathematical mechanisms by which it can take place (Friedman, Samuelson, and Leontief, to name a few).

Concepts from Environmental and Natural Resource Economics

As human activity increased, along with population redistribution and growth, and industrialization, the mobilization of resources became truly immense and the residuals from this process began to overwhelm natural sinks. This eventually led economists (trailing along after the public health professions) to begin asking how it was that beneficial economic activity could have such unforseen adverse effects and what to do about it. Many of the ideas that emerged are couched in terms of "market failure," i.e., why did prices fail to signal these effects so that individuals would alter their behavior to reduce residuals to the appropriate (optimal) level? Questions about preservation of un-priced or under-priced non-market resources such as species diversity were approached later in adaptations of the same paradigm. The basic concepts follow.

Externalities

It turns out that, in violation of one of the basic tenets, the consequences of individual choice do not all fall on the individual. Say that I decide to cut down a tree and pay someone $100 to do it. The price to me is $100. The cost includes the $100, but also the loss of habitat for birds and other creatures, soil conservation, shade, reduced heat island effect (this is an urban tree), and the aesthetic virtues the tree held for neighbors and passers-by. When I decide I am willing to pay $100 to remove the tree, I am not considering in my decision these external costs that fall on others. It may be optimizing behavior for me, but it won't be optimal for society. This is a small example, but one that is close to home. This is the classic explanation for pollution. Bodies of water and air are often commonly held and will be over-exploited for waste reception because none of us pays a price for their use equal to the damage that our use does. This kind of case was eloquently explained by Hardin (1968). Absent proscriptive regulation, the price to dump industrial waste in a river is zero; the case is similar with clean but warm water from a power plant outfall into a body of water. Ecosystem changes are external to the decision maker because they do not impact him directly in any material way and they are un-priced and, therefore, not part of his decision. A firm that is maximizing profit will select the least costly production process, as measured in prices that must be paid. There is, therefore, a tendency toward processes where costs are externalized. There is no self-correcting tendency in this case because there is no feedback mechanism, for example, a stream becomes so polluted that it can no longer be used for industrial processing, as was the case in the Ruhr Basin some years ago. The un-priced loss of soil microbes, urban airshed, habitat, CO_2 sinks, etc., continues because they are outside of the price system.

Public Goods

These are the flip side of externalities. They are *under* produced because the benefits of their production cannot be entirely captured by their producer(s). If available to anyone, they are available to many, without exclusion, and they are non-rival in consumption (your use does not affect mine). The traditional example is a lighthouse. Any individual who takes action to "produce" or protect such an asset knows that others will also benefit while they bear all of the costs. Other examples include clean air and species preservation. This is also referred to as the free-rider effect. The result is underproduction, stalwart altruists being the exception, not the rule.

Un-priced and Under-priced Goods

Goods or services that have value but that are un-priced will be overused and underproduced. This is really a generalization of the first two concepts. In essence, such goods and services are partially or completely outside of the price system that makes our economic world go around and will not be accounted for by the only feedback mechanism recognized by most economic models—market prices. Many, or most, of the biological and geochemical foundations of individual ecosystems are un-priced. On the grandest scale, the global ecosystem is un-priced. Narrowing it down, the complex mechanisms and sinks that "produce" stable climate (in terms of human time) are un-priced. Narrowing it again, the value of expanded wetlands is at best under-priced, except perhaps by duck clubs.

Scarcity

It is not traditional to include scarcity *per se* as a notion from the environmental and natural resource sub-disciplines of economics. Perversely, it is the recognition that all resources are scarce and that human wants can, therefore, not all be met, that drives the theory that scarcity can best be ameliorated by individual responses to price signals. Changes in relative prices then, at least in theory, signal increasing or decreasing scarcity. It is natural resource and environmental economists, however, who have focused debate on whether or not we in fact have any adequate methods to measure the degree of scarcity of natural resources and environmental services and what the implications may be of any inadequacies. Recent mathematical and conceptual constructs (Hall and Hall 1984; Hall et al. 1992) have demonstrated that increasing scarcity is not always reflected in prices and that other measures are not always good indicators either. The manifestation of scarcity depends on the way in which a resource becomes scarce and on the physical nature of the resource itself. This is in part because un-priced or under-priced goods and services are critical inputs to production and processing of the natural resources we use, and the value of such inputs is, therefore, not reflected in prices of the resources whose production and use they support. Consequently, the price of the resource could actually fall, indicating *decreased* scarcity, even as the most essential inputs to its production and processing became more scarce. Consider the case of a ton of coal mined in Wyoming. Looking at the price trend for that coal will tell you nothing, or in fact mislead you, about the economic scarcity of Wyoming coal. The value of grassland habitat lost to strip mining will be under-priced since reclamation requirements will not restore it fully, the water used is under-priced because of federal subsidies to water projects, the air and water quality impacts of the mining operation will be un-priced, and so on. So, as clean air and water, and grassland become scarcer in Wyoming, the price of coal will not reflect this and coal will be assumed to be no more scarce, or even less scarce while the associated support structure necessary to mine it is rapidly depleted. This logic can be extrapolated to consider what happened to California agriculture during the 1987-1992 drought, etc.

Incorporating These Concepts in Economic Analysis

Economists, in concert with federal agencies such as the Environmental Protection Agency and the Department of the Interior, have put tremendous effort into trying to determine how to incorporate un-priced and under-priced environmental and natural resource assets into traditional economic models such as cost-benefit analyses. They do this, for example, by trying to elicit information about what people are willing to pay to preserve, protect, or enhance some environmental characteristic or factor. The logic of this is that market prices actually paid should reflect what society is willing to pay for a product because that is what is actually paid. So, if you can get people to reveal what a non-priced "product" is worth, you have a surrogate for price, and the economic models will work just fine.

Most of what is now traditional environmental economic analysis is aimed at forcing ecological reality into fairly standard economic models with the inevitable conclusion that once correct price signals flow through the system, optimal use of environmental services and natural resources will result. This comes about either by changing individual behavior directly, or changing it indirectly through legislation and regulation that explicitly correct for these problems by mandating or limiting particular choices. Examples include the average vehicle fuel economy standards and the prohibition on dumping used motor oil in a storm drain.

There are economists who propose, and are developing, measures of value that explicitly incorporate the effects of changing natural assets. (See, for example, Pearce et al. 1989). Price can be a complex concept involving both monetary and physical measures of value. This approach falls under the general rubric of environmental accounting and is a promising concept within the broader field of sustainable development. Ultimately, the influence of resource availability and quality (including environmental amenities) on an economic measure of welfare—sustainable income—would be measured, producing a clear link between ecological and economic models.

Human Behavior in Ecological Models

The similarities between economic and ecological models have been elucidated elsewhere (see Rapport and Turner 1977) and will not be dwelt on here. Suffice it to say that there are parallels in consumption, production, specialization, and exchange. What is more crucial to the issue of how the human factor can be accounted for in basic ecosystem research is how the models are different. The models reflect reality as it is understood by the respective practitioners and, therefore, basic differences must be taken into account. Economic models can be taken as reflecting the significant dominant human values of our society, especially as expressed in the public policies that manage our common non-market resources. Such policies in turn influence human behavior, in essence creating reality. Humans are, in fact, inseparable from nature, but dominant western culture and its social and economic paradigms miss this point (White 1967).

This is where the metaphor of the iceberg and the Titanic comes in. The iceberg, largely submerged and out of sight, is a threat to the existence of the Titanic which looms large. Our industrialized economy and the human behavior that drives it are perceived to be the iceberg. The impacts of economic activity and growth on the essential natural systems that the economy depends on are submerged.

Economic models work on the tenets set forth earlier and, to the extent to which these models either accurately reflect values or determine behavior because of their use in policy formation, these tenets become reality. Understanding the human factor is, therefore, based on assumptions that individual maximization optimizes the system, that the past is past, that very simple signals, i.e., prices, encapsulating extraordinarily complex information are sufficient to maintain this optimization process, that human values are captured in such signals which are adequate measures of value, that each economic actor knows what he is doing, and that growth is inevitable, unlimited, and desirable. Many of these assumptions run directly counter to observations from ecological models. Individual maximization will not likely optimize a natural system (Norgaard 1987). Optimizing and maximizing are not the same thing (the mockingbirds who have converted my shrub to a nursery are not always building bigger nests). Growth within a system is not unlimited. Feedback mechanisms rely on multiple kinds of signals. Interactions are often quite direct. Everything adds up. And so on.

Perhaps most fundamentally, economists assume that the basic concepts can be generalized over the human universe, with no consideration of context (see Vayda, Chapter 6, this volume). The outcomes will differ slightly with culture and resource availability, but the assumptions, mechanisms, and objectives are the same. This is in diametric opposition to ecologists' discovery that context is crucial and even basic concepts may not travel well (diversity begets stability, for example) (Norgaard 1989).

As our political-economic structure has developed, these assumptions embedded in our behavioral models have almost become articles of faith to the body politic. Econometric modeling is more and more relied upon to provide policy guidance on everything from the level of optimal taxation to the value of the Valdez oil spill. Consequently, since large public expenditures, along with an array of taxes, subsidies, and direct regulations, result from such advice and since these resulting policies themselves are determinants of economic behavior, the behavior predicted by the models becomes a self-fulfilling, self-perpetuating prophecy. At this point, whether the assumptions are logical or empirically verifiable becomes irrelevant because behavior is altered by the predictions, even if the assumptions are invalid.

This is why the human factor, at least as represented by economic behavior predicted or measured by economic models, should for now be treated as an independent variable in ecological models. This might not remain true and will not be true in all cases. One can easily find examples of local communities making political choices that diverge from what the models recommend, but this is not easy to find at the national level. We have set ourselves up in the role of the iceberg, the values embedded in the assumptions of the model drive policy and, therefore, behavior, and so we drift along. This is undoubtedly entirely unrealistic as a sustainable role. Any reasonable view of the First and Second Laws of Thermodynamics makes it transparently clear that this cannot go on indefinitely. Yet, the assumption that rising prices signal scarcity, thus creating an incentive to invent alternatives and to substitute something less scarce, underlies the notion that economic growth is a concrete example of the realness of perpetual motion.

As long as policy-makers rely on the illusory precision of economic models to guide them, and society continues to view maximization of material goods as desirable, human economic behavior should realistically be treated as an independent variable in ecological models. This will be true until values change, non-price signals are conveyed in a meaningful way, or the iceberg and the Titanic swap roles—that is when some catastrophic ecological threshold is reached.

Implications for Ecological Models

First of all, the picture is not entirely as bleak as it might appear. There are many excellent economists working hard to point out the fallacies of the dominant economic thinking and to construct useful new paradigms (Norgaard 1987; Boulding 1966; Daly 1984 among others). Others build models that reach policy conclusions at odds with the results of the dominant models, leading to gradual acceptance of alternative views and assumptions, ultimately changing policies and thence behavior (Chapman and Drennan 1990; Pearce et al. 1989; Fisher and Hanneman 1986; Hall 1990; Hall et al. 1992 among many). It is noteworthy that virtually all of the economists in this cadre have embraced direct involvement with their colleagues in the natural, physical, and other social sciences and that they rely explicitly on integrated understanding of the social, biological, and physical systems of the problems they study.

Effective change often takes place first at the local level where much ecosystem research is carried out. Consequently, the most useful way to incorporate the human economic factor into ecosystem studies is to ask what human behavior is predicted to be (housing development, road construction, water diversion, whatever) and then analyze how that predicted behavior will alter the path of an ecosystem's development either directly or indirectly.

The importance to the system of the differences made by the human element can then be evaluated. This in turn ties back to the evolving field of environmental accounting. For example, I tell you that the human population in a region will increase by 35%, leading to a 40% increase in demand for park services. You then tell me what this implies for the integrity of an urban fringe wilderness park. Between us, we know a lot about what the changing human factor means about the quality of the area

60   Jane V. Hall

and the future usefulness of the park to humans as wilderness. Then we know something that can alter human behavior as this information becomes feedback to these humans, probably via the political system. Ehrlich (1981) identifies this as a dual task for ecologists; to predict the consequences of various courses of action and to communicate the results to the public. Without the human factor, the results won't mean much.

So, intellectually and in concrete cases, change in economic thinking is taking place, however slowly. This is a propitious time to begin developing ways to explicitly incorporate the economic manifestations of human behavior into ecological models.

Recommended Readings

Costanza, R., ed. (1991). *Ecological Economics: The Science and Management of Sustainability*. Columbia University Press, New York.

Cropper, M. and W. Oates. (1992). Environmental economics: a survey. *J. Econ. Lit.* 30:675-740.

Daly, H. (1991). *Steady State Economics*. Island Press, Washington, D.C.

Pearce, D. and R. Turner. (1990). *Economics of Natural Resources and the Environment*. Johns Hopkins Press, Baltimore.

6
Ecosystems and Human Actions

Andrew P. Vayda

A central purpose of this book is to provide a forum on the question of "how human influences can be better incorporated into ecological studies." In this chapter, the answer that I want to give to the question is, very simply, that we need to make concrete human actions and their concrete environmental effects our primary objects of study, and that we need to proceed in our research by progressively relating these to factors that can explain them, without *a priori* constraints on the factors that may be included. In the latter part of the chapter, some studies, including Indonesian research in which I have been involved, will illustrate the kind of approach I am advocating. First, however, I want to discuss certain other approaches because, although I regard them as extremely problematic, those approaches may be closer to what many biologists expect from a social scientist and it may, therefore, be useful to indicate what is problematic about them.

Emphasizing Concepts and Values About the Environment

On the assumption that human behavior affecting ecosystems is governed by basic conceptualizations or values concerning nature or the environment in general, social scientists concerned with human influences on ecosystems might be expected to accord priority in their research to identifying these conceptualizations or values and showing how behavior conforms to them. Such expectations and the assumption behind them are fueled by arguments which have been presented by some scholars. A well-known example, first put forward in 1967, is Lynn White's view that western cultures, in contrast to eastern ones without roots in Judeo-Christian religion, have an ideal of the conquest and exploitation of nature for human ends, and that this is what has produced "our ecologic crisis" (L. White 1967). More recently, other scholars have distinguished what they regard as basic interpretations

61

of nature—or "myths," as the ecologist Holling (1978, 1986; see also Chapter 2, this volume) has called them—and have been using these to explain such concrete, environment-related actions as a multinational household-products corporation's introduction of a new toilet-bowl deodorizer in West Germany and the subsequent protest of that introduction by members of the German Green Party (Schwarz and Thompson 1990). The myths of nature which social scientists Schwarz and Thompson (1990) refer to are four: views of nature as *benign*, always returning to equilibrium no matter what knocks it receives; nature as *ephemeral*, subject to collapse from the least jolt; nature as *perverse/tolerant*, unperturbed by most events but vulnerable in face of certain unusual occurrences; and, finally, nature as *capricious*, operating randomly, thwarting attempts at rational environmental policymaking and management, and allowing us only to cope with erratically occurring events (cf. Thompson 1988; Thompson et al. 1990). The fact that the multinational corporation failed to anticipate the reaction of the Greens and went ahead with its deodorizer introduction is explained by Schwarz and Thompson (1990) by referring to the corporation's adherence to the myth of nature as perverse/tolerant while the Greens held to the view of nature as ephemeral. According to the latter view, paradichlorobenzene, a waxy material which cements the active ingredients (perfumes and detergents) into a solid block which gradually dissolves as the bowl is flushed, is dangerous simply because its not having been previously introduced to the water cycle makes for uncertainty about its toxicity and biodegradability. According, however, to the corporation's nature-as-perverse/tolerant view, its original product (which, incidentally, was eventually replaced in response to the Green attacks) is innocuous and the Greens' failure to recognize that nature could handle it is irrational.

The foregoing is presented by Schwarz and Thompson not as an illustration of actions which they sought to understand and then determined empirically to be underlain by particular views of nature. Rather they *started* with just the sort of assumptions referred to earlier, namely that there are basic conceptualizations of nature which are few in number and essentially timeless and that, once identified, they can be used over and over again to make sense, as in the toilet-bowl deodorizer case, of the strategies adopted and actions taken (Schwarz and Thompson 1990:85).

What's wrong with all this? Since a few social scientists and perhaps a greater number of physical and biological scientists still subscribe to the old positivist notions (Boyd et al. 1991) that the same, so-called materialist models of explanation used in classical physics and chemistry must be appropriate to all science, including social science, and that, accordingly, ideas cannot be used at all to explain behavior, let me say right off and in intentionally peremptory fashion that I see approaches of the conceptualizations-of-nature type as problematic not because they involve using ideas to explain behavior, but rather because of *which* ideas they use and *how* they use them (cf. Vayda 1990). With respect to the old positivist notions, I forbear here from flogging what seems to me an almost dead horse and instead simply affirm my agreement with the philosopher John Searle, that it is possible to regard the world as consisting of material particles and of systems composed of material particles and still to recognize that subjective mental states, which are properties of brains, function causally in producing human and animal behavior (Searle 1984, 1991; for examples of persisting anthropological opposition to using ideas to explain behavior, see the "cultural materialist" manifestos by Price 1982 and Harris 1987).

About other objections than the positivist-materialist one to emphasizing basic conceptualizations concerning nature, we cannot be so quickly dismissive. A necessary anthropological objection is that it is ethnocentric to assume that such conceptualizations, known to us from our own culture, are to be found among every people. Anthropological accounts indicate that they are to be found among *some* people but possibly not among others. Consider just two examples. On the one hand, Koyukon Indians of northwestern interior Alaska are said to conceive of nature as conscious and consisting of entities with various degrees of spiritual power; all these entities, in the Koyukon view, must be shown respect (Nelson 1983). By contrast, Ponam Islanders, a maritime people of Manus Province in Papua New Guinea, are said not to have any unitary conception of nature; instead, they see different species in their environment as existing more or less independently and, in the case of fish and shellfish, being simply "things to be caught, eaten, or traded" (Carrier 1982, 1987).

But even when basic conceptualizations or values concerning nature are shown to exist among a people, what their efficacy is in causing actions with significant environmental consequences remains a question. That they seldom affect "more than a fraction of the total range of environmental behaviour" was stated by the geographer Yi-Fu Tuan (1968:177) more than twenty years ago in commenting on Lynn White's thesis about the Judeo-Christian tradition as a source of environmental spoilation. The main consideration on which Tuan's conclusion is based is that our actions are often guided not by our ideals or very broad conceptualizations, but rather by ideas about achieving more specific or more immediate objectives. Thus, Tuan notes that ideas of quiescence toward nature did not keep Chinese from deforesting vast areas as trees were burned to deprive dangerous animals of their hiding places and were felled for such purposes as making charcoal and obtaining the timber which was the predominant building material of old Chinese cities (Tuan 1968, 1970). Moreover, different ideas may be involved under different circumstances or at different times, in connection with the same objectives. For example, in a recent research project in which I took part, it was found that some Central Javanese farmers, during the early stages of ricefield infestation by insect pests, acted on the basis of practical ideas about chemical control and then, as infestation became more severe and the measures taken proved unavailing, switched to a course of inaction, recommended to them by resuscitated old Javanese ideas that the pests in outbreaks were brought forth from her realm by Nyai Loro Kidul, the goddess of the Southern Ocean, and that one could do nothing but wait for the pests to leave after sating themselves on the crop. White, like other essentialist thinkers, may be committing the fallacy of assuming that ideas or ideals which are widespread and/or persistent are *ipso facto* efficacious. They may, of course, be efficacious at times, but it must be regarded an empirical question whether *they* rather than some more idiosyncratic or transient ideas are the basis on which some particular actions are taken at particular times or in particular contexts (Vayda 1990; for further discussion of context-dependent variability in the ideas or cognitive attitudes guiding actions, see Bratman 1992 and Thomason 1987).

"Besides the more glaring contradictions of professed ideal and actual practice, there exist also the unsuspected ironies," says Tuan in his 1968 article (p. 188), and this remark serves to introduce a remaining important point to be made against emphasizing ideals or values about nature to explain human actions. The point is that many—perhaps most—of the environmental consequences of human actions are unintended. Possible examples are countless. For purposes of illustration, let me refer to just a few, including what I think would qualify as *subtle* effects.

My first set of examples is drawn from an article by Richard Mack in the March 1990 *Natural History* on weed dispersal through the agency of the 19th century mail-order seed business in the United States. Mack notes that the typical customers at whom seed catalogs were directed in the early 1800s were self-reliant farmers, growing plants for food (for themselves and their livestock) as well as for medications and for fiber for homemade clothes. However, in the boom decades after the Civil War, seed catalogs increasingly offered ornamental plants as more and more Americans no longer grew their own food and had more leisure time and discretionary income for lawns and ornamental gardens, and for such hobbies as collecting dried and pressed botanical specimens. Using elaborately illustrated, multicolored catalogs hundreds of pages thick, seed merchants vied with one another to offer the most diverse and exotic plants. These included what have become some of the most persistent weeds in many parts of the United States, e.g., baby's breath, Japanese honeysuckle, bachelor's button, and the water hyacinth.

Consider just a few details about the last-named, which is indigenous to the Amazon Basin and attracted considerable interest at the 1884 International Cotton Exhibition in New Orleans. By the early 1890s, nursery owners, lavishly praising the water hyacinth's easy care and its beauty in ponds, were selling it through their mail-order catalogs for as much as a dollar a plant, equivalent in today's prices to about eleven dollars. Neither the merchants nor the customers knew at this time about the plant's ability to run wild outside its home range in South America, where natural enemies kept it in check. Its floatability meant that currents could carry it miles down-stream, and when, for example, caught in the paddle wheels of a river steamer, it could travel long distances upstream, too. Since each plant can replicate itself hundreds of times each year, the spread of the water hyacinth was spectacular. In just a few years, many navigable rivers were reduced to streams by "impenetrable tangles of living and dead plants" (Mack 1990), which, moreover, became breeding grounds in some regions for the insect vectors of malaria and encephalitis. The plant is still a serious problem today in parts of the southeastern United States. A satisfactory explanation of all this lies not in the basic values or ideals which seed-sellers and seed-buyers held about nature but rather in the contexts in which they acted, the intentions and knowledge (or lack of knowledge) with which they did so, and the reproductive and other characteristics of the water hyacinth itself.

I want to give just one other example of evidently ecosystem-disrupting unintended effects. The example is the greater decline of warblers, wrens, nightingales, and other farmland birds in Britain than in France in recent decades. The factors pointed to by research on this include differences in national agricultural policies whereby there were incentives in Britain and not in France for the consolidation of farms and the attendant destruction of the hedgerows which are favored songbird habitats (Mills 1983). Insofar as this constitutes "unexpected action at a distance" (Chapter 1, this volume), I suggest it qualifies as a "subtle human effect."

In addition to evidently ecosystem-disrupting unintended effects, there often are what seem to be ecosystem-*maintaining* ones. Thus, French farmers are reported to tend their hedgerows not for the sake of keeping birds in the ecosystem but rather for having sources of firewood, boundaries around their land, and windbreaks to prevent erosion (Mills 1983). Such ecosystem-maintaining unintended effects are to be found in the pre-modern as well as the modern world. For example, Bulmer (1982), in an excellent review of alleged "traditional conservation practices" in Papua New Guinea, found no evidence of conservation ideologies behind the practices. Rather, he found that the traditional practices which had conservation effects were performed in pursuit of short-term yields. Thus, it may be good conserva-tional practice for shifting cultivators to let the land lie fallow for 15 or 25 or 40 years, but, notes Bulmer, the object for Papua New Guinea shifting cultivators was *not* conservation. It was getting that "next good crop for a reasonable minimum of effort." The cultivators knew from experience the approximate fallow period required both for an undiminished crop and for less effort than would be entailed if older secondary growth were to be hacked down. Similarly, Bulmer (1982:63) notes that: "Erosion barriers constructed along the contours of hillside gardens were not primarily in the interests of long-term soil conservation, but to stop that year's sweet potato or taro crop from sliding down the hill."

From Bulmer's review, as well as from my own New Guinea research which has resulted in findings like Bulmer's, I draw support then for assigning priority in research not to basic values or ideals about nature but, rather, to concrete human actions and their concrete environmental effects.

Studying Humans as Components of Predefined Systems

The second problematic approach which I want to discuss is studying human beings as components of ecosystems whose boundaries are defined or demarcated *in advance* of the research. In this and the next section, I incorporate material from some earlier articles of mine (Vayda 1986, 1988).

First I want to refer to studies in which human actions are regarded as having roles in the self-regulation of ecosystems and as thereby contributing to what is seen, whether dimly or clearly, as ecosystem stability or persistence. Without repeating the criticisms that I have made elsewhere (Vayda 1986) concerning the ontological assumptions underlying such studies, let me simply note that their concern with the self-regulation and stability or persistence of ecosystems sets the studies apart from others in which humans are regarded as ecosystem components in the sense of being organisms whose locale-specific interactions with organisms of other species and with things in the abiotic environment are included among the objects of study by ecosystem investigators.

The one example I will give here is what has become a classic of ecological anthropology, Roy Rappaport's *Pigs for the Ancestors* (1968). The concern with systemic self-regulation, i.e., with respect to maintaining within certain ranges such variables as the size and composition of both human and pig populations in the ecosystem, is made explicit by Rappaport at the beginning of his book. And I know from having been in the field with him that he assumed from the start that the ritual actions he was observing had roles in ecosystemic self-regulation. Indeed he became interested in the ritual actions precisely because of this assumption. Examples of the kind of claims made in his book are that: 1) certain ceremonies are initiated among New Guinea's Maring people when the relation between them and their pigs has changed from one of cooperation or commensalism to a competitive relation; and 2) the initiation of these ceremonies and their culmination in massive slaughters of pigs are to be regarded as systemic (or ecosystemic) self-regulation. Although I was at one time sympathetic to Rappaport's interpretations and, indeed, wrote the foreword to the original edition of his book in 1968, I made a number of critical points in my review (Vayda 1986) of the book's enlarged 1984 edition. Two of these are appropriate to note here.

The first point is that Rappaport gives us no adequate criteria for what constitutes behavior or responses by a higher-level unit like an ecosystem or a population or a society. What are actually observed by us as ecological anthropologists are the behavior of human beings and their interactions with specific components of their environments. If some of the observed behavior is to be regarded as, for example, societal functioning or ecosystemic response, we need to have rules or criteria for so regarding it. Even if the behavior in question could be shown to contribute to something definable as the persistence of a higher-level unit, the contribution could happen fortuitously rather than as a result of teleological control by the unit. In other words, apparent benefits accruing to higher-level units from behavior are no warrant for regarding the units as the agents of the behavior or as being capable of agency and goal-orientation (see Richerson 1977 and Williams 1966 for fuller discussion of similar points).

A second and related point is the difference between being, by virtue of location, included in a unit and being regulated or controlled by it. In the "Epilogue" of the 1984 edition of *Pigs for the Ancestors*, Rappaport concedes that organisms of different species included in ecosystems may have come together accidentally and may have initially not been subject to strong systemic controls; however, he assumes that the coexistence of the species (including human beings) proceeds under stable conditions and that the constitutions of their ecosystems are accordingly "likely to become increasingly elaborate and coercive through time." By such assumptions, simply being included in ecosystems implies being to a greater or lesser degree controlled by them. Ignored in this view is the generality of disequilibrating processes whereby interspecies articulations are recurrently sundered (Pickett 1980; Wiens 1984). Ignored, too, are the special problems posed by the mobility of many species in ecosystems. Elton, writing more than sixty years ago in reaction to the notion that animal communities operate like clockwork, expressed the problems figuratively by saying that the clockwork idea could be sustained only if we were to accord each cogwheel "the right to arise and migrate and settle down in another clock" (Elton 1930:17). A more recent statement is the following by Wiens (1984:453-454) about difficulties resulting from mobility characteristics of birds:

"These characteristics produce a substantial flux of individuals in local populations, and spread the sources of biological limitation of populations, and the consequent determination of community structuring, over a large and undefined area. Because most local assemblages of birds contain mixtures of species that differ in migratory tendencies and pathways and in longevity and fecundity, the dynamics of any given local assemblage are likely to be driven by an amorphous complex of factors, the effects of which are likely to be different for almost every local assemblage. This uncertainty decreases the probability that any 'patterns' that seem apparent in such communities are in fact real."

The disarticulating processes and mobility characteristics that are being referred to in current debates among biologists, raise questions about the utility of higher-level units simply as analytic concepts, regardless of whether or not they are regarded as entities with teleological properties. Ontological claims for the units have, it should be noted, been given up by most biological ecologists (Simberloff 1980; Colinvaux 1976) and also by many ecological anthropologists (Moran 1984). This means that even for those calling themselves community ecologists, the key questions right now can concern the "existence, importance, looseness, transience, and contingency of interactions" among species (Abele et al. 1984). Rappaport is impervious to these developments in ecology and rejects as "rather muddled" and "mistaken" (Rappaport 1984, 1990) the arguments which McCay and I previously put forward for regarding the ecosystem not as an objectively real entity, but rather as an analytical concept for dealing with interactions of organisms of different species living together in restricted spaces (Vayda and McCay 1975).

In the case of people, it is surely clear that their interactions are not necessarily restricted to particular ecosystems. Rappaport is probably now in a minority among ecological anthropologists in making his ontological claims on behalf of ecosystems and similar "higher-level" units. Yet, there are *many* ecological anthropologists who, while disavowing notions of ecosystemic self-regulation as well as other reifications to be found in Rappaport's work, still retain a methodological bias toward studying bounded systems (cf. Boyden, Chapter 7, this volume). A good example is Roy Ellen (1982) who, while rejecting *a priori* assumptions about the closure and stability of systems, still adheres to the view that the way to understand and explain relationships or interactions between people and components of their environment lies in thoroughgoing studies of some larger system which has its boundaries demarcated before the start of the research and which presumably contains the people and/or environmental components that are of interest. As will be further discussed shortly, an alternative approach is proceeding empirically to put people-environment interactions into context without antecedent stipulation or demarcation of what that context is, but this is an alternative which Ellen does not explicitly consider. He does acknowledge that describing an ecosystem holistically, which calls for measuring and describing how energy and materials flow throughout the whole system, is a formidable task (Ellen 1982) and "can make extraordinary demands on the time, care, patience, and skill of the investigator and his back-up facilities" (Ellen 1982:113). However, except for noting that there are practical limits to the flows that ecological anthropologists can deal with (Ellen 1982), he expresses no qualms about the undertaking, and devotes many pages of his 1982 book to methods and procedures in ecosystem research. By contrast, there is very little in his book about any answers which such research has provided to significant questions in anthropology or human ecology. And when Ellen makes suggestions about questions which *might* be answered (for example, some of the questions in his Chapter 5 about people's diets and energy expenditure), it is by no means clear that answers could not be provided more expeditiously and completely by inquiries much more focused—and, at the same time, much less restricted in the pursuit of causes and effects through space and time—than the holistic study of ecosystems. In truth, Ellen gives no convincing reasons for such study.

Actions and Consequences in Context

Turning now to my preferred approach, namely, starting with concrete actions and/or their concrete consequences as our objects of study, I want to give first an example from the journal *Ecology*, to show that it carries some articles in which humans are importantly and appropriately featured. The article in question concerns the invasion of coral cays by alien flora. Having examined the composition of the flora on 10 coral cays in the southern Great Barrier Reef region, the authors, Chaloupka and Domm (1986), identified "a substantial invasion of the region by an alien flora comprising mainly annual and perennial agrestal weeds found commonly on the adjacent mainland." They were then able to show that the percentage of alien plant species recorded on a cay depends on the frequency of human visitors to that cay and is independent of cay size. Their studies and analyses point clearly to the importance of inadvertent human dispersal of alien diaspores onto the cays.

For further examples, I turn to our research related to deforestation in the Indonesian province of East Kalimantan on the island of Borneo. Initial objects of study in the research included such human actions as cutting down trees and such effects of human actions as gaps and clearings in the forest. These effects were of special interest to us because, partly as a result of collaboration with Steward Pickett and other ecologists (cf. Collins et al. 1985), we were very aware of the importance of gap or clearing size with respect to the ability of plants of particular woody species to re-establish themselves, either by invasion or by growing from coppices or seeds surviving in the soil. Finding considerable variation in gap or clearing size resulting from human actions, we then looked more closely at the actions in order to try to account for variations in the effects. This, in our research among Dayak shifting cultivators in the interior of East Kalimantan, involved us in looking at, among other things, not only such contextual factors as topographic, technological, and manpower constraints on the people's ability to make the clearings larger (Kartawinata et al. 1984), but also such intentions as making the swiddens of one's own household adjacent to those of neighbors, such purposes behind the intentions as having neighbors to lighten the burdens of trail maintenance and swidden fencing, and such beliefs or knowledge behind the intentions as that having one's own swiddens a part of a cluster will facilitate the burning of felled trees and slash and the protection of crops from monkeys and other pests from the forest (Mackie et al. 1987; on distinctions among beliefs, ideas, purposes, and intentions in explanations of actions, see Davidson 1980; A. White 1967). Appropriately, but not too successfully, we also sought to obtain data on the origins, determinants, or reinforcers of the beliefs (cf. Ruyle 1973 on causes of the "differential replication of ideas"; and Sperber 1985 on the "epidemiology" of mental representations). It is, of course, conceivable that some shifting cultivators will be shown by research to deliberately keep their clearings small because of knowledge of the relation between the size of gaps and the re-establishment of woody species, and because of a belief in the desirability of hastening the reversion of fields to forests. A point to be emphasized here is that to explain the actions of such shifting cultivators we cannot refer simply or directly to the generalizations relating gap size and forest recovery, but must instead have data on—and make reference to—the cultivators' knowledge or beliefs.

In the foregoing illustrations, research proceeds from concrete environmental effects to finding out more about the actions that have led to them and about the ideas behind the actions. I want, however, to note also the possibility of having findings about behavior serve as guides in deciding which studies concerning consequences should be pursued. Thus, findings about the preference of shifting cultivators for making swiddens in previously farmed sites in secondary forest rather than in primary forest sites which are more difficult to clear, have been cues for research on the botanical effects of recurrent felling and burning. The East Kalimantan research so far, involving comparisons of plant successions on sites varying in how far back in time their histories of being re-cultivated extend, suggests that such cyclical disturbances, as are associated with so-called traditional shifting cultivation, favor certain types of trees—namely, vigorous resprouters and pioneer, light-demanding trees which produce seeds abundantly (Mackie et al. 1987; Mackie 1986).

One final point I want to make concerns the elucidation of actions and consequences by means of putting them into context. In our East Kalimantan studies, this sometimes involved us in going far beyond the boundaries of a region, nation-state, or island (e.g., in looking at how demand from Hong Kong, Japan, North America, and Western Europe affected the extraction and movement of forest products from the Borneo interior) and at other times kept us within the boundaries of a single Dayak village and its land (e.g., in looking at the villagers' collection of forest products for local use in building, cooking, and medicine). The point that I have made elsewhere (Vayda 1983, 1988) and want to reiterate here is that putting actions and consequences into context needs to be an empirical procedure without antecedent stipulation or demarcation of what the context is (cf. Vayda et al. 1980; Hayek 1973). The rationale for this is that contexts often are transitory and do not correspond to the conventional wholes of social and ecological science. Contrary to a notion expressed by Rappaport (1984, 1990), giving up holistic concepts need not restrict us to "the study of decontextualized interactions between humans...and one or a very few nonhuman elements in their surroundings." In view of the generality of flux and instability in the world, it is specious for Rappaport and others (e.g., Geertz 1984) to make lofty appeals for attention to context which are attached to *a priori* judgments that the context is necessarily an ecosystem or a culture or some other predefined whole. Due attention to context in the study and explanation of actions and consequences may often mean having to deal with precisely the kind of factors and processes scanted or denied by holistic approaches: loose, transient, and contingent interactions; disarticulating processes such as are increasingly being discussed by ecologists (e.g., Pickett 1980; Wiens 1984); and the many movements of people, resources, and ideas across whatever boundaries ecosystems, societies, and cultures are thought to have.

Recommended Readings

Bulmer, R.N.H. (1982). Traditional conservation practices in Papua New Guinea. In: L. Morauta, J. Pernetta, and W. Heaney, eds. *Traditional Conservation in Papua New Guinea*, pp. 59-77. Institute of Applied Social and Economic Research, Boroko, Papua, New Guinea.

Mack, R.N. (1990). Catalog of woes. *Nat. History*, March, pp. 45-53.

7
The Human Component of Ecosystems

Stephen Boyden

One of the outstanding biological characteristics of the human organism is its aptitude for culture.[1] We can assume that this attribute was of considerable selective advantage under the conditions prevailing in the environment in which the early hominids evolved. Certainly, from the outset, culture brought about important changes in the relationships between our ancestors and the other components of the ecosystems in which they found themselves, and these impacts, in turn, then influenced humans themselves and their culture (Fig. 7.1). In fact, interplay between cultural and biological processes has been a consistent and highly significant feature of *all* human situations that have ever existed. It occurs in all ecosystems of which humans are a part or on which they have any influence.

For the tens of thousands of generations when all humans were hunter-gatherers, their main culturally-induced ecological impacts were those that resulted from the deliberate use of fire, which in some regions of the world had major ecological consequences—leading in the conversion of large areas of forest to grassland and an associated increase in the numbers of grazing animals (Jones 1969; Woodburn 1980; Williams, Chapter 3, this volume). In other respects, hunter-gatherers fitted into ecosystems in much the same way as any other omnivorous animal. However, culture-nature interplay took on a new significance with the advent of farming some 450-500 generations ago. With this development, the human aptitude for culture led to much greater changes in local ecosystems, involving widespread displacement and redistribution of animal and plant species and, in some areas, loss of soil fertility. It was also associated with an important increase in the human population.

---

[1] The word *culture* is used here in the abstract sense, referring to such aspects of human situations as knowledge of language, assumptions, understanding, values, world view, and technological know-how. The word *nature* is used for things and processes of a kind that existed before human culture came into being.

BIOSPHERE ⟶ HUMANS ⟶ CULTURE

Figure 7.1. Basic conceptual model.

The formation of the first townships and cities, around 250 generations ago, brought into being an entirely new kind of human ecosystem in which large populations were dependent on food provided by farmers living beyond the borders of the settlement. This development was also associated with major changes in environmental conditions and behavior patterns of humans themselves, with significant consequences for their patterns of health and disease, and for the causes of fear and sources of enjoyment among them (Boyden 1987; Boyden et al. 1990).

The industrial transition, beginning only about 8 generations ago, heralded a new phase in the ecology of the human species—a phase characterized by massive intensification of culture-nature interplay. One of the consequences has been a surge in population growth, so that there are about 1000 times as many people on Earth today as at the time when farming was first introduced. During this phase, there has been a great increase in physical impacts on the environment of a kind which were already apparent in earlier times due to such human activities as farming, mining, logging, and road building. The industrial era has also given rise to a new set of changes of enormous ecological consequence resulting from the release into the biosphere of vast quantities of chemical waste products of technological processes. For example, cultural developments have resulted in the human species as a whole producing about 12,000 times as much carbon dioxide each day as was then the case at the beginning of the domestic transition when our ancestors first took up farming (Boyden et al. 1990). The great majority of atmospheric scientists take the view that the increase in atmospheric carbon dioxide will lead to progressive global warming (Houghton et al. 1990). Ninety percent of this carbon dioxide is the result of the combustion of fossil fuels used as a source of energy for driving machines and other purposes. Eighty percent of it is produced by the developed countries, in some of which the per capita daily production of carbon dioxide is now between 50 and 100 kg (World Resources Institute 1991). An equally important development has been the beginning of the depletion of the ozone layer in the stratosphere as a result of the release of CFCs in immense quantities by the industrialized populations.

Thus, in our own lifetime, and for the first time in the history of life on Earth, a single species is now producing progressive ecological change at the level of the planet as a whole.

It is now well appreciated that the biosphere, as a system capable of supporting humankind, will not tolerate indefinitely the present pattern of resource and energy use and waste production by human society (World Commission on Environment and Development Staff 1987; Boyden 1992a). Apart from these global concerns, there is also widespread evidence of undesirable impacts of humans on ecosystems at local and regional levels (Brown 1991).

It stands to reason that the proper understanding of the ecology of any ecosystem of which humans are a part, from a backyard or small nature reserve to the biosphere as a whole, requires that account be taken of all the impacts of humans on the system. In fact, a *full* understanding requires that we identify not only the human activities (Vayda, Chapter 6, this volume) that are responsible for given changes (e.g., a certain manufacturing process), but also the abstract cultural variables which underlie the human activity in question. Figure 7.2 represents an extension of the basic conceptual model presented in Fig. 7.1, from which it differs in the following respects:

1) The diagram includes *biometabolism* and *technometabolism*. Biometabolism is the human population's inputs and outputs of materials and energy used in the biological metabolism of humans. Technometabolism is the inputs and outputs of materials and energy resulting from technological processes;

2) The diagram includes *artifacts*, which are the material products of human labor;

3) *Human activities* has been singled out as an important aspect or component of humans themselves; and

4) Culture has been separated into two sets of variables—namely, *culture* itself, which includes beliefs, assumptions, values, and understanding, and *cultural arrangements*, such as economic arrangements, legislation social hierarchies, and the political system. The dominant values of society, for example, are enormously significant ecologically, in that they have a major influence on economic arrangements which, in turn, affect human activities. These activities then influence the nature and rate of technological waste production (Fig. 7.2), and consequently the health and well-being of ecosystems and of humans themselves.

Just how far an analysis of an ecosystem should go in the direction of abstract cultural aspects of the system depends on the purpose of the investigation. It may be considered sufficient in some instances simply to identify the connections between significant ecological variables (e.g., the presence of heavy metals in soil) and the human activities (e.g., a certain industrial activity) which are responsible for this fact. Nevertheless, for a *full* understanding of the system one has to go further than this to reveal the underlying causes in the societal system of the activities which are responsi

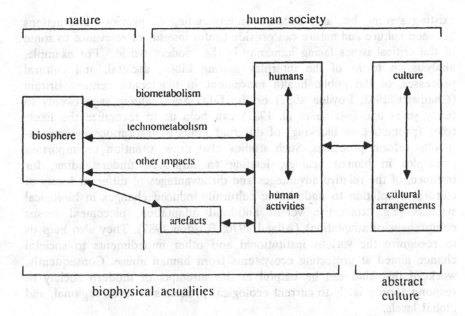

Figure 7.2. Conceptual model—extended version. Note: The term "biosphere" in this diagram refers to all aspects of the biosphere other than those incorporated under the general heading "human society." In reality, of course, humans and their artifacts are all part of the biosphere.

ble for significant ecological changes.

The connections between the economic arrangements of society and the rates of extinction of forms of wildlife is one obvious example (Marshall 1966; Swanson and Barbier 1992). Another is the influence of economic pressures (e.g., high interest rates), societal institutions, and dominant attitudes on land degradation in Australia, and it is very apparent in this instance that policies aimed at improving the ecological situtation can be successful only if they take account of the full range of interacting factors, biological, societal, and cultural (Chisholm and Dumsday 1987; Dovers 1993). And much has now been written, of course, about the relationships between the societal value systems and changes in the planet's ecosystems (e.g., Naess 1989; Boyden 1992b).

However, it has to be recognized that little intellectual effort has been effectively directed to the development of rigorous conceptual and method-ological frameworks for the study of the important interrelationships between these different classes of interacting variables. Let us hope that this serious deficiency will soon be rectified.

I would like to emphasize here the usefulness of taking account of the historical dimension in the studies on ecosystems of which humans are a part. This is not only because past impacts of culturally-inspired activities of humans are often important influences on the ecological dynamics of an existing system, but also because understanding of previous interactions between culture and nature can provide useful insights of relevance to some of the critical issues facing humanity in the modern world. For example, analysis, in terms of the interplay among biotic, societal, and cultural processes, of the public health movement in nineteenth century Britain (Chadwick 1843; Boyden 1987) or the DDT-*Silent Spring* saga twenty to thirty years ago (Marco et al. 1987) can help us to recognize the likely roles (promotive or blocking) of different kinds of institutions in environmental reform processes. Such studies also draw attention to important principles in human ecology, leading to improved understanding, for instance, of the relative advantages and disadvantages of different forms of cultural adaptation to undesirable culturally-induced changes in biological systems (e.g., corrective versus antidotal adaptation, piecemeal versus comprehensive adaptation) (Alland 1970; Boyden 1987). They also help us to recognize the various institutional and other impediments to societal change aimed at protecting ecosystems from human abuse. Consequently, work of this kind can be helpful in the attempts of modern society to respond appropriately to current ecological problems at local, regional, and global levels.

The study of ecosystems in a manner that takes account of the human component therefore requires an integrative conceptual framework which facilitates consideration of a range of very different kinds of variables, all the way from the most biophysical (e.g., soil characteristics, animal or plant populations), through societal (e.g., transportation), to the most abstract (e.g., dominant values), so that the full spectrum of causal interrelationships can be taken into account, if so desired.

For example, in a study of a local ecosystem, changes in flora might be found to be related to inputs of toxic chemicals into the system. It might then be considered desirable to identify the source of these chemicals in terms of *human activities* (e.g., a manufacturing process). It might also be considered appropriate to identify the societal group responsible for such activities. For a more complete picture, it would be necessary to go further and to establish the links between the human activities responsible for the ecological changes and, for example, the economic arrangements of society and such factors as cultural assumptions and values.

So far in this chapter, it has been assumed that the starting point for research is in the area of the natural sciences. Clearly, as we move to the right in the diagram (Fig. 7.2), we find ourselves dealing with variables normally studied in the social sciences and the humanities. Thus, a full or complete investigation, at least given the present structure of academia, requires information from those other areas of specialization. An effective integrative ecological study of an ecosystem containing humans therefore requires a team effort, with contributions from people from different academic disciplines.

However, this is not a plea for the "integration" of different academic disciplines or intellectual paradigms, or for interdisciplinarity *per se*, worthy as such notions may be. It is rather a call for an approach which recognizes the significance of the constant interplay that is taking place in all human situations between cultural and societal variables on the one hand and biological and physical variables on the other. The aim is to understand these interrelationships better and to achieve this improved understanding by inviting social scientists to provide the necessary information and insights relevant to the cultural and societal variables involved in this interaction.

The whole process, of course, can just as well be initiated at the other end of the spectrum. A social scientist interested in various options for changes in economic arrangements may wish to inquire into their likely impacts on human activities, human health, and the health of local ecosystems. This will require information from ecologists and biomedical scientists.

The above comments reflect the present structure of the world of learning with its historically determined pattern of academic disciplines. However, there is a real need for a new class of research workers and teachers whose main, full-time interest is the study of the patterns of interplay among biological and physical, human, societal, and cultural components of human situations and of the principles that govern this interplay and determine its outcome.

Improved understanding, not only among academics, but across the community at large, of human situations in terms of culture-nature interplay may well be a prerequisite for ecological sustainability and, hence, for the survival of the human species.

Recommended Readings

Boyden, S. (1987). *Western Civilization in Biological Perspective: Patterns in Biohistory.* Oxford University Press, Oxford.

Boyden, S., S. Dovers, and M. Shirlow. (1990). *Our Biosphere Under Threat: Ecological Realities and Australia's Opportunities.* Oxford University Press, Melbourne.

Jones, R. (1969). Fire-stick farming. *Aust. Nat. History* 16:224-228.

Woodburn, J. (1980). Hunters and gatherers today and reconstruction of the past. In: E. Gellner, ed. *Soviet West. Anthropol.* pp. 95-117. Duckworth, London.

World Commission on Environment and Development Staff. (1987). *Our Common Future.* Oxford University Press, Oxford.

Section II   Approaches to the Study of Humans as Components of Ecosystems

Several contributions to Section I have indicated that humans are parts of ecosystems in often unexpected or inconspicuous, but still important, ways. Section II will take as one of its goals the continuing illustration of how humans have acted and continue to act as components of both terrestrial and aquatic ecosystems. In the aquatic realm, lakes, rivers, and coastal zones are examined.

A second goal is to reinforce the awareness that social and economic decisions affect basic ecology. This leads to the third goal, that of showing how knowledge of ecological and human processes can be creatively combined to yield novel ecological insights.

The fourth goal of Section II is to indicate, using examples, how to detect subtle human effects and expose the ecological role of human populations on the areas they occupy or influence. We begin with Russell's (Chapter 8) refined analysis of the nature of subtlety of human effects and their detection.

Foster (Chapter 9) uses the history of the New England forest to examine the variety of causes and the continual flux of human roles in the ecology of a large region. Kitchell and Carpenter (Chapter 10), using midwestern U.S. lakes, show how humans are critical and increasingly more effective components of lake ecosystems they fish and manage. Richerson (Chapter 11) takes a longer term view to show the long history of human induced changes in a tropical lake, and human responses to those changes. The system has responded, both positively and negatively, to the effects of humans, alternately opening and closing opportunities for certain components of the ecosystem. Cole et al. (Chapter 12) focus on rivers and a strategy for predicting human impacts on the coarse scale, while showing the futility of a search for the pristine. Moving to the coastal zones, Castilla (Chapter 13) seeks to understand the place of humans in food webs. The combination of direct and indirect, or subtle, effects of humans in such communities stretches from ancient to modern times, and is a controlling factor in community organization.

Although air pollution is itself conspicuous, the effects of materials deposited from the atmosphere may be subtle and widespread (Aber, Chapter 14). That attention has only begun to shift from the effects of sulfur to the more subtle effects of nitrogen deposition is telling. The explicit problem of populated areas is taken up in two chapters (McDonnell et al., Chapter 15; Rapoport, Chapter 16). Again, settlement in urban areas is an example of a conspicuous human agency, but one whose effects have been largely ignored by ecology. The diversity of novel environments created by urbanization, and the variety of human interactions, values, and tolerances expressed in them, create a new arena for ecology. Gardner et al. (Chapter 17) end this part of the book by modeling a feature of all human dominated landscapes, fragmentation. Prediction, understanding, and specifying policy implications of landscape fragmentation are critical issues. The chapters in Section II show the power of historical ecology, of its combination with the disciplines that focus on humans, and of the investigation of new ecological arenas and scales. Together, these approaches expose the ubiquity and persistence of human roles in ecosystems, the social causes, and the policy implications of subtle human effects and human occupancy of landscapes.

8
Discovery of the Subtle

Emily W.B. Russell

Introduction

Astute observers have long recognized that human activities have indirect, unexpected, and often deleterious effects on natural systems. George Perkins Marsh was probably the first to consider explicitly the indirect and cumulative effects of human activities, observing in 1864 that "[w]e are never justified in assuming a force to be insignificant because its measure is unknown, or even because no physical effect can now be traced to its origin" (Marsh 1965:465). The critical importance of understanding the interconnections between human activities and the natural world is even greater today than it was in the mid-nineteenth century as the pace and pervasiveness of human influence intensifies. There are ever-increasing human activities with currently subtle but potentially devastating effects.

By analyzing the development of thought on several complex problems of subtle human impacts, we can discern patterns by which scientific research has provided important clues to actual and potential interactions. While research does not provide conclusive proof of many of the more remote cause and effect relationships, it suggests ways to communicate both the certainties and the uncertainties about these relationships in the contentious arena of public policy, as well as ways of using such issues to advance basic research.

The unintended consequences of human activities may be obvious or subtle. For example, obvious effects occur when fumes released from a copper smelter kill most of the vegetation, and probably animal life as well, in the vicinity of the smelter, or when the release of highly toxic waste into a body of water kills much of the life in the water. Many other unintended consequences are much more subtle, and thus difficult to predict, such as the eutrophication of lake waters by phosphate-rich runoff or the change in species distributions caused by modifying habitats.

Such subtle causation may take place because of indirect action, far away in time and/or place from the actual activity initiating the effect (see Vayda, Chapter 6, this volume), and may also be the result of interacting, even synergistic, factors. For example, major silting of the Adriatic Sea was attributed by Marsh to deforestation in the Alps and Apennines (Marsh 1965). Deforestation occurred far from the site of deposition of sediment and was driven by market forces also far removed from the local population. Erosion continued long after the timber was gone. Marsh made the connections between forest decimation, a drastic event that had occurred long in the past, and siltation, which was slow but continuing. One could argue, alternatively, that poor farming practices or changes in the weather caused the excessive erosion, or that changes in local currents caused the siltation. But by amassing copious evidence of the relationships between trees and soil stability, Marsh reached a conclusion which we accept today as fact, or close to it. He could not experimentally reproduce the sequence of events, but provided a plausible hypothesis to explain them, the inferences from which were all consistent with observations.

To clarify this process of subtle causation, I will classify several more current problems of subtle causation into four categories based on the difference between a subtle or obvious "cause," or initiating human activity, and a subtle or obvious "effect," or consequence for the environment (Table 8.1). In all of these cases I am using "cause" in the sense of the human activity that precipitates a series of effects, rather than as the cause of the human activity itself, which is another, and equally important, issue (see Turner and Meyer, Chapter 4, this volume):

1) *Obvious activity with an obvious effect.* An example would be production of SO_x and other pollutants by burning fossil fuels, and loss of fish populations in lakes;

2) *Obvious activity with a subtle effect.* As an example, repeated logging in northeastern North American forests has led to an increase in the amount of birch (*Betula* spp. depending on the location) in the forests which regenerated in the nineteenth and twentieth centuries (McAndrews 1988);

3) *Subtle activity with an obvious effect.* An example would be releases of small amounts of DDT into the environment, resulting in major declines in raptor populations (Cramp 1963); and

4) *Subtle activity with a subtle effect.* For example, the increase in CO_2 and other "greenhouse gases" and changes in global climate are both subtle and have subtle connections (Firor 1990).

Techniques to establish these relationships have required a variety of approaches, since most are not easily, if at all, amenable to experiment. These were the methods used by Marsh as well. His process of argument, from observations to hypotheses to inferences, which can be tested by observation, make up much of the science of studying subtle or remote

Table 8.1. Examples of different categories of human activities and their effects on ecosystems.

Category	Example
Obvious activity with an obvious effect	Production of SO_x and other pollutants by burning fossil fuels, resulting in loss of fish populations in lakes
Obvious activity with a subtle effect	Repeated logging leading to increase in amount of birch in regenerated North American forests
Subtle activity with an obvious effect	Small amounts of DDT released resulting in major declines in raptor population
Subtle activity with a subtle effect	Increase in CO_2 and other "greenhouse gasses" causing changes in global climate

interactions between causes and effects. His method used historical documentation and observations of the phenomena to develop hypotheses that explained the observed relationships. The hypotheses, in turn, suggested logical inferences which can be tested to falsify or corroborate the hypotheses (Medawar 1967). However, as with any hypothesis, one can at best only corroborate or fail to falsify, never conclusively to prove it (Popper 1959). In most of the cases of interest here, observations were used for testing inferences, supplemented in some cases by experimentation and modeling, to predict and generate more inferences. In the highly politically-charged atmosphere of environmental issues, this is a particularly important point to make clearly. Falsifying the hypotheses without strictly controlled experiments can be as difficult as corroborating them, so that competing hypotheses cannot be easily eliminated, and because of the social importance of the issues, it is very difficult to be objective about which corroboration is most convincing. Some of the approaches are similar to those used in epidemiology, where, obviously, one does not spread a disease to test a hypothesis, or in geology, where because of temporal and spatial scales, many hypotheses are not experimentally testable. The hypotheses in both of these disciplines are, however, falsifiable, and are removed from the realm of mythical explanations, which generally cannot be falsified (Egerton, Chapter 2, this volume).

Several examples will illustrate research that has been done on these topics, and show how a consensus has been reached on some. This consensus has most often been reached where generalizations based on many similar situations have led to more general hypotheses or theories, which can then be used for prediction.

Examples of Subtle Connections

Category 1: Obvious Activity With an Obvious Effect

Human-caused air pollution has been with us ever since humans learned how to control fire. Smoky caves undoubtedly had very high levels of carbon monoxide and other air pollutants. It was only when people began to congregate in large population centers where fossil fuels were burned, however, that major harmful effects on human health from burning fuels were recognized as a major problem. Additional deleterious chemicals were added to the air with increasing industrialization in the twentieth century. A major European conference in 1968, for example, investigated the effects of such air pollutants as SO_2, peroxyacylnitrates (PAN), hydrogen fluoride (HF), smoke, and O_3 on economically valuable plants and animals (Air Pollution 1969). Generally, neither the chemicals nor their effects were subtle, i.e., they were fairly obvious to the astute observer. However, by the early 1970s there were suspicions of other, subtle, effects with much more tenuous and convoluted trains of causation.

By 1972, in both Europe and North America, scientists had observed that rain was acidified beyond what would be expected by solution of CO_2 from the atmosphere (Likens et al. 1972). In the late 1960s and early 1970s, it was also clear that some lakes in Ontario and the Adirondack Mountain region were too acid to support fish populations (Haines 1986). While most fish do not thrive in water with pH < 5, the cause of the low pH was more difficult to unravel. Direct acidification of lakes by the addition of H_2SO_4 was a drastic, direct, experimental approach, but did not adequately mimic the complex effects of acidified precipitation reacting with watershed soils and vegetation before entering the lake (Haines 1986). Controls for "natural experiments" were difficult to find because of the wide diversity in atmospheric circulation patterns, emissions, and chemistry; watershed topography, geomorphology, soils, and vegetation; and lake size, morphometry, and biota, to name but a few of the variables. In addition, variables such as lake size, pH, and cation concentrations, which account for much variation in fish populations, appear to be so strongly correlated in many instances that the variability cannot be partitioned among them (Haines 1986). Historical approaches have yielded the most convincing evidence connecting changes in pH over decades: historically documented use of fossil fuels, and historical and current trends in fish populations (Davis and Stokes 1986; Haines 1986). These studies have relied on carefully established models which allow pH to be inferred from the relationships between diatom species and pH in local lakes. When these models are applied to diatom assemblages found in dated lake sediments, it is clear that a decrease in pH has occurred within the last 50 years in some previously acid lakes located in watersheds where there have been no major recent changes in land-use in both Scandinavia and some parts of northeastern North America (Davis

1987). This decrease in pH corresponds also with a decrease in dissolved organic matter, which would mean less organic matter to immobilize potentially toxic trivalent Al in the lake waters (Davis et al. 1985). This model supports the hypothesis that accelerated acidification of these lakes in the second half of the twentieth century has been responsible for contemporaneous declines in fish populations. Because other sources of acidification seem quite unlikely, the conclusion is that the acidification has been caused mainly by acid atmospheric deposition.

The complexity of getting from initiating cause to effect indicates a number of critical steps. First is the finding of correlations between an effect and a possible cause. Then multiple hypotheses, as exhaustive as possible, must be posited to explain the cause and be rigorously tested by determining the consequences of each. Considerable effort must be made to distinguish anthropogenic from natural causes (Davis and Stokes 1986), and to sort out the various anthropogenic causes, such as changes in land-use or atmospheric acid deposition, and interactions among them.

The tremendous amount of basic research on acid precipitation and its effect on aquatic and terrestrial systems (more than half a billion dollars on the major U.S. national effort, the National Acid Precipitation Assessment Program (NAPAP) alone) revealed some clearcut damage from acid deposition (Roberts 1991a). That it has failed to find conclusive answers to the potential for damage in general should come as no surprise given the magnitude and complexity of the potential problems. As Ellis Cowling observed, saying that "'you can say no symptoms were found, and we looked hard' but that is different from saying that no problem exists" (Roberts 1991b). This statement, which echoes George Perkins Marsh, is a conundrum with which scientists must come to grips in studying subtle environmental impacts—some are so subtle in the first stages that they are undetectable.

Category 2: Obvious Activity With a Subtle Effect

The destruction of forest vegetation by European colonists in eastern North America was an obvious event, comparable in its effect on the pollen record to a major climatic change (Davis 1965). The immediate and intended effect was obvious—the transformation of a primarily forested landscape to a primarily agricultural and product-oriented one (Williams, Chapter 3; Foster, Chapter 9, this volume). Many of the residual effects are, however, not so obvious. Many previously oak (*Quercus*)-dominated forests contain much more birch (*Betula*) than they did in the past (McAndrews 1988; Russell 1980). The difference is clear from pollen records from lakes, but not necessarily from observing the vegetation directly, because of the lack of a clear basis for comparison. The cause of the increase in *Betula* is not obvious unless one considers the uses that some of these forests were put to in the past, combined with the ecology of the dominant taxa. Many forest stands in northeastern North America that were not cleared for agriculture were used heavily to produce fuel for industrial processes such as iron smelting. The procedure used was to clearcut the forest in patches, making charcoal from the wood. The forest trees then sprouted back to make another stand to be cut in 15-30 years (Hough 1882). Although there was little system in this procedure, it probably resulted in patches of different-aged stands, with vigorous sprouters, such as chestnut (*Castanea dentata*) and oaks, along with species which could grow into the light, open canopy from seed, such as sweet birch (*B. lenta*). In the early twentieth century, another human influence, an introduced disease, decimated the chestnut trees—an obvious effect, but also with subtle consequences. The current high density of birch, which thrives in heavily disturbed forests, is most likely a residual effect of the past land-uses and death of chestnut trees (Good 1965; Russell 1979).

Interpretation of these subtle effects on forest composition requires analyses of past activities and of vegetational composition. If one incorporates only very recent human activities into a study, ignoring disturbance 60-100 years ago, these "echoes of the past" may lead to misinterpretation of the present. It appears that past land-use is reflected in present vegetation in subtle ways which can only be discerned by integrating historical and current studies (Russell 1980; Glitzenstein et al. 1990).

Category 3: Subtle Activity With an Obvious Effect

Ridding the world, or at least one's corner of it, of pests has been a concern of humans for centuries, if not millennia. Many substances used for this purpose have been either plant products which destroyed or repelled pests, or inorganic poisons. A main problem with the former is that they do not last, so must be reapplied often, and with both is that they are frequently lethal not only to pests but also to past predators and humans. In the nineteenth century, the development of chemical pesticides accelerated with advances in technology. The most commonly used chemical insecticides were the arsenicals, especially lead arsenate (Whorton 1974). While people recognized that arsenic was poisonous to humans in large enough doses, most accepted the contemporary wisdom that small doses were harmless, even beneficial in certain illnesses. While some doctors and others suggested that small doses could have cumulative effects that were not easily predicted from the effect of each small dose, their criticisms were generally ignored. While permissible amounts on fruit were limited by statute, small amounts were deemed safe for general consumption.

Invented in 1939, DDT seemed to be a new miracle pesticide. Its low toxicity to humans and other mammals, and persistence after application were ideal characteristics for a pesticide. Now, with one application, one could eliminate a pest without harming other animals. Shortly after World War II, government agencies (inspired by the military) quickly espoused its use, and major chemical corporations began large-scale production. But fairly soon it became obvious that insect pest problems were not over, as many insect species developed resistance to the chemical. Larger doses were needed, but it was recognized that much larger doses could be harmful to humans as well as insects.

Up until the early 1960s, most concern about the negative effects of pesticides related only to possible deleterious effects on humans or their domestic stock, and to potential (and actual) resistance of insects to the pesticide (Wickendon 1955). Some few were concerned about killing predatory insects which acted as natural controls on the herbivorous ones. But the publication in 1962 of Rachel Carson's *Silent Spring* brought together a tremendous amount of evidence that pesticides affected other non-target species also, in ways that were not predicted. She documented evidence of the concentration of DDT and its degradation products up the food chain, and of lingering amounts in phytoplankton many generations after the water in which they grew had undetectable (at the time) amounts of DDT. She observed that "our fate could be sealed twenty or more years before the development of symptoms" (Carson 1970) because of the slow accumulation of persistent pesticides in animal tissues. Earlier warnings of potential problems had been suppressed by both governmental agencies and companies dedicated to manufacturing the chemicals (Carson 1970; Graham 1970).

Others brought more evidence that these insecticides affected animals other than insects, especially birds. In England, Cramp contrasted the decline in raptors after 1955 to that during World War II (Cramp 1963). During World War II, the Royal Air Force had destroyed peregrine falcons which indiscriminately killed passenger pigeons as well as less useful prey (Graham 1970), but, subsequent to the war, the populations of peregrines recovered. The decline that started in 1955, however, was gradual and continued into the 1960s. It also correlated with areas where pesticides were in heavy use. Dead birds and undeveloped eggs generally had high concentrations of DDT. Cramp concluded that it was the pesticides that caused the decrease in peregrine populations, both directly and by affecting reproduction. These correlations, as well as the apparent concentration of DDT in higher levels of the food chain, were also documented in North America (Woodwell et al. 1967; Wurster 1968).

The hypotheses that very small concentrations of persistent pesticides in the environment could become much larger in animals high on the food chain, and that damage to reproduction could have serious consequences for populations did not go unchallenged, but the challenges were generally oblique and did not provide evidence that falsified the hypotheses. For example, a highly controversial article in *BioScience* in 1967 argued that the low concentrations of DDT in air, water, and soil, and the prevalence of other naturally occurring poisons, negated the argument about the harmfulness of DDT (McLean 1967). McLean also attributed concern about reproduction to a "[preoccupation] with the subject of sexual potency."

Basic physiology and ecology played critical roles in understanding the relationship between persistent pesticides and remote effects. What the persistent insecticide evidence showed was that materials that are tied up in less readily metabolized substances, such as fats, may accumulate, so that the concentration increases, and concentrations in the air or water that are vanishingly small may become dangerously high by the time they reach higher order consumers, such as carnivorous birds. Thus, knowledge of basic metabolic physiology and the ecology of foods webs supported a theory of cumulative concentration even when a pesticide was hard to detect in the environment.

Secondly, these pesticides were often not lethal even at the doses found higher in the food chain. Rather, they reduced reproductive rates and, therefore, exerted very strong but subtle effects on population sizes, such that lethality was not the only criterion for the dangers of the chemicals. The dangers posed to populations by reducing their reproductive rates have been found to be generally applicable in the study of endangered species.

Many people originally doubted the potentially dire consequences of seemingly minor problems. Careful documentation of inferred effects convinced most of the scientific community, and much of the general public of the validity of the predictions, and led to action to reverse the situation. Raptor populations began to recover, offering further support for the hypotheses explaining the decline, that is, strategies for environmental mitigation served as a test of the hypotheses concerning the causes of the decline. Accepting these hypotheses as theories, which can be applied in similar situations, is the next critical step in enabling us to deal constructively in time with such subtle consequences of human actions.

Category 4: Subtle Activity With a Subtle Effect

Increases in CO_2 in the air are an inevitable consequence of burning biomass, whether fossil or recent, more rapidly than it is regenerated by photosynthesis or sequestered in the oceans or sediments. This relationship has been recognized at least since the end of the nineteenth century, as has the potential effect of increased CO_2 on global temperature (Firor 1990). The increase of CO_2 in the atmosphere is not obvious without the aid of sensitive instruments collecting data over a long time period. Such a record from the Mauna Loa Observatory in Hawaii indicates both yearly fluctuations related to the annual cycle of photosynthesis, and an increase in average CO_2 from about 315 to 345 ppm from 1958 to 1986 (Firor 1990). The question we must ask is: what may the effects of this increase be?

The effects on climate would be overall global warming and changes in precipitation related to changing temperature and global circulation. The causal connection is at first glance fairly direct, as CO_2 and other "greenhouse gases," e.g., CH_4 and O_3, allow solar radiation, including high energy wavelengths, to penetrate the atmosphere, but absorb outgoing infrared radiation, thus causing an increase in the heat loading of the atmosphere. Complicating factors relate to global circulation patterns and specially to the effects of different categories of clouds on both incoming and outgoing radiation. The effects of the climatic changes on organisms are even more difficult to predict. While it is likely that species' ranges would shift poleward, there are complicating effects due to differential migration rates, changing precipitation, and seasonal patterns. While there have been changes in species distributions in the last 30 or so years, none of these can be directly attributed to climatic change. Many relate to changed land-use or other human activities, some as subtle as the prevalence of bird-feeding stations providing winter seed for birds that would otherwise migrate south in the winter. Rising sea level is another potential consequence of global warming which is difficult to differentiate from post-glacial rise in sea level or to detect where isostatic rebound is lowering local sea level.

Here we are dealing with effects so subtle that even given the most extreme models of the effect of CO_2 on global climate, we would be unable to disentangle them from normal climatic fluctuations over the short term (Karl et al. 1991), and major, on-going cultural modification of landscapes obscures any subtle changes in the patterns of distribution of species. Does this mean that there really is no problem, or that we should ignore the potential until we see some obvious effects? Theory indicates strongly that this is not a good approach. In this case of a very subtle (at the moment) effect from a subtle cause, theory has provided us with a strong prediction of future change. This prediction is not in the nature of a hypothesis that can be tested over the reasonable term by empirical evidence, either experimentally or by observation. The inferences can be tested, however, by using models based on empirical evidence and theories of global climate. These models yield predictions which can be tested only by allowing the present situation to continue. They predict, however, dire consequences of non-action, so if one acts on the predictions by controlling emissions, there will be no way of ever corroborating the hypothesis, or even of falsifying it. In such a situation, where consequences are presently subtle, to the point of being undetectable, we must rely on the validity of models and their basic assumptions to assess dangers of continuing the activity. As both Rachel Carson and George Perkins Marsh observed, we cannot dismiss the importance of effects that we cannot detect at this time. This is an area that requires the most convincing application of present theory to future unknown events.

Conclusion

Establishing cause and effect in situations where the connection between the two is subtle requires many different techniques. The results that are obtained can be convincing, but because the temporal and spatial scale of the problems generally makes experimental confirmation inappropriate, they will always be open to different interpretations. Experiments can suggest relationships, but must be supplemented by observation and general theories to expand beyond very limited predictability. Where clear hypotheses have been used to explain the phenomena, the resulting general theories have become accepted because the application of the theory to applied problems has produced predicted results.

The reasons for lack of clear concensus must be made clear when interpreting the results of studies of subtle consequences of human actions. Because these studies often affect controversial social or economic policies, they must include explicit recognition that, generally, we cannot be sure of the "truth," but that we are searching for wisdom to guide decision-making (Levi-Strauss 1978). Paul Sears was aware of this problem in 1935 when he said that in order to develop public awareness and affect policy on environmental issues, "it is not merely soil, nor plant, nor animal, nor weather which we need to know better, but chiefly man himself" (Sears 1988:196). This is as true today as it was in 1935—we must know the basic science better, but basic science alone cannot provide solutions to problems that we anticipate but often cannot even detect. By using scientific methods, we can anticipate future problems before the subtle effects become obvious, and by applying this knowledge to public policy, deflect the worst of the effects.

Recommended Readings

Carson, R. (1970). *Silent Spring*. Fawcett Publications, Greenwich, Connecticut. Reprint of 1962 edition.

Marsh, G.P. (1965). *Man and Nature*. Belknap Press of Harvard University, Cambridge, Massachusetts. Reprint of 1864 edition.

Medawar, P.R. (1967). *The Art of the Soluble*. Methuen and Co., Ltd., London.

Sears, P.B. (1988). *Deserts on the March*. Island Press, Washington, D.C. Reprint of 1935 edition.

Whorton, J. (1974). *Before Silent Spring*. Princeton University Press, Princeton, New Jersey.

9
Land-use History and Forest Transformations in Central New England

David R. Foster

Most of the northeastern U.S. has been extensively altered by human activity over the past 200-300 years. In New England, much of the landscape was deforested, farmed in diverse ways, and eventually allowed to reforest naturally. Any understanding of the forest vegetation and ecosystems of this broad region, therefore, requires close consideration of human impacts, past and present, and their continuing effects over the modern landscape (Cronon 1983). At the Harvard Forest, the effect of human history on structure and management of forests in central New England has been a focus of study for foresters (Fisher 1925, 1931; Spurr 1956), soil scientists (Gast 1937; Griffith et al. 1930), economists (Gould 1942; Black and Brinser 1952; Barraclough 1949), and ecologists (Fisher 1928, 1933; Raup and Carlson 1941). Today the Long-term Ecological Research program continues to compare and contrast the effects of human and natural disturbance processes (Foster and Smith 1991).

One major theme that emerges from these diverse studies is that at a range of scales, from the stand to the region, the forest vegetation of central New England has been in a constant state of flux in response to dynamic and ever-changing modes of human activity. Changes in the structure, composition, and pattern of forest during the last three centuries have produced a sequence of unique landscape mosaics that are distinctly different from the landscape during the pre-settlement period. The human impacts to the forest ecosystem have not only altered the type and arrangement of forests, but the manner in which the forests function and respond to natural processes. In turn, these changing conditions and responses may affect the research emphases that scientists have pursued in this region and the conclusions that have been drawn.

In the following discussion I examine forest change across different spatial scales resulting largely from the impact of European settlers. First, the history and underlying causes of human impacts are outlined. On a township to regional scale, changes in forest extent and pattern, as well as composition and structure, are examined from the pre-settlement period. The complexity and dynamics of these forest transformations are then investigated through the detailed history of pre-settlement and post-settlement disturbance processes and the dynamics of a single woodlot that was extant throughout the period of intensive agriculture.

Regional Characteristics of the New England Landscape

Physical and Biological Features

The New England states, excluding Maine, form a roughly rectangular area 250 by 450 km in size that extends north and west from the Atlantic Ocean. Physiographically, the region consists of six broad areas, the coastal lowland, central uplands, the Connecticut River Valley, and the White Mountains, Green Mountains, and Taconic Mountains. The geological substrate is comprised predominantly of acidic, nutrient-poor material with surficial deposits left by the last glaciation 10,000-13,000 years ago. In general, the soils are shallow and bedrock is extensively exposed.

Substantial variability in regional climate results from elevational and coastal-inland gradients. Average annual rainfall exceeds 100 cm and is evenly distributed through the year. Summer temperatures (July) average 22°C whereas winter averages drop to -4°C (January) in inland locations. Regional differences in growing season length exceed three weeks between the south coastal and northern locations.

The regional vegetation changes latitudinally, with local variation due to elevation in the Connecticut Valley and northern mountains (Fig. 9.1). Northern hardwoods-conifer forest extends southward into northern Massachusetts. Important hardwood species in this forest include sugar maple (*Acer saccharum*), beech (*Fagus grandifolia*), yellow birch (*Betula alleghaniensis*), paper birch (*Betula papyrifera*), and red maple (*Acer rubrum*). Among the conifers, red spruce (*Picea rubens*) and balsam fir (*Abies balsamea*) are common in the north, whereas hemlock (*Tsuga canadensis*) and white pine (*Pinus strobus*) increase to the south. Southern New England forests (Central Hardwoods) include more oak (*Quercus alba*, *Q. velutina*, *Q. rubra*), gray birch (*B. populifolia*), and hickory (*Carya ovata*, *C. cordiformis*), along with red maple and hemlock. A transition forest including elements of both the southern and northern community occurs across central Massachusetts, up the Connecticut River Valley, and through eastern New Hampshire. A distinctive vegetation of pitch pine (*Pinus rigida*) and

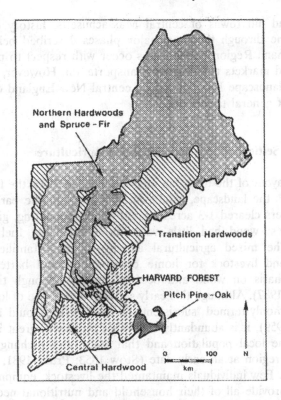

Figure 9.1   Forest vegetation of New England showing the Harvard Forest in Petersham, Massachusetts, and the outline of Worcester County, Massachusetts, USA.

oak (*Q. ilicifolia, Q. stellata*) species occurs on sandy soils across Cape Cod and inland on outwash deposits.

Post-settlement History of Central New England

The forest changes during the past 300 years are a response to a human history that was dynamic in population size, technology, mobility, economic structure, and intensity of interaction with the landscape. These factors, in turn, were largely driven by broader social, economic, and scientific developments operating externally to the township and region. The history of human activity and, therefore, of forest change appears to be much more a consequence of this external social environment than a result of changes in the physical environment (e.g., climate) or edaphic conditions (e.g., soil fertility) (Raup and Carlson 1941; Black and Westcott 1959). As summarized by Hugh Raup (1966), "The land itself did not change. Only the people's ideas changed in response to...ideas brought to bear from outside the region."

In the upland hill towns of central Massachusetts, history of land-use activity has gone through the four major phases described below for the town of Petersham. Regional differences occur with respect to proximity to urban areas and markets or access to transportation. However, similarities in human and landscape histories across central New England corroborate the existence of general trends.

1730 to 1790: Settlement and Low Intensity Agriculture

Following the layout of the township of Petersham in 1733, the first settlers dispersed across the landscape, establishing homes and the earliest farms. Individual farmers cleared 1-3 acres annually through cutting, girdling, and burning of trees; wood had little value other than for fuel and local construction. The mixed agricultural activity of most families included raising crops and livestock for home consumption and barter, with an increasing emphasis on wood and beef production through time (Jones 1991; Garrison 1987). Although this early settlement lifestyle of low-intensity farming is commonly termed "subsistent" (Bidwell 1916; Gould 1942; Black and Westcott 1959), it is abundantly clear that there was great interdependence within the local population and that trade and exchange occurred throughout the region at an early date (Stow 1853; Pruitt 1981; Baker and Patterson 1986). Few individuals maintained the livestock, equipment, skills, or acreage to provide all of their household and nutritional needs, rather, there was a continual trade of labor, services, and goods (Gates 1978; Garrison 1987). Roads were poor and yet cattle and sheep could be driven to regional markets. Trips to Boston, Hartford, Springfield, and Worcester are noted regularly in accounts of the day (Bogart 1948; Gates 1978; Rothenberg 1981; Baker and Patterson 1986).

1795 to 1860: Commercial Agriculture and Diversified Industry

The late eighteenth century brought a major transformation in the economic setting of central New England and in the focus of human activity (Bidwell 1916; Baker and Patterson 1986; Merchant 1989). With regional population growth, the development of villages and small urban areas, and improved transportation, the hill towns of central New England developed a market focus for agricultural products and small industrial goods (Davis 1933; Raup 1966; Jones 1991). Improvements in the turnpike and railroad system across Massachusetts allowed access to distant markets and created new local ones in the developing mill towns (Kimenker 1983). Farmers in towns like Petersham responded to these market opportunities by greatly increasing their production and by concentrating on a restricted number of products (Baker and Patterson 1986). Specialization in perishable goods (beef, cheese, and butter) and bulky items (hay, firewood, potash) involved the increasing use of marketmen and drovers and the development of a cash economy (Pabst 1941; Munyon 1978; Baker and Patterson 1986). Forest land was cleared extensively for agricultural development and forest products; at the peak of forest clearance around 1860, many of the hill towns in central New England supported forest on less than 25% of the uplands. Home industry produced such wares as straw hats and shoes (Gould 1942), whereas local workshops manufactured wooden wares and leather goods, and processed farm products (Brown 1895). The population peaked in many hill towns near the middle of the nineteenth century during a period of relative prosperity and stability (Doherty 1977).

1865 to 1935: Agricultural Decline and Commercial Logging

The development of regional urban centers and the expansion of valley towns, based on industrial development presented an attractive alternative to the rural life for many villagers. This, plus agricultural competition from the midwest and general changes in social structure brought on by the Civil War and industrialization, prompted many families to sell their hill farms in the late 1800s (Chase 1890; Peters 1890; Brown 1895). Lands were consolidated by neighboring farmers (Torbert 1935; Gates 1978) or were purchased by outsiders on speculation or as country estates (Lutz 1938; Mann 1889; Currier 1891). Marginal farmland was neglected and allowed to reforest naturally. In the 50-year period before 1900, more than 9 million acres of farmland was reforested in New England with more than one million of these located in Massachusetts (Barraclough and Gould 1955). By 1920, over 30% of Massachusetts villages had lower populations than in 1870 (Black and Brinser 1952). Agricultural activity became geographically concentrated around the manufacturing and urban centers and focused on dairy products, wood, and hay (Davis 1933; Pabst 1941; Black and Brinser 1952).

In Petersham and much of central New England, reforestation of abandoned agricultural fields with even-aged stands of white pine or hardwoods including chestnut stimulated the development of a thriving, though relatively short-lived, timber industry (Raup 1966; Barraclough and Gould 1955). Producing lumber, container materials (boxes, crates, barrels), and furniture, this industry prompted broad-scale cutting in the late 1880s (Gould 1942), peaked in 1910 (Raup 1966; Cook 1961; Black and Westcott 1952), and declined through 1930. Much of the economic return from this activity went to out-of-town landowners and lumber mill operators (Lutz 1938).

1935 to 1990: Modern Period

Through the 1920s and beyond, the once-thriving woodworking industries of central New England declined as new forms of packaging replaced wood containers and as the quantity and quality of old field white pine (*Pinus strobus*) decreased (Fisher 1925). Farming activity lessened with declining milk prices and the rising labor costs and tax burden on farmers (Gould 1942). Regionally, there occurred an increasing shift to part-time farming, continued abandonment and coalescence of farms, and intensification of effort and output on the remaining agricultural lands (Wheeler and Black 1954; Black and Brinser 1952). This process is epitomized in the town of Petersham where only a single family continued farming in 1991, but on rented land.

Commencing in the 1930s and continuing after World War II, most of the hill towns of central New England experienced an increase in population based on a steady influx of older couples and commuters deriving their income in neighboring towns or distant cities. This new population is less connected to and less dependent on the land, which is being worked less intensively than at any time in the preceding 250 years. The extensive network of former secondary roads and laneways has been largely abandoned and the nearly continuous forest is aging. Local firewood cutting, low-intensity timber management, high-grading of quality trees, and small-scale clearing of new house lots are the major direct impact on the forest. Commercially, the current population is largely removed from the land.

Land-use Impacts on the Forest

The history of human activity in central New England has resulted in a series of major transformations of the forest vegetation across the landscape and at individual sites. On a township or regional basis, changes have occurred in terms of the extent and pattern of forest cover and the structure and composition of the extant forests.

Forest Cover and Pattern

Information on the pattern and geography of land clearance in New England awaits the painstaking assessment of individual tax and property records, which has been accomplished for only limited areas to date (Raup and Carlson 1941). These data and maps will be essential for addressing a series of important questions: 1) what physical, logistical, and social factors influenced the pattern of land clearance—e.g., were soils, forest type, and agricultural concerns more important than accessibility and transportation,

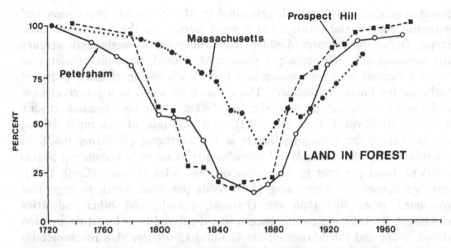

Figure 9.2. Changes in the percentage of the land area covered by forest during the historical period in the state of Massachusetts, town of Petersham, and the Prospect Hill tract of the Harvard Forest. Information is derived from the following sources: Dickson and McAfee (1988), MacConnell (1975), Rane (1908), and Baldwin (1942) for Massachusetts; Raup and Carlson (1941), Anonymous (1959), MacConnell and Niedzwiedz (1974), Cook (1917), and Rane (1908) for Petersham; and from Foster (1991) and Spurr (1950) for Prospect Hill.

or was land clearance an exploratory process determined more by the geography of ownership patterns and haphazard processes; and 2) did clearance involve a general process of expansion of agricultural areas from the center of town and centrally located farms, or did it occur through progressive fragmentation resulting from dispersed activities?

The information that is available suggests that settlers arriving to Petersham in the 1730s found scattered Indian clearings along the broad ridge running north-south through the center of town (Coolidge 1948; see Botts 1934; Bogart 1948; Donahue 1983 for other information on the use of Indian fields and villages in central New England). The availability of Indian sites and the preference by settlers for high ground which is usually less stony, better drained, and more fertile than valleys (Frothingham 1912; Black and Brinser 1952), led to the establishment of Petersham along this ridge. Forest clearance proceeded at a slow though increasing pace through the late 1700s (Fig. 9.2). The average farmer cleared less than 5 acres of land per year and, during this period of low-intensity agriculture, the overall clearing rate for Petersham was approximately 130 acres per year (E.M. Gould, unpublished data). Although the specific pattern of deforestation is unknown, most farms were located along the central ridge and a secondary ridge to the east, and by 1790 only three farms were found west of the village (Coolidge 1948). The general pattern appears to follow that of dispersed settlement and clearing on favorable lands without the development of a compact village center (Clark 1970).

Through the middle of the nineteenth century, the demand for firewood, forest products, and farmland, generated by the increasing population and extensive commercial activity, had a great impact on the forest and landscape. Deforestation rates doubled from those of the eighteenth century (to approximately 260 acres per year; E.M. Gould, unpublished data) as farmers cleared land for pasture and hay. Much of the clearing provided fuelwood for home and industry. The reliance on wood as a source of heat and energy continued through the late 1800s, when coal became readily available (Baldwin 1942; Cook 1961), and was one of the main driving forces altering the remaining forests in the northeast (Williams 1982). In the mid-1840s, the railroads in Massachusetts alone were consuming 54,000 cords of wood per year to fuel trains on 560 miles of track (Cook 1961). Average households were using 15-20 cords per year, which in aggregate consumed more fuel than the charcoal, potash, and other industries combined. By 1850, statewide concern was displayed over forest decimation (Cook 1961) and the Massachusetts General Court issued a recommendation that forest plantations be established in order to address future wood requirements for agriculture, manufacturing, and home fuel (Massachusetts General Court 1846).

For the last 150 years, which encompasses farm abandonment and reforestation, federal and local census records and maps do provide good information on the rate and pattern of forest change for towns like Petersham. A rough map of roads, houses, and woodlands for the township in 1830 depicts the landscape at the height of deforestation and intensive farming activity (Fig. 9.3). The remaining forest areas are indicated as isolated woodlands tens to hundreds of acres in size. Although this map may ignore very small woodlots or recently cut areas, it documents that large forested areas persisted in swamps, rocky and wet lowlands, and steep, ledgy areas. The distribution of stone walls, most of which date from the late 1700s to early 1800s, provides an indicator of farming intensity and, indirectly, of uncleared areas (Fig. 9.3). The densest areas of stonewalls correspond well with the most fertile agricultural soils, whereas areas of poor soils overlap with the woodland areas on the 1830s map. Remaining woodlots were repeatedly cut and were often grazed to the point of being wooded pasture (Cline et al. 1938).

Reforestation, occurring upon the neglect or abandonment of pasture lands, commenced as early as the 1840s in the town of Petersham, and somewhat later across central New England (Fig. 9.2). Abandonment rates increased sharply through the late 1800s, when nearly 300 acres were reforested annually, slowed from 1900-1940 (approximately 35 acres per year), and have been decreasing since that time as nearly the entire township has become covered with forest (MacConnell and Niedzwiedz 1974; E.M. Gould, unpublished data; Fig. 9.4). The spatial pattern of reforestation has involved the gradual expansion of individual woodlots and progressive coalescence of adjoining forest lands (Fig. 9.3). The driving factor behind this process is apparently that the abandonment of agricultural

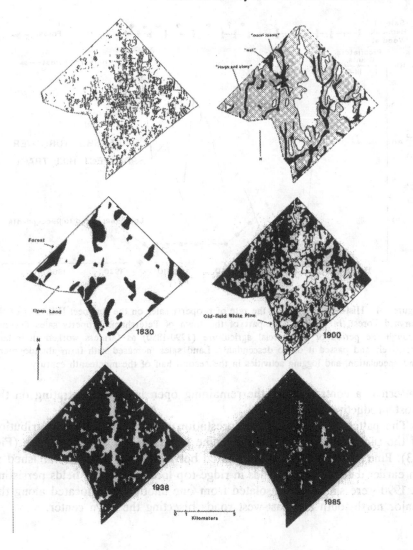

Figure 9.3. The township of Petersham, Massachusetts, depicting agricultural suitability (white, shaded, and black), stonewalls, and forest cover for the periods 1830, 1900, 1938, and 1985. In the forest map for 1900, shaded areas indicate agricultural land that was abandoned between 1870 and 1900 that developed forest of white pine. Stone walls and agricultural land during the abandonment period (mid 1800s to present) are concentrated in areas of more productive soil. Maps are compiled from the atlas of Worcester County (1830), Worcester County Land Use Project (1930), and analysis of aerial photographs for 1985.

lands proceeded across marginal areas in valleys, slopes, and poorly drained sites adjacent to existing woodlands, and at the back part of individual farm properties. Thus, reforestation appears to progress upslope from steep or poorly drained areas to lands of gentle, upland relief, from back lots toward roadways, and from poorer to better agricultural lands. The overall

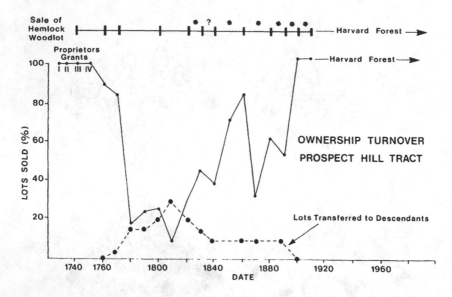

Figure 9.4   Historical changes in the rate of property sales on the Prospect Hill tract of the Harvard Forest, in the northern part of the town of Petersham. Property sales declined through the period of commercial agriculture (1790-1850) as farmers worked their land intensively and passed it on to descendants. Land sales increased with farm abandonment, land speculation, and logging activities in the second half of the nineteenth century.

pattern is a contraction of the remaining open land, concentrating on the most productive ridge-top sites.

The pattern of progressive reforestation can be seen in the distribution of the successional, old-field white pine stands in the 1900 landscape (Fig. 9.3). Pine stands are generally situated between older forests established at an earlier date, and open fields in ridge-top locations. Open fields persisting in 1990 were small, often isolated from one another, and located along the major north-south and east-west roads bisecting the town center.

Compositional and Structural Changes in the Forests

Pre-settlement Environment and Forest Composition

Based on palynological studies and reconstructions of individual forest histories, it appears that stand-regenerating and broad-scale disturbance processes, such as intense fire or catastrophic windstorms, occurred relatively infrequently in the pre-settlement landscape of central New England (Stephens 1955; Foster and Zebryk 1993). The overall abundance of charcoal in lake and swamp sediments in this region is low (P. Schoonmaker, T. Zebryk, D. Gaudreau, pers. comm.) and suggests an intermediate occurrence of fire between that of the frequently burned coastal regions of eastern Massachusetts (Patterson and Backman 1988; Winkler 1985) and locations in the western hills of the state where no fires have been detected in the previous 1000 years (Backman 1984). Detailed investigation of one lowland site at the Harvard Forest recorded eight fires in the last 7800 years, indicating that fire is an important though infrequent disturbance (Foster and Zebryk 1993).

Evaluation of the disturbance regime of windstorms relies on the single study by Stephens (1955), although an assessment of the impact of historical storms to the New England landscape is in progress. (E. Boose, pers. comm.). Stephens' study indicated that there were major storms in the 1500s and early 1600s as well as the historical storms in 1788, 1815, and 1938 (Smith 1946). These storms would have had highly heterogeneous impacts on the landscape, depending on their intensity, wind direction, and the interplay of landscape physiography and structural and compositional characteristics of the vegetation (Foster and Boose 1992).

Human activity, notably Indian clearing and possibly burning for fields and villages, created local openings in the forest. However, these impacts would have been minor on a regional scale, for example, not showing up on regional pollen diagrams (Patterson and Sassaman 1988). Structurally, therefore, the pre-settlement forest landscape was comprised largely of older growth and uneven-aged stands with occasional openings or even-aged patches resulting from natural and human disturbance. A combination of gap dynamics and patch formation, resulting from chronic background mortality and broader-scale damage, would create a mosaic landscape structure.

Three sources of information provide an insight on the composition of the pre-settlement forest. Analysis of observations and relative abundance data contained in the descriptions by Peter Whitney (1793) of the 62 townships in Worcester County describe phytogeographical trends that are indistinguishable from those of today. Forest vegetation types corresponding to central hardwoods, transition hardwoods, and northern hardwoods (cf. Westveld 1956) are arranged on a southeast to northwest axis across Worcester County and follow the elevational and latitudinal gradient (Fig. 9.5; Foster 1992; Raup and Carlson 1941). Central hardwood communities were dominated by hickory (*Carya*) and oak (*Quercus*) species, chestnut (*Castanea dentata*), and white pine. Transition hardwood forests included these taxa (with less hickory) plus greater amounts of red maple (*Acer rubrum*), birch (*Betula* spp.), ash (*Fraxinus americana*), beech (*Fagus grandifolia*), hemlock (*Tsuga canadensis*), and pitch pine (*Pinus rigida*). In the few northern towns represented by northern hardwood forest, hickory,

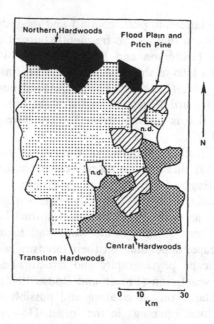

Figure 9.5. Map of Worcester County, Massachusetts, depicting the major forest vegetation zones as independently derived from an analysis of descriptions by Peter Whitney (1793) for the mid eighteenth century. Modified from Foster (1992). The Flood Plain and Pitch Pine forest is a variant of the Transition Hardwood Forest. Compare this historical map with the modern vegetation of Worcester County outlined in the New England map by Westveld (1956) in Fig. 9.1.

chestnut, and pitch pine were unimportant, and sugar maple (*Acer saccharum*), yellow birch (*Betula leutea*), and white birch (*B. papyrifera*) were common. The forests of Petersham were described by Whitney as dominated by oak across the uplands with birch, beech, maple, ash, elm (*Ulmus* spp.), and hemlock common in lowlands. Interestingly, he described chestnut as a species that was increasing on the uplands as a result of settlement activities.

Corroboration of Whitney's description of Petersham is provided by witness tree data collected during the laying out of original lot boundaries in the town (G. Whitney, unpublished data). Species comprising the 384 trees sampled include oak (32%, mixture of black [*Q. velutina*], white [*Q. alba*], and red [*Q. borealis*] oak), pine (21%, primarily, white pine), chestnut (9%), maple (9%), and hemlock (9%) with lesser amounts of beech, birch, ash, poplar (*Populus spp.*), and hickory.

Finally, pollen data from a lowland swamp in the center of the Harvard Forest provide information on the composition of the regional vegetation (Fig. 9.6; Zebryk 1991). Pre-settlement sediments contain high levels of beech, pine, oak, birch, and hemlock with lesser amounts of red maple, sugar maple, and chestnut. The three sources of information concerning

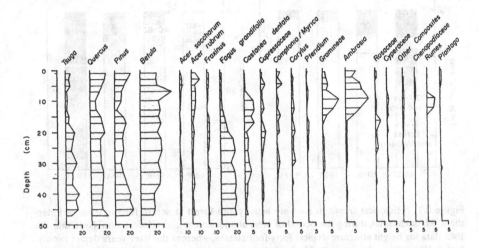

Figure 9.6. Pollen diagram from the Black Gum Swamp in the Harvard Forest (Petersham, Massachusetts) depicting changes in the regional vegetation of central New England. The presettlement vegetation is represented by the lower portion of the diagram, initial settlement activity (ca. 1750) is indicated by the rise in *Gramineae* and weed species (*Ambrosia, Rosaceae, Compositea*) at 30-32 cm, and the chestnut blight (ca. 1910-15) occurs between 7 and 10 cm. Data from Zebryk (1991) and modified from Foster et al. (1992).

these early forests thereby suggest that the broad vegetational distinction of this area as transitional hardwood forest was applicable (cf. Raup and Carlson 1941) and that the landscape may have been comprised of oak, chestnut, and pine forests on the uplands, forests of beech, birch, maple, and ash with hemlock in the valleys and lowlands, and swamps dominated by red maple or by spruce (*Picea*), larch (*Larix*), and black ash (*Fraxinus nigra*). Beech, sugar maple, chestnut, and hemlock were more abundant in early forests than at present (Fig. 9.6; Foster and Zebryk 1993).

Post-settlement Forests

With the arrival of European settlers, structural changes in these forests were initiated with progressive forest clearance accompanied by selective logging of specific size categories and species of tree. Grazing of livestock in forests and the overgrowth of neglected pastures led to the development of a new category of vegetation—the wooded pasture or brush pasture, much of which was included in census records under the large category of unimproved land (G. Whitney, pers. comm.). Due to grazing and repeated

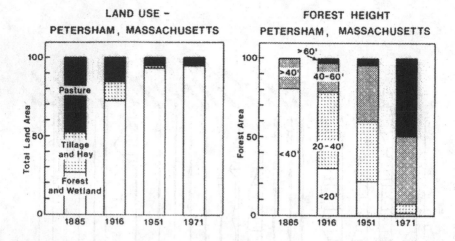

Figure 9.7. Historical trends in land-use activity and forest structure for the town of Peter-sham, Massachusetts, during the period of farm abandonment and reforestation. Note that the 1885 data for height structure depicts forty-foot classes, whereas the later years depict twenty-foot height classes. As the township became increasingly covered with forest, there occurred a progressive aging and height increase in the extant forest. Sources include Cook (1917), Rane (1908), and MacConnell and Niedzwiedz (1974).

cutting, woodlots and forest remnants must have resembled sprout woods rather than the original older growth forests (Barraclough 1949; Donahue 1983; see Russell, Chapter 8, this volume, for similar conclusions in a different region).

Essentially, no historical records describe the composition of the forests through the mid nineteenth century with the exception of occasional references in journals, town histories, and letters. Thus, we have no details on the spatial arrangement and specific structural and compositional characteristics of this changing vegetation. However, the palynological record depicts the major regional changes which include a striking decrease in northern hardwood species (beech and sugar maple), hemlock, pine, and oak and a sharp increase in chestnut (Fig. 9.6). Specific agricultural activities reflected in the pollen spectra include hay production (*Gramineae*), grazing (*Comptonia, Pteridium, Rosaceae*), and tilled and marginal lands (*Chenopodiaceae, Ambrosia, Rumex,* and *Plantago*).

Land-use practices since the mid-1800s have produced a mosaic of essentially even-aged patches of forest (Fisher and Terry 1920; Fig. 9.7). Although the process of land abandonment generally occurred through neglect often accompanied by grazing, the constituents of the resulting forest canopy generally dated to within 20 years of abandonment (Fisher 1931). The establishment of even-aged white pines in former fields initiated a lumber industry in the 1870s that operated through the use of portable sawmills and a practice of clear-cutting (Ahern 1929; Fisher 1931; Cline et al. 1938). Advanced regeneration of hardwoods below the pines was generally cut or damaged during logging or was burned with the slash. Thus, an even-aged forest of hardwood sprout generally followed the pines (Thoreau 1962). Overall, the development of even-aged forests through the early part of the twentieth century can be ascribed to these two major processes (Barra-clough 1949): 1) repeated culling of woodlands; and 2) reversion of agricultural land to woodland.

The 1938 hurricane added the latest broad-scale structuring to the forest landscape. On exposed sites, conifer stands over 30 years old were com-pletely windthrown (Black and Brinser 1952) and hardwoods, though less susceptible, were severely damaged (Foster 1988; Foster and Boose 1992). Across Petersham, approximately 50% of the forest suffered extensive windthrow. Salvage operations throughout New England involved clear-cutting of severely damaged areas and burning of slash, thereby accentuating the conversion of conifer stands to even-aged sprout hardwoods (NETSA 1943).

Forestry practices since the 1930s have progressively emphasized a combi-nation of patch cutting and selective removal. Thus, through forest develop-ment, small-scale natural disturbance, and reduced harvesting intensity, the forests are aging and becoming more diverse in terms of within-and-between stand age-structure (Fig. 9.6; MacConnell and Niedzwiedz 1974). The predominant forests in the landscape, however, are 50-80 years old with the majority post-dating the 1938 hurricane.

Overall through this century, major changes in the composition of forest vegetation reflect the dominant influence of cutting practices and the 1938 hurricane, progressive aging of the forest and the introduction of major pathogens (Cline et al. 1938). Timber harvesting and the selective impact of the hurricane have resulted in a declining importance of white pine. Other species that seeded into fields and abandoned areas, such as gray birch (Betula populifolia), aspen (Populus grandidentata), pin cherry (Prunus pensylvanica), and red maple (Lutz 1928), peaked in abundance in the first part of the century and have decreased as the forest has aged. Meanwhile, chestnut, which appears to be the species that increased most prolifically through the agricultural and logging period, has been converted to an understory sprout sapling by the chestnut blight (Paillet 1982). The recent trend in forest composition has been for a progressive increase in longer-lived and shade tolerant hardwoods (e.g., sugar maple, beech, yellow birch, and red oak) and especially hemlock (Foster 1992). Large, older growth hemlock are primarily restricted to areas that remained wooded throughout the settlement period. However, hemlock is common in understory situa-tions and is gradually increasing in abundance across the landscape.

Detailed History of a Woodlot Forest

In order to assess the magnitude of transformation of the forest vegetation resulting from human activity, it is instructive to examine the composition and dynamics of the remnant forests that persisted through the settlement period. Many of these forests are currently dominated by hemlock, pine, and northern hardwood species 75-125 years old. As a result of the relatively large size of the individual trees, the presence of typical "climax" species (hemlock, beech, red oak, yellow birch; cf. Nichols 1913), and the absence of evidence of recent human activities, certain of these stands appear to provide possible examples of the original forest cover of the area. In fact, considerable attention has been given to such forests in the past in an effort to understand the pre-settlement vegetation (Raup and Carlson 1941; Spurr 1956).

One such forest in northern Petersham was recently examined using paleo-ecological and reconstruction techniques in order to describe the dynamics, structure, and composition of the original vegetation, and to follow its history through the settlement period described above. The forest is a lowland stand of hemlock with white pine, red spruce, red oak, beech, and red maple. The area was part of an extensive woodlot that appears on the 1830 map of forests in Petersham and it occupies the central part of the Prospect Hill tract of the Harvard Forest (Foster et al. 1992).

Palynological investigation of sediments in a small hollow in the center of the stand indicate that over the past 7000 years the local vegetation has been dominated by hemlock, pine, and hardwoods (Fig. 9.8, T. Zebryk, unpublished data; Foster and Zebryk 1993). Spruce increased in abundance approximately 2000 years ago and chestnut migrated into the region and became a component of the local vegetation approximately 3000 years ago.

However, despite the long-term dominance of the site by a small number of species, the vegetation has undergone dynamic changes in response to fire (7650, 6650, 6150, 4700, 1900 B.P.), the hemlock decline (4700 B.P.) and human activity (250 B.P.). Following each of the disturbance episodes, the abundance of hemlock and associated species has declined and then recovered (Fig. 9.8). The nature of the successional sequence preceding this recovery has depended strongly on the type of disturbance and the pool of available species.

From 8000-2000 B.P., fire (and the hemlock decline which was coincident with a fire at 4700 B.P.) initiated a decline in hemlock and northern hardwoods, whereas birch, oak, and pine increased for a 300-1000 year interval after the disturbance. Following the arrival of chestnut to the region, it became a minor component of the forest and then increased prolifically after a fire at 1900 B.P. Chestnut was able to replace pine, birch, and oak as a successional component following disturbance, due to its tremendous sprouting capacity and growth rate. In the period extending

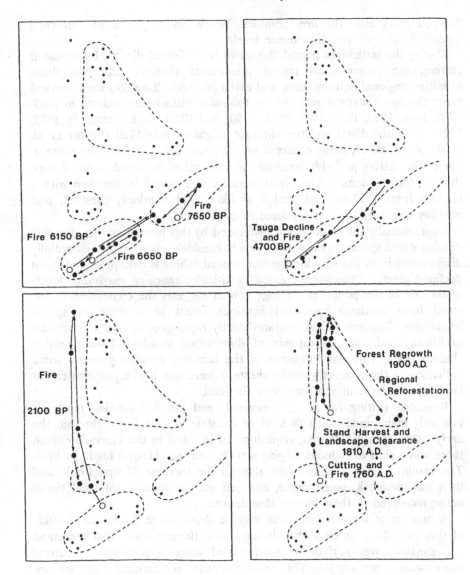

Figure 9.8.   Position on the first two DECORANA axes of pollen stratigraphic samples for the Hemlock Woodlot in the Prospect Hill tract of the Harvard Forest. Enclosed areas contain samples representative of major pollen zones and inferred vegetation types. Areas indicate the direction of sample movement through time from oldest (>9,000 B.P.) to modern. Progressive change in sample positions from 9,000-7,500 B.P. reflects species migration in response to environmental change (A). The period 7500-200 years marks a time of relatively stable vegetation of hemlock and hardwoods (B), with chestnut and spruce increasing approximately 2000 B.P. (C). The distinctive position of the post-settlement samples is indicative of the unique assemblages of pollen deposited at this time and is strongly controlled by the presence of native and introduced weed species (D). Modified from Foster and Zebryk (1992).

300-500 years after the fire, hemlock, beech, and sugar maple returned essentially to their pre-disturbance levels.

During the settlement period, the stand was affected directly by repeated cutting; the palynological record documents changes reflecting these activities, regional deforestation, and extensive agriculture. During historical times the forest was subjected to the following disturbance: cutting in 1765, 1790, 1830, 1870, 1893, 1900, 1913, 1940, and 1957; wind damage in 1815, 1921, 1938, and 1941; and the chestnut blight in 1914-1920 (Foster et al. 1992). Apparent clearcuts occurred in 1765 and 1870, whereas the remainder of the cutting probably involved the removal of firewood or small logs, through forest thinning. The forest woodlot was owned in common with a tannery from 1839 onward, and it is likely that hemlock, chestnut, and possibly oak were used as a source of tannic acid.

Vegetationally the stand was transformed by this human activity. Former dominant tree species in the forest—beech, hemlock, sugar maple—essentially disappeared from the very local pollen record, whereas oak, pine, and birch declined greatly. Chestnut increased prolifically, reaching maximum levels of 60-80% of the pollen assemblage, which suggests the conversion of the stand from northern hardwoods-hemlock forest to a stand of sprout hardwoods. Chestnut has an extraordinarily rapid growth rate and sprouts prolifically and, thus, is capable of dominating stands following cutting (Paillet 1982). This transformation of the hemlock woodlot provides some indication of the process whereby chestnut increased in the post-settlement landscape throughout southern New England.

Repeated cutting for poles, cordwood, and possibly tanning materials evidently maintained the high level of chestnut in this forest through the early 1900s. Shortly following acquisition of the land by the Harvard Forest, there was a shift to reduced logging activity and the chestnut blight in 1913. The resulting decline of chestnut allowed the increase of first birch, and then oak, hemlock, pine, spruce, and red maple. Sugar maple and beech never recovered to their former abundance.

A number of major conclusions may be drawn from the intensive study of this one site. It is clear that during pre-settlement times that the stand was dynamic over periods of hundreds of years in response to natural disturbance, especially fire. The post-disturbance vegetational sequence was dependent on the pool of available species; for example, the successional sequence changed greatly upon the immigration of chestnut. In all cases the species assemblage reverted largely to that which was present before the disturbance. Settlement activities generated a similar initial response in the vegetation as chestnut dominated the pollen rain. However, the repeated cutting of the stand and the broad-scale deforestation of the landscape created conditions that produced a unique vegetation in which hemlock is now abundant and beech and sugar maple have been essentially eliminated. Thus, as the modern landscape has become reforested and the hemlock stand is aging, the vegetation is developing along a trajectory that is distinct from any that has occurred in the past.

Conclusion

The examination of the woodlot stand at the Harvard Forest provides convincing evidence of the long-term dynamics of forests in this region and the different quality of the disturbance processes resulting from human activity. During the course of the last 300 years, the stand has been modified in long-lasting ways that differ considerably from the consequences of repeated fires and pathogen effects during pre-settlement time. Elsewhere in the landscape the transformations were even more severe; entire forests were replaced by open fields that were grazed, cut, or tilled. The entire pre-settlement flora including buried seed populations was eradicated locally from these sites. It is quite clear that although the major geographic trends in forest vegetation still hold, that the individual forests, their composition, structure, and arrangement in the landscape, are distinctly different from the primary woods, let alone the natural vegetation (Whitney 1992; Whitney and Foster 1990).

The post-settlement changes in this landscape have involved shifts in the extent and pattern of the forests, and apparent constant alteration in the structure and composition characteristics of the extant stands at any given time. These forest dynamics were driven by human activities embedded in an ever-changing cultural context. The forest picture has not been static because the human forces were and are continually changing.

Many questions remain. There is great uncertainty regarding the nature and arrangement of the pre-settlement forests and the influence of aboriginal populations and natural disturbance processes on them. The extent of change in forest cover throughout the post-settlement period is well-known but we know little about loss of species, changes in community complexity, and alterations in the relative importance of species. Of particular interest is the manner in which changes in community and ecosystem characteristics of these forests may have long-lasting impacts in their functional aspects. We know that the aggregate changes that have occurred affect the manner in which broad-scale and local disturbance processes operate (Foster and Boose 1992; Sipe 1990). These changes may have very important consequences for our interpretation of fundamental ecological attributes as well as management considerations (Metropolitan District Commission 1991). Preliminary studies suggest that land-use changes may have long-lasting impacts on the way in which modern forest soils process organic matter and nitrogen compounds (J. Aber, pers. comm.). This, in turn, could have far-reaching impacts on forest productivity, biosphere-atmosphere exchange, and ultimately on the global atmosphere (Steudler et al. 1989; Aber et al. 1989). To sort out these questions will require continual interdisciplinary efforts among social, physical, and biological scientists.

Recommended Readings

Aber, J., K. Nadelhoffer, P. Steudler, and J. Melillo. (1989). Nitrogen saturation in northern forest ecosystems. *BioScience* 39:378-386.

Cronon, W. (1983). *Changes in the Land. Indians, Colonists, and the Ecology of New England*. Hill and Wang, New York.

Merchant, C. (1989). *Ecological Revolutions—Nature, Gender, and Science in New England*. University of North Carolina Press, Chapel Hill.

Patterson, W.A. and A.E. Backman. (1988). Fire and disease history of forests. In: B. Huntley and T. Webb III, eds. *Vegetation History*, pp. 603-622. Kluwer, The Hague.

Raup, H.M. (1966). The view from John Sanderson's farm. *Forest Hist.* 10:2-11.

10
Variability in Lake Ecosystems: Complex Responses by the Apical Predator

James F. Kitchell and Stephen R. Carpenter

Introduction

Ecologists are typically among the last to appear on the shores of lakes already feeling the effects of humans. Deforestation, agriculture, fisheries, and urbanization have usually preceded limnological research. We should expect that most aquatic systems accessible to humans have already experienced substantial and profound changes before the first scientific evidence becomes available. This generalization is especially true for lakes whose fresh water is vital as a potable resource and whose drainage basin includes soils of agricultural value. The common result is expressed in cultural eutrophication (Likens 1972; Schindler 1977, 1981).

Less obvious than the increase in nutrient loading which causes eutrophication are the effects of humans on lake food webs. These can have equally profound effects on the basic productivity and variability of lake ecosystems (Carpenter 1988). Human exploitation of fish populations is virtually ubiquitous. Enhancements through stocking or introduction of exotics or both are also widespread. A growing body of evidence and experience demonstrates that management of food webs serves as a means for enhancing water quality (Shapiro et al. 1975; Kitchell 1992). Similarly, inappropriate or insufficient management can produce or amplify water quality problems (Gulati et al. 1990). Thus, management of anglers and the fish populations they alter, presents a special challenge of growing ecological importance (Shapiro 1990).

In a more general sense, there is a growing interest in and need for understanding of the mechanisms and the magnitude of ecosystem responses to fishery management and its effects on the nutrient loading that regulates basic algal productivity. The issues of ecosystem resilience and resistance are highly relevant in dealing with the multiple stressors placed on lentic food webs (DeAngelis et al. 1989; DeAngelis 1991).

111

Fishery exploitation pressures and enhancement activities differ among systems and tend to change rapidly due to the adaptive response of anglers. As apical predators, humans learn rapidly, are highly mobile, and can operate with devastating efficiency. In the larger ecological context, a trophic cascade derives from human actions and passes through all levels of local food webs. In some cases, effects are profound and persistent. In others they are transient or not readily apparent. In most cases, they are not anticipated.

While unexpected or poorly anticipated responses occur at many levels of manipulated food webs, changes in the zooplankton community are a predictable and common denominator. Zooplankton community dynamics are highly responsive to predator effects and have been successfully used as indirect and integrative sources of insights for entire ecosystems (Kitchell and Kitchell 1980; Kerfoot and Sih 1987; Mills et al. 1987; Carpenter and Kitchell 1988). Our analyses use the historical records of zooplankton as a basis for evaluation of food web responses (Leavitt et al. 1989).

This chapter describes lake ecosystem responses derived from: 1) large-scale change in lake productivity and its consequent effects on the trophic cascade; and 2) observations made as fisheries management activities inadvertently changed in food web interactions. One example is taken from the history of Lake Mendota where cultural eutrophication and its reversal are reasonably well documented and where a more recent, intensive effort has focused on management of food web interactions. Responses included increased variability in water quality characteristics; some were desirable while others were not. A second example is taken from the results of intensive salmon stocking of Lake Michigan where a highly successful sport fishery provided billion-dollar benefits. Salmon and the anglers that exploit them caused a cascade of effects expressed throughout the trophic network. Recent events in Lake Michigan suggest that this artificial system is undergoing extensive and generally negative changes.

Background

Lake Mendota in Wisconsin is among the most studied lakes in the world. Its general limnology and the history of its cultural eutrophication are well documented (Brock 1985). Evidence of the reversal of eutrophication and attempts to improve water quality through management of Lake Mendota's food web are detailed in a more recent volume (Kitchell 1992). Similarly, the history of Lake Michigan's food web is well described (Kitchell and Carpenter 1987; Kitchell et al. 1988a; Hartig et al. 1991). The following provides a brief overview of major dynamics pertinent to the analyses we have performed.

Lake Mendota

The plow was put to the rich prairie soils of Lake Mendota's catchment basin as Europeans settled into the region of Madison during the early 1800s. Nutrients released through the resultant increase in soil erosion inevitably found their way to the lake. They were enhanced by sewage from the growing villages and towns throughout its 200-square mile drainage basin and the increased use of manure on frozen fields. Each spring, runoff brought a nutrient load strongly correlated with the rapidly growing populations of humans and dairy cattle. The lake responded with massive increases in algal productivity and the consequent oxygen depletion of its hypolimnetic waters (Hasler 1947; Brock 1985).

Initiated in the 1960s, a program of diversion and treatment of sewage plus some improvement in agricultural practices, reduced the nutrient load. Unlike some of the most dramatic reversals of eutrophication (Schindler 1987), Mendota's century of sediment-stored nutrients and six-year hydraulic flushing time have slowed the manifestations of reduced nutrient load. Nevertheless, over the course of recent decades, Lake Mendota has evidenced gradual recovery (Lathrop 1992; Lathrop and Carpenter 1992).

Food web structure in Lake Mendota changed in concert with both the enhanced productivity and the shoreline development brought by growth of the City of Madison and neighboring communities. Loss of bordering wetlands reduced the spawning marshes for many fish species. Exploitation of fishes increased in correspondence with the growing human populations. Oxygen depletion of the hypolimnion precipitated major summerkills of cold water fishes (Magnuson and Lathrop 1992). Exotic species invaded and flourished (Nichols et al. 1992).

As detailed elsewhere (Vanni et al. 1990; Kitchell 1992), recent changes in management practices now emphasize reestablishment of the large, piscivorous fishes, reduced populations of zooplanktivorous fishes, and, therefore, maintenance of a plankton community dominated by large herbivores such as *Daphnia pulicaria*. These practices are intended to augment the reduction in nutrient load and, thereby, reduce both the frequency and magnitude of noxious algal blooms.

Lake Michigan

Although cultural eutrophication has been and continues to be an issue of importance in Lake Michigan, most of its manifestations have been expressed in nearshore water quality and substantially diluted by the immense volume of the lake. Of more general interest are the dramatic dynamics of the biological communities and the effects of humans on the species interactions expressed in a food web. Fish communities and fisheries of Lake Michigan have been in constant flux since the early part of this century (Kitchell and Crowder 1986). Intense fishery exploitation coupled with invasions by exotic species, notably sea lamprey (*Petromyzon marinus*) and alewife (*Alosa pseudoharengus*), resulted in the decline of most native piscivores. Local extinctions were common. With few piscivores present, alewife became the dominant planktivore, replacing a suite of native species—including several endemic ciscoes (*Coregonus* spp.)—that had served as the forage base for native piscivores and supported major commercial fisheries. Alewife populations burgeoned to the extent that massive die-offs were common, which resulted in major economic costs due to clogged water intakes and fouled beaches (Eck and Wells 1987).

Development of sea lamprey control measures during the 1950s and 1960s allowed reintroduction of top predators such as the native lake trout (*Salvelinus namaycush*) (Kitchell and Crowder 1986). During the 1960s, management agencies began stocking Pacific salmonids such as coho (*Oncorhynchus kisutch*) and chinook salmon (*O. tshawytscha*), rainbow trout (*O. mykiss*), and the European brown trout (*Salmo trutta*) in an attempt to create a biological control for the overly-abundant alewife (Scavia et al. 1988).

While management through salmonid stocking was first and most intensively developed for Lake Michigan, similar practices have since evolved in each of the five Laurentian Great Lakes. The assemblage of stocked salmonids now supports a sport fishery estimated to yield economic benefits approaching $3-4 billion (U.S.) per year in the Great Lakes region (Hartig et al. 1991). The economies of many coastal communities are now heavily dependent on these fisheries and the support services they demand.

Central to the maintenance of current fishery values in Lake Michigan is the alewife population that supports the stocked salmon and trout. Over the period of the 1980s, alewife populations in Lake Michigan declined approximately five-fold due, in large part, to the sustained, high rates of predator stocking (Stewart and Ibarra 1991). At present, there is a substantial and growing concern about the sustainability of this artificial system.

Analyses of the Human Role

Our goal in this analysis was to develop a general approach that would allow analyses across the full range of change in environmental conditions known from the periods of agricultural-industrial, and fishery influence compared with those that preceded them. Changes in the zooplankton serve as a continuous indicator of food web interactions derived from selective predation. They also provide an estimate of changes in the basic productivity that passes up through trophic levels. Accordingly, we employed data derived from the ecological archive of zooplankton remains present in recent sediments. Detailed descriptions of the paleolimnological methods for

Lake Michigan cores are presented in Kitchell and Carpenter (1987), while those for Lake Mendota are described in Kitchell and Sanford (1992).

Cores from Lake Mendota included a diversity of zooplankton remains. We concentrated on two indicators of trophic status and food web interactions. The rate of loading of total pelagic cladocerans served as a general indicator of system productivity. Among the cladocerans, the relative and absolute abundance of large *Daphnia pulicaria* served as an indicator of the intensity of predation by zooplanktivorous fishes as it did among the preferred prey for abundant pelagic fishes such as the yellow perch (*Perca flavescens*) and the lake herring or cisco (*Coregonus artedi*).

Evidence of food web interactions in Lake Michigan focused on the morphological response of one of the most common zooplankton, *Bosmina longirostris*, whose fossils are well preserved and abundant. *Bosmina* exhibits dramatic changes in the length of the mucron spine at the base of its carapace (Kerfoot 1974). The mucron functions as an anti-predatory device induced in embryos as a response by their brooding mother to the odor of certain large predaceous copepods. Spine length corresponds directly to the intensity of predation and, therefore, is generally indicative of the abundance of both the predators of *Bosmina* and other large-bodied zooplankton such as the herbivorous Cladocera (e.g., *Daphnia* spp.). Mucron length, then, serves as an indicator of the relative abundance of fishes which are selectively predaceous on large zooplankton and, by extension, of the relative abundance of large piscivores that prey on the zooplanktivorous fishes. When mucrons are short, large copepods (and *Daphnia*) are rare due to intense predation by size-selective zooplanktivores such as the alewife. When *Bosmina* mucrons are long, the food web is strongly influenced by abundant piscivores which prey heavily on alewife (Kitchell and Carpenter 1987).

Mucron length of *Bosmina* may seem a modest representation of the complexity embedded in a food web. It is, however, the only quantitative evidence of predator-prey interactions available from the paleolimnological record in Lake Michigan. Until other or better indicators are found, it is the best evidence.

Statistical Analyses

Our initial question centered on the mechanisms and magnitude of human influence on food webs in two lakes of distinctive histories. We wished to evaluate and discriminate the patterns or trends ascribed to altered basic productivity from those due to changes at the top of the food web. In a more general sense, we sought to evaluate the relative stability of responses due to both the enrichment of increased nutrient loading and the cascade of effects that derived from the apical predators—large fishes and the fisheries that preyed upon them.

Ecologists have used varied approaches which require very different kinds of data, to assess stability and related properties in natural systems (DeAngelis 1991). We have used a very simple approach appropriate for univariate time series such as those commonly derived from paleolimnology. We separated each time series into signal and noise components using the following model which is a simple version of the Kalman filter (Mehra 1979).

$$s(t+1) = \phi s(t) + w(t) \tag{10.1}$$

$$y(t) = s(t) + v(t) \tag{10.2}$$

The system state, s, is the value of the paleolimnological variate being analyzed at a specified time, t or $t+1$. System state is measured by observing y, subject to measurement error, v. System state depends on the immediate past state through the coefficient ϕ, and on shocks to the system, w. An important advantage of this model is that two independent sources of variability are recognized: error due to measurements (v) and error due to unpredictability of the system (w). In this analysis, we viewed w as a measure of noise or variability in system state that cannot be explained by past states or measurement error. As a measure of signal, we used the system state, s.

At any given time, $t+1$, we can obtain two independent estimates of system state, $s(t+1)$: that given by past system performance (eq. 10.1) and that given by the most recent measurement (by using $y(t+1)$ in eq. 10.2). The Kalman filter is a mechanism for combining these two estimates to yield a weighted average with minimal variance (Shumway 1988). Details are given by Mehra (1979) and Shumway (1988). As a measure of noise, we used

$$w(t+1) = s(t+1 \mid t+1) - \phi s(t \mid t) \tag{10.3}$$

where $s(t+1 \mid t+1)$ and $s(t \mid t)$ are the Kalman filter estimates of system state based on all measurements up to time $t+1$ and time t, respectively. Values of ϕ, $s(t+1 \mid t+1)$, and $s(t \mid t)$ were calculated using the program of Shumway (1988). Plots of signal that follow are time series of s. Because w can be positive or negative, we plotted the analogue of the standard deviation as a measure of noise: $(w^2)^{0.5}$.

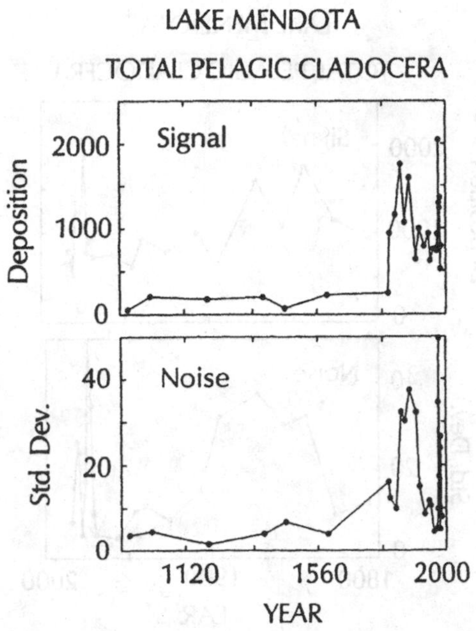

LAKE MENDOTA

TOTAL PELAGIC CLADOCERA

Figure 10.1. Long-term records of deposition rates (individuals per square cm per year) for total pelagic cladocera in deep sediments of Lake Mendota, Wisconsin, USA. "Signal" refers to rates of input and corresponds to productivity. "Noise" refers to variability. See text for details of statistical analyses.

The Changing Human Role in Lake Ecosystems

Lake Mendota

Deposition rates for all members of the pelagic cladoceran assemblage in Lake Mendota indicate an extended period of low and relatively stable productivity punctuated by a strong response to the advent of agriculture in the drainage basin during the early 1800s. System productivity increased rapidly then and by nearly an order of magnitude (Fig. 10.1). Cores from Lake Mendota show an abrupt color change from buff-colored marl at greater depths to black sediments above the time that roughly corresponds to the first appearance of ragweed pollen (Kitchell and Sanford 1992). Lake Mendota either developed or experienced increased hypolimnetic oxygen depletion during the summers that followed appearance of European agriculture.

The initial surge in productivity was followed by a gradual and continuous reduction to roughly half that of the response developed when soils were first disturbed. Similarly, the variability of deposition rates was initially high, then reduced during much of the early part of the twentieth century (Fig. 10.2). We interpret this response as an initial major change in trophic status

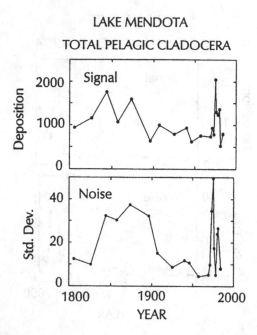

Figure 10.2.   Fine-scale resolution of data from Figure 10.1 for the period of 1800 to 1985. Other descriptions as in Figure 10.1.

of the lake followed by a stabilization as the rate of growth of agricultural development was reduced.

Although the paleolimnological record includes nearly ten centuries of evidence, the maximum, minimum, and most highly variable deposition rates occurred during the latter decades of this century (Fig. 10.2). Those extremes correspond with pulses of nutrient loading due to urbanization and a series of dramatic food web changes associated with fish community dynamics. The latter are better represented by the responses of *Daphnia pulicaria*, the largest of the zooplankton (Fig. 10.3).

Large *Daphnia* were a modest component of the total pelagic cladoceran fauna in the early history of Lake Mendota and remained at relatively low levels even during the period of agriculturally enhanced productivity. We interpret the low abundance of large *Daphnia* as strong evidence of a well-developed population of size-selective zooplanktivorous fishes whose functional and numerical responses paralleled the overall increase in system productivity. In other words, pelagic food web structure changed very little during the initial stages of eutrophication because increased nutrient loading simply translated into greater populations and biomass at higher trophic levels.

Daphnia pulicaria rapidly increased to dominance shortly after the beginning of this century. During this time, hypolimnetic oxygen depletion intensi-

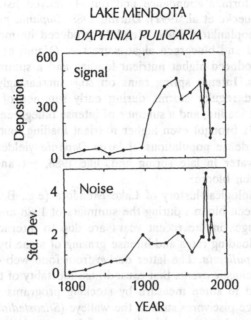

LAKE MENDOTA
DAPHNIA PULICARIA

Figure 10.3. Deposition rates of *Daphnia pulicaria* in Lake Mendota over the period of the past two centuries. Other descriptions as in Figure 10.1.

fied during summer months to the extent that cold water fishes were severely stressed by the restriction of anoxia below and high temperatures above their refuge on the thermocline. Local newspapers documented periodic summerkills of the abundant lake herring. Truckloads of dead fish were removed from local beaches.

Large *Daphnia* responded to the catastrophic reduction of one of its major predators and increased nearly tenfold in abundance. *Daphnia pulicaria* remained abundant throughout the middle of this century, then declined to lower levels through the last 20 years. This change corresponded with the improvement in water quality that followed sewage diversion in the 1960s and with the reappearance of strong year classes of cisco in the early 1970s (Magnuson and Lathrop 1992). Strong year classes and period mass mortalities of cisco continued through the 1980s (Rudstam et al. 1992). *Daphnia pulicaria* abundance varied hugely and, in general, as the reciprocal of its major predator (Vanni et al. 1990). Water quality as evidenced by bluegreen algal blooms showed similarly high variability. The interactions of food web structure and nutrient loading effects were dramatic (Kitchell and Carpenter 1992).

Recent events in Lake Mendota allowed a more complete and specific documentation of the mechanisms inferred from historical records. A major cisco die-off during the late summer of 1987 reduced the population to less than 10% of its former abundance and quickly halved fish predation on large *Daphnia* (Luecke et al. 1992). During 1988, *Daphnia pulicaria* heavily dominated the zooplankton, Lake Mendota produced its most remarkable clear-water period and bluegreen algae were rare (Vanni et al. 1990). The spring of 1989 produced higher nutrient loading and a summer of periodic bluegreen blooms. Intense spring rains on soils increasingly disturbed by urban growth and regular storms during early summer of 1990 brought increased nutrient loading and a summer of intense bluegreen algae blooms. The spring of 1991 brought even higher nutrient loading, but early thermal stratification and dense populations of large *Daphnia* yielded an extended period of clear water in late spring and, like 1988, yet another summer without a bluegreen bloom.

Given the limnological history of Lake Mendota (e.g., Brock 1985), the absence of bluegreen blooms during the summers of 1988 and 1991 is quite remarkable. Changes in the recent years are due to interactions between reduced nutrient loading rates and intense grazing of algae by large populations of *Daphnia pulicaria*. The latter derive from food web effects initially established by stochastic events that caused high mortality of the cisco. They are now sustained in some measure by stocking programs that enhanced populations of large piscivores such as the walleye (*Stizostedion vitreum*) and northern pike (*Esox lucius*), and by intense fishery regulations designed to protect these predators on the young and juvenile stages of potential planktivores (Johnson and Staggs 1992).

The fishery management practices of this large-scale and ongoing predator enhancement serve as a means for improving water quality through food web interactions. However, response by the next higher trophic level now seems likely to minimize ecologically effective expression of their effect. In keeping with its obligation to constituents, the Wisconsin Department of Natural Resources made public its plans for increase in predator stocks. Local newspapers, sporting goods dealers, and telephone "Fishing Hotlines" promoted the Lake Mendota fishery. Anglers responded to the publicized prospect with a four fold increase in fishing effort (Johnson and Staggs 1992). Mortality due to fishery exploitation increased by about the same amount. Survivorship of the stocked cohorts has and will likely continue to decline as the technological and communication capacities of the apical predator—humans—respond with remarkable speed and efficiency (Kitchell and Carpenter 1992).

Lake Mendota will probably continue to exhibit the highly variable responses witnessed over recent years. Only vigilance in preventing increases (or further reductions) in either the nutrient loading or increased protection for the large piscivores, or both, will increase the probability of continued improvements in water quality. Uncontrollable weather effects will confound the nutrient-food web interactions but, on average, the lake's water quality will continue to improve in response to the current management regime.

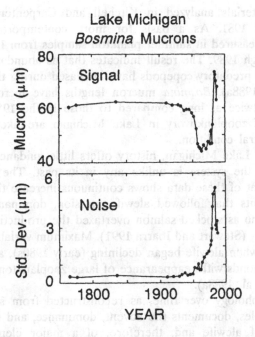

Figure 10.4. Length of the mucron spine on carapaces of *Bosmina longirostris* recovered from a deep sediment core of Lake Michigan, USA. Other descriptions as in Figure 10.1.

Lake Michigan

The changes in exploitation of fisheries and intense predator-prey interactions through time have produced very different community structures in Lake Michigan. An ancestral system dominated by lake trout and native *Coregonus* spp. was replaced by one dominated by Pacific salmon derived from hatcheries and dependent on the exotic alewife as the primary prey. Which of the possible food web structures is most likely to represent a stable configuration within the productive capacity of Lake Michigan? We addressed this question by examining *Bosmina* morphology as a quantitative indicator of the intensity of predator-prey interactions and food web structure.

The mean length of the mucron on *Bosmina* carapaces decreased dramatically at a core depth representing sediments deposited around 1960, the time of rapid increase in alewife abundance (Fig. 10.4). Prior to 1960, zooplanktivory in Lake Michigan was relatively low, abundance of predatory copepods was high, and long-spined *Bosmina* morphs dominated. The increase in alewife abundance after 1960 reduced abundance of predatory copepods through size-selective predation and the short-spined morph of *Bosmina* became abundant.

The core materials analyzed in Kitchell and Carpenter (1987) were collected before 1981. As a basis for more contemporary evaluation, *Bosmina* were measured in summer plankton samples from Lake Michigan from 1978 through 1989. The result indicates that as abundance of alewife has declined and predatory copepods have increased during the early 1980s (Kitchell et al. 1988a), *Bosmina* mucron lengths have increased and are currently about twice as long compared to those in the 1950s. Thus, the current levels of zooplanktivory in Lake Michigan are likely lower than under any ancestral condition.

In the case of Lake Michigan, history offers little guidance because the current state of the system is unlike any in its past. The variability or "noise" component of these data shows continuous increase throughout the sequence of events that followed alewife invasion, dominance, and their subsequent decline as stocked salmon overtaxed the productive capacity of their primary prey (Stewart and Ibarra 1991). Maximum variability appeared during the time when alewife began declining (early 1980s), and the recent reduction corresponds with reappearance of large zooplankton (Scavia et al. 1988; Kitchell et al. 1988a).

Bosmina morphology over time, as reconstructed from sediments and open-water samples, documents the advent, dominance, and recent decline in the impact of alewife and, therefore, of a major element in Lake Michigan's pelagic food web. It also appears that *Bosmina* morphology indicates alternate steady-state behavior of the predator-prey system. The morphological response exhibited in Lake Michigan was neither gradational nor linearly related to alewife populations, suggesting a combination of switching and depensatory mechanisms often observed when systems change state (Kitchell et al. 1988b; Walters and Holling 1991).

The current pelagic food web of Lake Michigan is artificially maintained by predator stocking. It also demonstrates intense variability and, as a logical extension, lessened predictability. As in the case of Lake Mendota, the apical predator—humans—has immense technological capacity and exhibits rapid responses. The early summer of 1989 produced massive mortalities of salmon infected with bacterial kidney disease. Tens of thousands of adult salmon carcasses appeared on Lake Michigan beaches. Catch rates in the sport fisheries dropped to half or less. Many anglers abandoned Lake Michigan for the greater prospects offered by Lake Huron or Lake Ontario.

Some fishery biologists reasoned that the dead salmon succumbed to an otherwise common and innocuous infection because they were nutritionally stressed by a shortage of the now rare alewife. They reasoned that the appropriate management action was to reduce predation on alewife by stocking fewer salmon. Another group reasoned that the high salmon mortality had reduced predation levels and that the appropriate management response was to heed the anglers' call for even more salmon in compensation for those that had died. This issue remains unresolved.

Humans as Components of Aquatic Ecosystems

When faced with variable and unpredictable ecosystem behavior, ecologists often ask whether the variability originates externally (environmental forcing) or internally (unstable or chaotic system dynamics). For Lake Mendota and Lake Michigan, our analyses provide examples of both. Lake Mendota is especially interesting because two distinct episodes of variability appear to have different causes. Perturbations of the watershed prompted high variability from about 1840 to 1910. Variability declined as the system stabilized from about 1910 to 1960. From 1960 to the present, the lake has again experienced high variability, this time attributable to food web dynamics. These episodes of high variability stand in marked contrast to the steady state conditions that prevailed from about 950 to 1800.

Lake Michigan also demonstrates two major periods of instability. Following centuries of minimal variation, the first major change appeared with the collapse of native fish populations and invasion of the exotic alewife. A two-decade period of relative stasis followed, then the populations of stocked predators forced the decline of alewife and sustained the high levels of piscivory evident in the most recent period of intense food web interactions. Although deforestation and agricultural development were fully manifest in the early history of Lake Michigan, neither produced evidence of changes in system productivity. The pelagic food web responded dramatically and twice to changes in species composition. Both were due to human activities that promoted invasion and, in the case of stocked salmonids, maintenance of exotic species.

If we broaden our definition of the system to include humans, then most of the variability can be attributed to internal dynamics rather than external forcing. The perturbations that caused major episodes of variability were land use change in the watersheds and intervention in the fish populations: exploitation and enhancement of stocks, introduction of exotics, and extirpation of native species. At the scale of our analyses, effects of climatic fluctuations pale before those of human perturbations of the watersheds and the lake communities (Schindler 1987).

One major conclusion of our analyses is that the rapid and massive response capacity of anglers calls for an ecosystem view and management system not yet developed by our research and resource management institutions (Cullen 1990). At the scale of populations and communities, fisheries are very efficient species and size-selective predators. At the scale of ecosystems, fisheries can be major perturbations to the trophic network. In both Lake Mendota and Lake Michigan, major decisions about fishery management are imminent, and they will have profound long-term effects on food web structure and the behavior of ecosystem-scale properties (Kitchell et al. 1988b; Carpenter et al. 1991). Clearly, the ecological and political power of organized angling interests should be counted as a variable of growing importance to aquatic ecologists and a factor of direct relevance to lake managers.

Our results have sobering implications for attempts to restore freshwater ecosystems (National Research Council 1991). Despite decades of restoration effort, Lake Mendota and Lake Michigan are far more variable and unpredictable now than in the past. Both lakes are vulnerable to new perturbations through development in the watersheds, species introductions, and shifts in fisheries management policy. While both lakes now boast improved water quality, productive fisheries, and some recovery of their original biotic integrity, neither could be regarded as a stable ecosystem. Until we learn how to restore lakes effectively, both water quality and fisheries management will require continual adjustment in response to largely unpredictable fluctuations and crises. An insufficiently explored alternative would be to promote the view that these systems should not be expected to be stable, that strong management actions have been and will be a requisite of the future, and that the opportunities for creative, informative actions are increasing (see Jordan, Chapter 21, this volume).

Acknowledgments

We thank Pat Sanford and Sharon Barta for their patience and dedication in collecting the fundamental paleolimnological data. Linda Holthaus wrought the final draft of this manuscript from an assemblage of pieces. Support for this study derived from the University of Wisconsin Sea Grant Institute (to J.F.K.), a grant to J.F.K. and S.R.C. from the Federal Aid to Fisheries Restoration program administered through the Wisconsin Department of Natural Resources, and a grant from the Andrew W. Mellon Foundation to S.R.C.

Recommended Readings

Carpenter, S.R. and J.F. Kitchell. (1988). Consumer control of lake productivity. *BioScience* 38:764-769.

Cullen, P. (1990). The turbulent boundary between water science and water management. *Freshwater Biol.* 24:201-209.

Kitchell, J.F., ed. (1992). *Food Web Management: A Case Study of Lake Mendota*. Springer-Verlag, New York.

Kitchell, J.F., and S.R. Carpenter. (1987). Piscivores, planktivores, fossils, and phorbins. In: W.C. Kerfoot and A. Sih, eds. *Predation: Direct and Indirect Impacts on Aquatic Communities*, pp. 132-146. University Press of New England. Hanover, New Hampshire.

Walters, C.J. and C.S. Holling. (1991). Large-scale management experiments and learning by doing. *Ecology* 71:2060-2068.

11

Humans as a Component of the Lake Titicaca Ecosystem: A Model System for the Study of Environmental Deterioration

Peter J. Richerson

Introduction

A deep understanding of the processes involved in anthropogenic environmental deterioration will require more than just the study of contemporary human environmental effects, as important as such studies are. There is much to be learned from comparative human ecology. Over the past few millennia, human societies have been typified by "cycles" of growth and decay (Turner and Meyer, Chapter 4, this volume). Ecosystems have been subjected to repeated episodes of high and low anthropogenic stress, giving rise to the replication that is required to extract dependable knowledge from the noisy world. The role of environmental deterioration in the cycles of civilized societies is also of considerable interest in it own right. There are a series of exciting problems in the area of environmental deterioration/societal collapse that would especially repay relatively large-scale interdisciplinary investigations by ecologists and social scientists. The purpose of this paper is to outline the problems that such a project might address, and how they might be tackled. My argument is illustrated by a specific system, Lake Titicaca, that might have suffered from subtle effects of anthropogenic environmental deterioration and also recorded effects on adjacent systems in its sediments. It provides an example of the kind of natural laboratory where the role of environmental deterioration by and on human societies can be explored in a broad and comparative perspective.

The concept of environmental deterioration is usually associated with human effects, even by professionals. However, the evolutionist R.A. Fisher (1958) gave it a very much broader definition. According to Fisher's

125

concept, most environmental changes will be disadvantageous to existing populations because selection will have adjusted their behavior to current and past environments. Changing environments will impose a new selective load on populations as they adapt to the change. Fisher was most intrigued by cases in which populations directly or indirectly cause the deterioration of their own environment. Take the evolution of a predator. Any increase in the efficiency of predators deteriorates the environment for prey, and selection is likely to make them larger, fleeter, more evasive, etc., deteriorating the environment for predators in turn. In more recent evolutionary biology, much of what Fisher meant by environmental deterioration has been collected under Ehrlich and Raven's (1964) term "coevolution." I like the older concept of deterioration because it underlines a certain similarity between human effects on environment and a much broader class of phenomena. We can use such concepts to facilitate the application of the more general principles to the special case of humans.

Dominant populations are most likely to be numerous enough and important enough to ecosystem structure and function to generate wide ranging deterioration affecting themselves and other populations. Indeed, a large capacity for environmental deterioration might be a good way to define dominance. Contemporary humans are an excellent, if extreme, example of a dominant, deteriorating species.

We can also associate the concept of deterioration with a sign, defining positive deterioration as changes in the environment that increase rather than decrease a population, given current adaptations. For the growth of a given population or a given physical environmental change, the sign of the deteriorating effect will, in general, vary from population to population, often in unpredictable ways due to indirect effects.

Thinking about environmental deterioration in an evolutionary context is important. In general, we cannot understand environmental deterioration from a purely ecological point of view because environmental deterioration will generate selective pressures. As Bradshaw's (Bradshaw and McNeilly 1981) classic work demonstrates, adaptive responses may be swift. Positive deterioration is liable to generate pressures to adapt even as populations grow. Human capital investment in land clearance and improvement is a good example, but natural examples are perhaps common as well (Wilson 1980). Evolutionary concepts are especially important in the human case because cultural evolution is so dynamic. Failure to account for the dynamics of technical innovation greatly weakened early analyses of human-caused environmental deterioration (Boyd 1973). As Paul Ehrlich's recent loss of his bet with Julian Simon on the trend of costs of basic materials illustrates, it is easy for those who know that environmental limits must exist in the long run to underestimate the power of technical innovation and other cultural adaptations to overcome limits in the shorter run.

From the most general point of view, humans are a fairly typical dominant or keystone organism. What we do matters more to ecosystem structure and function than what most other species in ecosystems do. This point of view should allow us to assimilate human populations into the general theory of ecology, rather than leave us standing as awkward outsiders. To be sure, we are certainly not typical in some respects. Our high levels of cooperation result in an ability to effect massive positive deterioration (as far as our own species is concerned). Our capacity for rapid cultural change allows us to adapt to negative deterioration and to sustain population growth for long periods in the face of what would otherwise be catastrophic deterioriation. (The positive deterioration we create for ourselves *is* catastrophically negative for many other species, of course.)

A companion concept to environmental deterioration is societal collapse. One apparently common feature of human populations during the last few thousand years is that they fluctuate dramatically. Societies rise to prominence, then decline, often both in the sense of political and of demographic collapse. Examples abound. A small selection of archaeologically or historically well attested cases include the Maya (Lowe 1985), Bronze Age, Iron Age, and Classical Europe generally (Renfrew 1973; McNeill 1982), China (UN 1973), Easter Island, and some other Polynesian islands (Kirch 1984), Mesopotamia (Jacobsen and Adams 1958) and Tiwanaku on the shores of Lake Titicaca (Browman 1981).

The realization that state collapse is common rather than restricted to a few famous cases like the Maya and Rome, has prompted considerable attention to the processes that might be involved. Yoffee and Cowgill (1988) is a useful collection of papers on the topic, and Conrad and Demarest (1984) and Lowe (1985) provide reviews of the theoretical debates surrounding the issue. Four types of hypotheses are advanced to account for various cases: 1) exogenous environmental shocks or changes such as a series of dry years, long term climate deterioration, or the introduction of new diseases; 2) endogenous environmental change due to anthropogenic changes such as soil salinization or deforestation and erosion; 3) exogenous political or economic changes, such as the rise of pastoral nomad confederations or the shifting of critical trade systems; and 4) endogenous political or economic changes, such as a stress on ideological and political legitimacy when elite manipulation of the economy fails to keep up with population growth, or when the limits to imperial conquest lead to an inability to reward the military establishment. Most explanations do not have a convincing account of why collapses rather than stagnation or graceful decline are apparently the rule. Exceptions to this generalization are Renfrew's and Conrad and Demarest's accounts, which depend upon endogenous political mechanisms to produce catastrophic collapse. The complexity of real cases seems to have outrun both theory and data. As Turner and Meyer summarize the current state of knowledge in Chapter 4 (this volume), all the simple monocausal explanations have been adequately tested and rejected in well-studied cases such as the Maya. The whole problem is subtle.

It is certainly not uncontroversial that either endogenous or exogenous environmental deterioration are more important than endogenous or exogenous political processes in generating the typical collapses of the historical record. Current political debates fracture along the same fault lines as scholars' arguments. Various flavors of the conventional Right and Left fear foreign enemies or internal decadence as the greatest threat. Typical environmentalists fear endogenous environmental change. Fear of exogenous environmental change has attracted curiously little ideological enthusiasm in the modern world, although placation of vengeful gods who represent the impersonal forces of nature is a powerful theme in many religions. The climatologist Flohn (1979) hypothesizes that the Holocene explosion of human societies is entirely due to a statistical anomaly, the absence of large volcanic irruptions during the last 10,000 years. On his view, the shift to warmer mean temperatures from glacial conditions is not so important as the fact that most of the Pleistocene past is characterized by large excursions of climate on the scale of centuries to millennia, caused by such irruptions.

A better general understanding of the phenomenon of societal collapse and the role of environmental deterioration therein would be most enlightening for our contemporary situation. The modern world system has generated the most spectacular boom, and could generate the most spectacular collapse, in human history. Nevertheless, it seems likely that the same or similar processes might be responsible for ancient and future fluctuations in human density and political elaboration. The collapse of ancient states is a scale model of what might happen to us.

The collapse of states is also interesting from the point of view of understanding subtle human impacts in ecosystems. State collapse must often turn human impacts on and off, often repeatedly in the same place. Peak population densities are often impressive even by modern standards, and nadirs seem to plunge by factors of two to ten. These repeated patterns provide natural replication that will ultimately allow powerful meta-analytical designs to tease out subtle effects. Collapses are a challenging but practical and interesting subject for critical empirical investigation.

Ecological Background

Lakes are excellent ecosystems to use as detectors of subtle human influences. Lakes are measurably influenced by events in their watersheds. Their resources are often more or less heavily used by human populations. The planktonic community that dominates lakes of any size is relatively simple in structure and relatively uniform horizontally compared to typical terrestrial systems. Many events concerning the watershed and plankton are recorded in the sediments (Kitchell and Carpenter, Chapter 10, this volume). Sediments accumulate at a rate on the order of 1 mm·yr⁻¹, so cores spanning millennia of human history are relatively easy to obtain. A lake's measurable response to deteriorating effects will tend to be averaged over the whole basin, and temporal resolution is limited by natural noise and the bioturbation of sediments; but at a certain spatial and temporal scale, lakes are perhaps the best systems for studying environmental deterioration. The idea that lakes are "reservoirs of history" is a classical bit of limnology (Hutchinson et al. 1970).

Lake Titicaca is a large (8,100 km², 273 m max. depth), tropical (16° S), alpine (3,800 m above sea level) lake (Fig. 11.1). The lake is cool (epilimnetic temperatures = 11-15°C), monomictic (isothermal for a few weeks in July and August), and moderately productive (1.13 gC·m⁻²day⁻¹, mean of 49 months of measurement). Typical chlorophyll-a concentrations are 2 µg·l⁻¹. For general limnological information, see Richerson et al. (1977, 1986) and Dejoux and Iltis (1992).

Although the lake is large, its basin has been heavily populated for a few thousand years (see below), and regional-scale subtle effects are possible—perhaps even likely. It is liable to be susceptible to both "top down" and "bottom up" deteriorating effects (Carpenter et al. 1985; Harris 1986). The system was nitrogen-limited during the best-studied years (1973, 1981-1982), as shown by a number of methods (Wurtsbaugh et al. 1985; Vincent et al. 1984; Vincent et al. 1985). The nitrogen budget appears to act as a keystone ecosystemic process. Geochemical and biotic effects are likely to be amplified and ramified throughout the system if they affect the dynamics of this nutrient. The largest loss of fixed nitrogen is through denitrification due to weak seasonal mixing, and the largest gain is nitrogen fixation by cyanobacteria, which is also highly episodic for reasons that we do not currently understand. Changes in basin geochemistry will influence nutrient budgets and, hence, whether nitrogen or some other nutrient is limiting, potentially affecting the overall productivity of the lake and the structure of food webs from the bottom up.

Changes in the abundance of species on higher trophic levels can ramify downwards and would profoundly affect the system if, say, the tendency for episodes of nitrogen fixation were affected ultimately by zooplankton grazing. In late 1982, large populations of *Daphnia pulex* dramatically altered nutrient limitation of the phytoplankton (Vincent et al. 1984). Most likely, *Daphnia* were abundant because zooplanktivorous fish populations were low, although data on this point were not collected. Limited evidence thus suggests that the Titicaca pelagic ecosystem is tightly coupled (Pimm 1982), and can exhibit top-down effects and perhaps endogenous, complex,

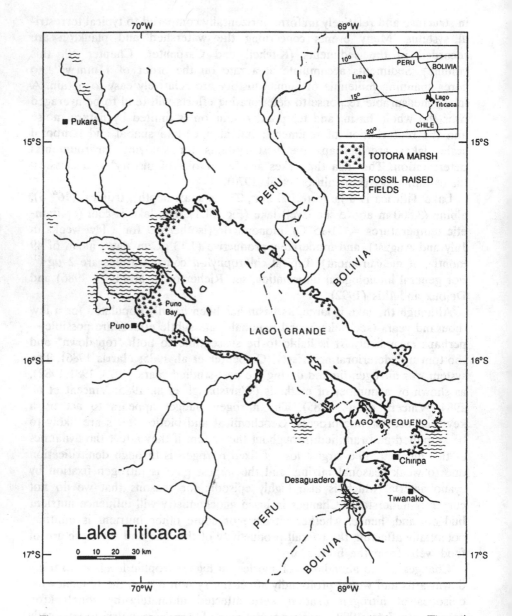

Figure 11.1. A map of Lake Titicaca showing the locations mentioned in the text. The main areas of raised fields and of totora marsh are also indicated. Map by Sharon Okada.

biotically driven dynamics (Steele and Henderson 1984; Murdoch and McCauley 1985).

The lake is host to a large "flock" of endemic fishes in the cyprinodont genus *Orestias*. Thirty species have been described from the lake or its immediate environs (Parenti 1984; Alfaro et al. 1982; Lauzanne 1982). As with the classic cichlid radiations in the African Great Lakes (Fryer and Illes 1972), *Orestias* species have diverged to fill a "superalimital" variety of niches including (to judge from morphology) littoral sit-and-wait predators, specialized shell-crushers, epibenthic predators, open-water clupeid mimic planktivores, and black bass-like piscivores. There has been little ecological study of these fishes (but see Vaux et al. 1988). Since 1950, two successful introductions have occurred, rainbow trout (*Onchyrincus gairdneri*) and a giant silversides (*Basilichthys bonariensis*). Coincidentally or not, the large piscivorous *Orestios cuvieri* has not been collected in recent years and is believed to be extinct.

Our rather short time series of data is sufficient to indicate that the Titicaca ecosystem is quite variable (Richerson et al. 1986; Richerson and Carney 1987). The seasonal cycle is muted in this tropical system, and most biological variables do not even show statistically significant repeatable 12-month patterns. However, there is considerable noisy variation within years, and yearly averages of most parameters are quite variable. The historical record indicates that our short sample of years underestimates the variation that will be seen on the decade time scale. For example, phytoplankton data from the 1937 Percy Sladen Trust expedition (Tutin 1940) documented a phytoplankton assemblage dominated by *Botryococcus braunii*, a species that has hardly been seen in our samples, although it was apparently a dominant in 1961 as well (Ueno 1967). Similarly, it is striking that the dominant planktivorous fish in Titicaca in the 1970s and 1980s, *Orestias ispi* (Lauzanne 1982), was not represented in the various moderately extensive collections obtained in the lake from the mid nineteenth century onward. It was first collected in abundance by two different collectors in 1979! Several other potential open-water planktivores have been described from the lake from previous collections.

Longer time scale variation is visible in cores, but the cores raised from the lake have not yet received complete analysis (Ybert 1987; Wirrman and de Oliveria 1987). A 1500-year-long ice core from the Quelccaya Ice Cap (150 km NW of the lake) records the climate of the general region (Thompson et al. 1988). Much of this variation is undoubtedly due to natural variations in weather, climate, and biota. Human effects must be detected against this background of natural variation, which is why the replication provided by societal collapse is potentially so important.

Human Ecological Background

Ravines (1982) and Browman (1981) provide convenient summaries of the history of human habitation in the basin. The Titicaca basin probably had no human population until the terminal Pleistocene arrival of human hunters ca. 11,000 B.P. (uncorrected radiocarbon dates). Browman (1974) argues that domestication of camelids and guinea pigs occurred very early, leading to a long period of pastoral economy from about 6,000 B.P. until shortly after 4,000 B.P., when pottery and evidence of settled villages first occurs in the local record. The environment of the high, cool, semi-arid plains around the lake is a difficult one for crop agriculture, but it is unusual to find well-developed animal pastoralism preceding plant agriculture. Analysis of plant remains (phytoliths) from sites dating in the 3,200-3,000 B.P. period shows that a complex agrarian system had developed by this period. Plants under cultivation included all of those that are the special hallmarks of tropical Andean agriculture: potatoes, the amaranth "grains," quinoa and cañihua, legumes including a domesticated lupine and common *Phaseolus* beans, and several lesser-known crops (Browman 1981). There is also evidence for the exploitation of lakeside resources such as *Scirpus*, *Juncus*, submerged macrophytes, and larger algae. Thus, it seems that by about 3,500 B.P. a mode of subsistence, mixing animal herding, cultivation, and exploitation of lake resources, was established. Evidence for large-scale trade also dates to this period. Both luxury goods (gems, metals, seashell, and lowland forest hallucinogens) and staples were carried by llama caravan in long-distance and regional trade networks. This is still the basic subsistence pattern in the basin. Major new domesticates were introduced in the Columbian period, including sheep, cattle, barley, and fava beans. Some modernization has occurred in the twentieth century, for example, motor transport has almost entirely replaced the llama caravan.

A series of distinctive cultural phases have been based on this agropastoral economy, according to Browman. The first of these are termed the Condori (3350-2850 B.P.) and Llusco (2850-2650 B.P.) Phases. Their influence spread over the entire Titicaca basin and over 200 km southeast to Lake Poopo. Ceramics appeared, and the wide geographical extent of these cultures testifies to the development of efficient llama caravan transportation. The Classic Chiripa Phase (2600-2150 B.P.) is marked by the development of distinctive polities in the northern part of the basin, centered at Pucara, and the southern part centered at Chiripa. The archaeologically impressive development during this period was the advent of monumental religious constructions, reflecting the development of an elaborate local tradition and including the first underground temples. Such projects require organization on a considerable scale. Reasoning from analogy with better-known cases, monumental architecture signifies the evolution of large-scale political systems ranging from advanced chiefdoms to the earliest form of states. The Polynesian trajectory of political evolution is particularly well worked out using a combination of ethnographic and archaeological data (Kirch 1984). In the Classic Chiripa and succeeding Kalasasaya Phase (2150-1875 B.P.), trade networks expanded in scale and sophistication of organization. Local communities were linked by regional trade, apparently conducted by entrepreneurs. A considerable local special-ization plus trade likely resulted in a more efficient economic system in the extremely heterogeneous Andean environment. Trade in craft specializations like pottery and textiles, less linked to environmental variation, also became important and also presumably led to a more efficient economy.

From about 1600-1900 B.P., the Titicaca basin and a large surrounding area was dominated by the city of Tiwanaku, situated on the southeast shore of the Lago Pequeño sub-basin of Lake Titicaca. Browman (1981) reviews evidence that the cultural unit associated with Tiwanaku was the Aymara language, which is still spoken in a wide area centered on the lake. Browman (1981) interprets the Tiwanaku "empire" as an economic federa-tion linked by trade and religion, in contrast to the Tiwanaku-influenced Wari state further north, which appears to have been more centralized. (The contrast between the centralized, but Hellenized Roman Empire and the decentralized Greek culture area comes to mind.) The dominance of Tiwanaku was apparently based in part on the development of a large, dense population supported by an intensive system of raised field agriculture (Kolata 1986). Denevan (1970) noted about 80,000 ha of raised fields around the shores of Lake Titicaca. Erickson (1988) presents evidence that the full extent of raised fields could be double this figure. Kolata (1986) reviews evidence that the 7,000 ha of raised field in the immediate vicinity of Tiwanaku could have supported from 2.9 to 7.9 people per ha. If all raised fields around the lake were in use simultaneously, this system alone could have supported something on the order of 250,000 to 650,000 people in the basin. In 1980, the population of Puno Department in Peru was 871,000 (INE 1990-1), most of whom live within the basin. The contempo-rary population on the Bolivian side is harder to calculate, as political and basin boundaries do not coincide, but an approximate total for the whole basin is 1-1.25 million. These people are supported largely by fairly traditional agricultural systems not including intensive use of the raised field zone. It is possible that the Tiwanaku period saw population peaks not yet reached in the modern period in the basin.

By about 800 B.P., the Tiwanaku trade system had collapsed and the city was abandoned. The collapse of Tiwanaku, unlike that of the related Wari state to the north (Anders 1986), was not catastrophic, but rather a slow eclipse lasting centuries. Kolata (1986) argues that a fairly sophisticated political system was necessary to manage the raised fields, as they depended upon massive public works such as channelized rivers. The raised field system was abandoned as the urban society it supported slid into obscurity and the requisite managerial institutions disappeared. No further use of this technology was made until experimental fields were revived by the archaeol-ogist Clark Erickson (1988) based on his excavations. The trade system, likewise, may have increased carrying capacity at the cost of a dependence on vulnerable political institutions.

Following the decline of Tiwanaku, various small Aymara "kingdoms" dominated the political landscape until the lake region was conquered by the Inca Empire about A.D. 1460. The Spanish conquest followed in 1532, causing severe depopulation followed by recovery. The Spanish conquest is a clear example of an exogenous political stress, but the main demographic harm was probably done by introduced diseases (Crosby 1972). The current population of the basin is growing at about 2% per year, but about half of the natural increase is lost by migration to the coast (INR 1990-1).

Thus, the archaeology and history of the region suggest that human uses of the Titicaca ecosystem can be divided into three phases, each of considerable duration: a hunting-and-gathering phase, a pastoral phase, and an agropastoral phase. The last of these includes episodes of higher human densities based upon the deployment of intensive strategies requiring state-level political institutions for their management, and lower ones during times of political disintegration. A detailed demographic history, which can be reconstructed to a reasonable level of detail for pottery-using populations (Marquardt 1978), has not yet been done in the Titicaca basin. However, the apparent buildup to the Tiwanaku maximum, decline to post-Columbian minimum, and subsequent recovery to contemporary levels seems secure enough, and more replications of such fluctuations will probably be uncovered.

Human Effects on the Lake Ecosystem

Human uses of the basin's resources, of course, reflect the economic adaptations people are currently using. None of the historic or current uses of these resources has been sufficient to cause gross disturbances of a lake of Titicaca's size, but there are a number of routes via which subtle effects are likely to have affected the lake. By the same token, human dependence on lacustrine resources is quite high, and autochthonous or anthropogenic deterioration of the system may well have feedback effects on human populations. No such effects are currently well documented, but we can combine information on lake ecosystem properties with what we know about the human ecology of the situation to construct hypotheses regarding likely subtle effects. Browman (1991) argues that the marginality of the Titicaca environment for agriculture leads to a marked sensitivity of human populations in the Basin to exogenous physical deterioration. He notes a number of associations between climate change and cultural change, so this type of hypothesis is important to keep in mind as well.

The hunting and pastoral phases are likely to have had their most profound effects on the lake through effects on the terrestrial watershed. The extinctions of big game, correlated with the arrival of hunters and possibly caused by them (Martin and Klein 1984), may have caused sufficient vegetation changes to have subtle effects on nutrient budgets. For example, human hunting might reduce game densities substantially, encouraging denser plant cover. Lower erosion rates should reduce phosphorus loading, provoking an oligotrophic tendency in the lake, while higher evapotranspiration might reduce the lake's water budget. Herding populations would eventually have intensified animal exploitation, promoting higher grazer biomass. Herders in marginal environments often seem to maximize herd standing crop as a risk-management strategy. Flannery et al. (1989) consider the risk-management complexities of herding camelids in a Peruvian area ecologically similar to the Titicaca area. Pastoralism might have significantly increased phosphorus loading due to higher erosion rates, driving the lake toward the nitrogen-limited condition observed today.

The evidence is that agropastoral populations came to make extensive use of lake and lakeside resources as well as converting significant areas of watershed to farmland, especially near the lake. Coupled with high densities in certain periods, the effects of this mode of subsistence are most likely to be detectable. Several patterns of use could have exerted substantial changes on the lake ecosystem.

Fishing and Fish Introductions

The lake is currently the focus of a significant artisanal fishery conducted almost entirely by residents of lakeside communities (Orlove 1986; Orlove et al. 1992; Alfaro et al. 1982). Fresh and dried fish are marketed in local towns, but there is also a long-distance trade in dried fish. Orlove et al. (1992) report results of a survey of fishermen indicating a total catch of about 8,000 tonne·yr^{-1} on the Peruvian shore of the lake (say 12,000 tonne for the whole lake). The fishery even today is lightly capitalized. Most fishermen use reed boats or other small craft suitable only for exploitation of the littoral zone. It is likely that littoral stocks are heavily fished or even over-fished, whereas the pelagic zooplanktivores are probably lightly exploited. Thus the direct effects of fishing are likely to have only subtle top-down effects. The current pelagic fishery (ca. 1,000 tonne·yr^{-1}) is harvested well below its presumed maximum sustainable yield of perhaps 20,000-50,000 tonne·yr^{-1}. If this fishery is (or was) ever developed, significant top-down effects such as those described by Kitchell and Carpenter (Chapter 10, this volume) are likely.

The successful introduction of rainbow trout and the giant silversides may have had more profound effects on pelagic populations, but the limited state of our knowledge of the population ecology of Titicaca fishes permits only speculation about what these might have been. Large but unsystematic collections from 1937 and 1979 differed substantially in species present and their abundances, but aside from the fact that the now dominant zooplanktivore, *O. ispi*, was not collected until the 1970s, it is difficult to know if these differences represent real variations, much less their causes. The apparent extinction of O. cuvieri is attributed to introductions, but even on this point, direct evidence is lacking. As one of the largest non-cichlid flocks of endemic fishes, study and protection of the Titicaca Orestias would be well justified on the grounds of biodiversity preservation.

According to Kent (1987), the intensive exploitation of fish and waterfowl was an early component of the agropastoral adaptation, so it is likely that fish and trade in fish has been more or less in proportion to population density for the past 3,500 years. It is conceivable that intensification of fishing was part of the general pattern of intensification of Tiwanaku or other large scale polities of the past. Offshore fishing in small reed boats is quite dangerous (Orlove, pers. comm.), but fleets of larger vessels could have been organized to overcome this limitation. Exploitation of waterfowl for eggs or meat, or perhaps disruptive guano harvest, is also a conceivable source of top-down effects. Archaeologists and paleoecologists should be alert for these possibilities.

Littoral Vegetation

Other littoral resources are extensively exploited by lakeside residents. The most important activity is the harvest of macrophytes, including bulrushes (*Scirpus totora*), locally called "totora," and a variety of submerged macrophytes (*Myriophyllum, Elodea,* and *Potamogeton*), locally termed "llachu." This community is described by Collot (1981) and its exploitation analyzed by Levieil and Orlove (1991) and by Levieil (1987). Many uses are made of totora; llachu is fed to livestock. Macrophyte beds are intensively managed. Totora beds are considered privately owned by local communities, although not by the national governments. Totora is carefully harvested to maintain production and is planted or transplanted when necessary (e.g., when lake levels rise or fall). Thus, the entire littoral community is heavily influenced by human activities. Subtle effects on the whole system are probable through anthropogenic effects on the population and nutrient dynamics of this community. Past intensity of these effects presumably varied substantially due to changes in population and degree of intensification. Industrial-scale exploitation of totora is contemplated from time to time, but would engender fierce conflict with local communities.

Raised Fields

The large areas of raised fields around the shores of the lake could certainly have had subtle impacts on the lake ecosystem by altering nutrient budgets. One of the possible functions of raised fields is nutrient retention (Erickson and Candler 1989). H. Carney et al. (1993) are investigating the hypothesis that raised field systems retard the flow of phosphorus into the littoral zone and act as sites of active nitrogen fixation. It is possible that extensive use of raised fields could shift the lake, or at least the smaller sub-basins, from an N-limited to a P-limited system.

Another potential effect of raised fields is alteration of major ion chemistry in the lake. The water budget of Titicaca, like many large tropical lakes, is dominated by evaporation (Carmouze et al. 1977). The current total dissolved solids concentration in the lake is close to 800 mg·l^{-1}. Richerson et al. (1977) present evidence that the lake is salinizing under current hydrological conditions. Carmouze et al. (1977) argue instead that the lake's salt budget is balanced by a large exfiltration of water. Given that the current salt content is somewhat marginal for agriculture, and that significant changes could occur with a time scale of centuries, either exogenous or anthropogenic salinization of the lake and low-lying raised field areas is a distinct possibility. In a persistent high-evaporation regime, the lake would lose its outlet and begin to shrink. Preliminary reports on cores taken from the shallow Lago Pequeño sub-basin indicates that the lake lost its outlet from 7700 to 3650 years ago, fell perhaps 50 m, and reached salinities approximating seawater (Wirrman and de Oliveira 1987). Much less dramatic deterioration would compromise the raised-field system.

Upland Cultivation

The cultivation of hillside terrain in the Titicaca basin is currently intensive up to perhaps 4,000 m (200 m above the lake elevation), and domestic animals heavily graze virtually the whole basin (Browman 1987). Manipulation of vegetation through grazing and farming is likely to have had effects on nutrient budgets and on hydrology and major ion chemistry. Intensive use of the watershed is likely to increase erosion and enlarge phosphorus budgets. Farmers may enhance basin nitrogen fixation by cultivating large areas of legumes, increasing available nitrogen in runoff. Fuel collection and heavy grazing will tend to increase water runoff by reducing evapotranspiration. Irrigation schemes and other water conservation strategies (Flores 1987), on the other hand, may reduce water supplies to the lake and lead to salinization. There are no well-documented ancient irrigation systems in the Titicaca basin, although they are common elsewhere in the Andes. Very little irrigation is used in the area today. Under current climatic conditions, rainfall agriculture is quite practical, though irrigation would extend growing seasons and provide some hedge against drought. It is conceivable that it was part of past intensification strategies.

Urban Runoff

Point-source pollution from urban agglomerations is likely to have had effects in the past, at least on a local scale. The city of Puno's sewage now grossly pollutes the inner harbor area (ca. 8 km², Northcote et al. 1989). Industrial activities are modest, though a mine on the Bolivian side of the lake is reputed to pollute badly one stream.

Testing Hypotheses

Modern environmentalists subscribe to the hypothesis that anthropogenic environmental deterioration is a serious threat to human societies and populations. People of other political persuasions hold other hypotheses about the main risks and opportunities that face us. As we have seen, students of past societies have entertained a very similar set of hypotheses to explain their decline. Ecologists and evolutionary biologists have become impressed with the inconstancy of natural systems in the face of biotic and abiotic deterioration of the environment (Egerton, Chapter 2, this volume).

The main thesis of this chapter is that ecologists and human ecologists are theoretically in a position to appreciate these phenomena in much greater detail than only a few years ago. However, the problems involved are complex and subtle. For example, the four basic kinds of hypotheses to account for societal collapse are certainly not mutually exclusive. In the case of the collapse of ancient societies, the failure of simple explanations to account for any given case, much less all cases, suggests that several interacting processes often contribute to given examples. Similarly, different cases might be due to rather different combinations of effects. The simplest hypothesis consistent with current knowledge is that every society at every point in its history is subject to a number of positive and negative deterioration effects, involving exogenous and endogenous environmental and political effects. Suitably scaled, these effects might merely need to be summed to predict the current trajectory of the society. Any of a large number of combinations of negative deteriorating effects might outweigh the positive ones to produce declines. What is the relative importance of the various deteriorating effects? Is this highly variable from case to case, or are there relatively standard syndromes of anthropogenic deterioration and societal collapse? Is there any evidence for complex interactions between deteriorating processes?

Even the simplest hypothesis is impossible to test with a standard hypothetico-deductive approach; more sophisticated strategies are required (Quinn and Dunham 1983; Richerson and Boyd 1987). In this case, it seems to me that there is no substitute for a series of relatively large-scale multidisciplinary projects exploiting favorable situations like the existence of the lacustrine Titicaca ecosystem, with its potential to register regional effects in an environment where a fluctuating succession of human societies have prospered and declined for several millennia. In such a system, archaeologists can assemble long and detailed records of human population and economic change, and paleoecologists, long records of environmental change. Living human populations are affecting the lake ecosystem in much the same way as past ones did. The frustrating ambiguities of the always-limited historical record can be supplemented by investigations on the living system (see Likens 1989 for an analysis of approaches). Past research ranging across this whole spectrum demonstrates the feasibility of exploiting each of these four components at Titicaca. Of course, a number of similar sites needs to be investigated to explore the range of variation across systems and to gain a comparative picture of environmental deterioration (see Cole et al. 1991 for analysis of approaches). In the absence of a planned, coordinated series of investigations, however, it will take many decades before a synthetic understanding is achieved.

A coordinated set of investigations on one ecosystem like Lake Titicaca might do the job in a decade and cost $1-2 million per year. Even assuming that a fairly large number of similar sites would have to be investigated to get a solid comparative perspective on these questions, the sums involved are quite small by "big science" standards, although large by the traditions of funding of such efforts as the International Biosphere and Man and the Biosphere programs (MAB). Total U.S. MAB funding is about $1 million per year; individual projects receive $50-150 thousand dollars per year (pers. comm. U.S. MAB office). This level of funding is not commensurate with the importance of the problem or with the potential for accomplishing useful results.

Conclusions

The question of the importance of anthropogenic environmental deterioration is one of the most widely debated issues of our time. It is arguably one of the most important factors in explaining the dynamics of human societies for at least the last 10,000 years. Nevertheless, the complexity and subtleties of the issue are such that little can be said about them with any real confidence by a suitably cautious scientist. In systems like Lake Titicaca, the general problem of environmental deterioration translates into a relatively straightforward research program in which systematic application of contemporary ecological, paleoecological, human ecological, and archaeo-logical investigations would yield great insights.

Recommended Readings

Conrad, G.W. and A.A. Demarest. (1984). *Religion and Empire: The Dynamics of Aztec and Inca Expansionism.* Cambridge University Press, Cambridge.

Erickson, C.L. and K.L. Candler. (1989). Raised fields and sustainable agriculture in the lake Titicaca basin of Peru. In: J.O. Browder, ed. *Fragile Lands of Latin America*, pp. 230-248. Westview Press, Boulder, Colorado.

Lowe, J.W.G. (1985). *The Dynamics of Apocalypse: A Systems Simulation of the Mayan Collapse.* University of New Mexico Press, Albuquerque, New Mexico.

Richerson, P.J. and R. Boyd. (1987). Simple models of complex phenomena: the case of cultural evolution. In: J. Dupré, ed. *The Latest on the Best: Essays on Evolution and Optimality*, pp. 27-52. MIT Press, Cambridge, Massachusetts.

Richerson, P.J., C. Widmer, and T. Kittel. (1977). *The Limnology of Lake Titicaca (Peru-Bolivia), A Large, High Altitude Tropical Lake.* (Institute of Ecology Publ. #14), University of California-Davis, Davis, California.

Yoffee, N. and G.L. Cowgill. (1988). *The Collapse of Ancient States and Civilizations.* University of Arizona Press, Tucson, Arizona.

12
Nitrogen Loading of Rivers as a Human-Driven Process

Jonathan J. Cole, Benjamin L. Peierls,
Nina F. Caraco, and Michael L. Pace

Introduction

The ecology of areas populated by humans is viewed in a number of ways in this book. In this chapter, we examine the biogeochemistry of nitrogen in the major rivers of the world as a function of the number of humans inhabiting the watersheds of these rivers. This study should be of interest for several reasons. First, it is an ecological study that includes, quite explicitly, the humans within the system. Further, the coastal margins tend to have the greatest concentration of urban areas and human population density. For example, even in the United States, a country with a great deal of interior relative to coastal zone, 53% of residents live within 50 miles of the coast (Schubel and Bell 1991). The rivers that pass through these coastal zones are a major source of nutrients, including nitrogen, to coastal waters. Nitrogen (N) is an essential plant nutrient in aquatic systems and is considered a limiting factor for primary productivity in coastal marine systems (Caraco et al. 1987; Howarth 1988). Significant increases of riverine or estuarine N, therefore, can lead to undesirable consequences such as nuisance algal blooms, anoxia or hypoxia, or loss of native or economically important species (D'Elia 1987).

Rivers receive nitrogen from watershed runoff, atmospheric deposition, and point sources. Human populations can increase these N inputs to rivers by increasing industrial and automobile emissions to the atmosphere, increasing application of fertilizer, increasing sewage discharge, or by decreasing watershed retention through watershed disturbance (Figure 12.1; Likens et al. 1977). Increasing eutrophication of the coastal zone has been linked to increasing discharges of nutrients, especially nitrogen, from the

Figure 12.1. Conceptual diagram for the nitrogen economy of a human-dominated watershed.

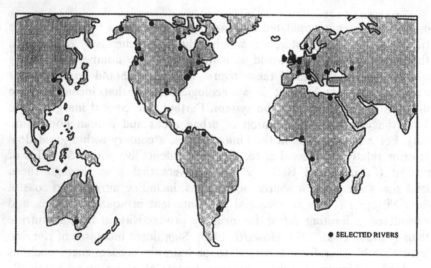

Figure 12.2. Map showing the locations of rivers in our data set.

atmosphere, point sources, and upland areas (Walsh et al. 1981; Fisher and Oppenheimer 1991). Here we present data from 42 major rivers that indicate human population density in watersheds is strongly related to both riverine nitrate concentration and nitrate export. Our results demonstrate that sewage loading and atmospheric deposition, both controlled by human populations, correspond closely to the magnitude of river nitrate exported to coastal areas. We suggest that projected growth of world human population will lead to continued increases of nitrate in rivers and estuaries.

Sources of Data and Analysis

To determine the relationship between river N and human population density, we summarized data from a variety of sources on nitrate concentrations, watershed characteristics, human population densities, and nitrate deposition for 42 major rivers of the world that empty to the ocean (Tables 12.1-3; Fig. 12.2). We selected only large rivers for which we could obtain data on discharge as well as seasonally-averaged nitrate concentrations.

Particularly relevant to the goals of this book is the extreme difficulty we had in obtaining high-quality information on human populations. Population statistics are compiled according to political boundaries rather than ecological ones. One cannot look up, for example, the number of people inhabiting the watershed of the Amazon River. Rather, we had to piece together watershed population data from what we could garner on a political basis, and our estimates of human population are correspondingly crude.

The rivers and their watersheds selected for our data set are representative of the world's rivers and account for 37% of the total freshwater discharge to the ocean. The mean runoff for the selected rivers ($13.2 \, \mathrm{l \cdot s^{-1} \cdot km^{-2}}$) is very close to an estimate of mean runoff for rivers of the world ($11.8 \, \mathrm{l \cdot s^{-1} \cdot km^{-2}}$ [Meybeck 1982]). Discharge for the 42 rivers ranged from 7 to 175,000 $\mathrm{m^3 \cdot s^{-1}}$, spanning more than four orders of magnitude. Human population densities range from a low of 0.15 individuals $\mathrm{km^{-2}}$ in the Mackenzie River basin to 400 individuals $\mathrm{km^{-2}}$ in the Thames. The mean human population density for our data set (40 individuals $\mathrm{km^{-2}}$, weighted by watershed area) is comparable to the mean for the land area of the entire planet (34 individuals $\mathrm{km^{-2}}$ [Population Reference Bureau (PRB) 1989]). Our mean river nitrate concentration, (19.6 μM, weighted by discharge), is somewhat higher than other estimates of mean river nitrate for the world (7 μM [Meybeck 1982]), probably due to the large number of European rivers in our data set.

The Pattern and Its Meaning

A log-log regression of nitrate concentration on human population density (Fig. 12.3) shows a highly significant ($p < .00001$) relationship explaining 76% of the variation in nitrate concentration in rivers of the world (Table 12.5). Because humans likely represent an input of N, we would expect population to be more closely related to nitrate export than to nitrate concentration. Nitrate export is predictable from population density by a similar model

Table 12.1. Characteristics of rivers. Data used in our analyses of river nitrogen and human populations. Systems are listed alphabetically by river name. Discharge, runoff, and nitrate concentrations are our estimates of annual average values. Export is the product of runoff and mean concentration and, thus, does not take into account any relationship between concentration and discharge. "PPT NO_3" is the concentration of nitrate in wet precipitation at sites located as near to the watersheds (w.s.) as we could find existing data for. Deposition is the product of nitrate concentration and precipitation amount. Sources for these data are listed in the "REF No." column (Table 12.3); these numbers refer to the citations listed by the same reference number in Table 12.4.   ND refers to No Data available.

SYSTEM/LOCATION	DISCHARGE $(m^3 s^{-1})$	RUN-OFF $(l\ s^{-1} \cdot k^{-1} \cdot m^{-2})$	W.S. AREA (km^2)	POP. DENS. $(indv \cdot km^{-2})$
Adige/Italy	2.23e+02	18.3	1.22e+04	102.0
Amazon/S America	1.75e+05	24.8	7.05e+06	1.0
Caragh/Ireland	7.29e+00	45.6	1.60e+02	7.15
Columbia/USA	7.90e+03	11.8	6.7e+05	10.0
Danube/Rumania	6.5e+03	8.1	8.05e+05	90.0
Delaware/USA	3.36e+02	19.1	1.76e+04	100.0
Fraser/B.C., Canada	3.55e+03	16.1	2.2e+05	2.0
Ganges/India	1.6e+04	14.9	1.07e+06	300.0
Glåma/Norway	7.06e+02	16.9	4.177e+04	12.0
Huanghe(Yellow)/China	1.47e+03	2.0	7.5e+05	200.0
Hudson/NY, USA	5.6e+02	16.1	3.47e+04	150.0
Kazan and Back/Canada	1.9e+03	6.1	3.12e+05	0.4
Mackenzie/Canada	1.06e+04	5.9	1.787e+06	0.15
Magdalena/Columbia	7.50e+03	31.3	2.4e+05	30.0
Mekong/SE Asia	1.5e+04	19.2	7.83e+05	43.0
Mersey/England	2.1e+01	17.5	1.2e+03	200.0
Meuse/Neth.,Belgium	3.17e+02	9.1	3.49e+04	250.0
Mississippi/USA	1.61e+04	5.0	3.220e+06	30.0

Table 12.1 (cont.)

SYSTEM/LOCATION	DISCHARGE $(m^3 \cdot s^{-1})$	RUN-OFF $(l\,s^{-1}\cdot k^{-1}\cdot m^{-2})$	W.S. AREA (km^2)	POP. DENS. $(indv\cdot km^{-2})$
Murray-Darling/Australia	3.182e + 02	0.3	1.073e + 06	1.5
Nelson/Manitoba, Canada	2.37e + 03	2.2	1.07e + 06	2.0
Niger/W Africa	7.0e + 03	6.2	1.125e + 0	20.0
Nile/NE Africa	9.5e + 02	0.3	2.96c + 06	50.0
Orange/S Africa	1.7e + 02	0.2	1.02c + 06	20.0
Orinoco/Venezuela	3.390e + 04	33.9	1.0e + 06	2.0
Paraná/Argentina	1.59e + 04	5.7	2.8e + 06	10.0
Po/Italy	1.47e + 03	22.0	6.67e + 04	232.0
Stikine/Canada, USA	1.1e + 03	22.0	5.0c + 04	1.0
St. Lawrence/Canada, USA	1.07e + 04	10.4	1.025c + 06	15.0
Susquehanna/USA	1.1e + 03	15.1	7.3c + 04	100.0
Tees/England	5e + 01	27.7	1.806e + 03	100.0
Thames/England	7.8e + 01	7.8	9.95e + 03	400.0
Tiber/Italy	2.3e + 02	13.5	1.7e + 04	262.0
Uruguay/S America	3.85e + 03	10.5	3.65e + 05	10.0
Vistula/Poland	1.1e + 03	5.5	2.0c + 05	120.0
Volga/USSR	8.2e + 03	6.1	1.35e + 06	50.0
Yangtze/China	2.9e + 04	15.4	1.9e + 06	200.0
Yukon/Canada	6.18e + 03	7.4	8.31e + 05	0.4
Zaire/Zaire	3.973e + 04	10.4	3.82e + 06	11.7
Zambezi/SE Africa	3.2e + 03	2.5	1.3c + 06	15.0
Average	1.0e + 04	13.2	9.4e + 05	85.0
World Average		11.8		33.0

Table 12.2   Nitrate chemistry of rivers. Conventions as in Table 12.1.

SYSTEM/LOCATION	(μM) NO$_3$	(μmol NO$_3^-$·s^{-1} km^{-2}) EXPORT	DEPOSITION
Adige/Italy	67.0	1224.7	1237.5
Amazon/S America	3.0	74.5	120.6
Caragh/Ireland	3.6	164.0	86.5
Columbia/USA	26.6	313.6	62.8
Danube/Rumania	46.0	371.4	826.4
Delaware/USA	61.0	1167.2	851.7
Fraser/B.C., Canada	6.4	103.3	739.7
Ganges/India	91.3	1361.4	294.3
Glåma/Norway	24.0	405.7	975.0
Huanghe(Yellow)/China	139.0	272.6	286.4
Hudson/NY, USA	47.8	771.4	851.7
Kazan and Back/Canada	1.1	6.7	60.9
Mackenzie/Canada	5.7	33.8	73.9
Magdalena/Columbia	17.0	531.3	87.5
Mekong/SE Asia	17.0	325.7	334.1
Mersey/England	156.0	2730.0	919.4
Meuse/Neth., Belgium	230.0	2089.1	742.3
Mississippi/USA	63.0	315.0	691.7
Murray-Darling/Australia	15.0	4.4	74.8
Nelson/Manitoba,Canada	5.0	11.1	248.6
Niger/W Africa	7.0	43.6	555.2
Nile/NE Africa	20.0	6.4	50.9
Orange/S Africa	50.0	8.3	ND
Orinoco/Venezuela	6.0	203.4	92.5
Paraná/Argentina	14.2	80.6	ND
Po/Italy	102.0	2247.3	1237.5
Rhine/Europe	286.0	3395.6	1647.9

Table 12.2 (cont.)

SYSTEM/LOCATION	(μM) NO_3	(μmol $NO_3^- \cdot s^{-1}$ km^{-2}) EXPORT	DEPOSITION
Rhône/France	57.2	1012.9	695.9
Shannon/Ireland	54.0	727.7	252.8
Stikine/Canada, USA	6.1	134.2	76.8
St. Lawrence/Canada, USA	16.0	167.0	673.2
Susquehanna/USA	66.0	994.5	821.5
Tees/England	75.0	2076.5	608.7
Thames/England	520.0	4076.4	1125.1
Tiber/Italy	100.0	1352.9	1237.5
Uruguay/S America	29.0	305.9	ND
Vistula/Poland	70.5	387.8	832.8
Volga/USSR	30.0	182.2	151.8
Yangtze/China	58.17	897.0	370.5
Yukon/Canada	9.3	69.2	ND
Zaire/Zaire	6.0	62.4	467.2
Zambezi/SE Africa	9.3	22.9	ND
Average	62.3		
World Average	7.0		

Table 12.3 Precipitation. Conventions as in Table 12.1.

SYSTEM/LOCATION	(µM) PPT NO_3^-	(cm yr^{-1}) PPT	REF No.
Adige/Italy	46.0	84.8	26,36,38
Amazon/S America	2.1	181.1	2,8,14,36,45
Caragh/Ireland	2.6	104.9	18,36
Columbia/USA	2.0	99.1	24,35,36,38
Danube/Rumania	45.0	57.9	22,24,36,38
Delaware/USA	25.0	107.4	2,24,27,35,36,39
Fraser/B.C.,Canada	16.0	145.8	24,27,35,36,38,39
Ganges/India	5.8	160.0	20,24,35,36,39,45
Glåma/Norway	45.0	68.3	32,36,38,39
Huanghe(Yellow)/China	28.0	32.3	13,27,35,39
Hudson/NY, USA	25.0	107.4	2,17,23,24,36,37,45
Kazan and Back/Canada	7.0	27.4	24,27,35,36,38
Mackenzie/Canada	7.0	33.3	5,30,36,38,39
Magdalena/Columbia	2.6	106.2	2,24,27,35,36
Mekong/SE Asia	7.57	139.2	24,25,36
Mersey/England	28.9	100.3	21,35,39
Meuse/Neth.,Belgium	36.0	65.0	2,9,24,36,40
Mississippi/USA	19.0	114.8	2,20,27,36
Murray-Darling/Australia	4.4	53.6	2,36,39,41
Nelson/Manitoba, Canada	21.0	37.3	27,35,36,38,39
Niger/W Africa	9.64	181.6	24,35,36,42
Nile/NE Africa	10.2	15.7	24,27,31,33,35,36
Orange/S Africa	ND	ND	6,16
Orinoco/Venezuela	3.0	97.3	2,24,27,36,45
Paraná/Argentina	ND	ND	3,20,35
Po/Italy	46.0	84.8	26,36,38
Rhine/Europe	60.0	86.6	9,12,27,35,36,38

Table 12.3 (cont.)

SYSTEM/LOCATION	(μM) PPT NO$_3^-$	(cm yr^{-1}) PPT	REF No.
Rhône/France	30.0	73.2	20,27,35,36,38
Shannon/Ireland	8.6	92.7	19,36,38
Stikine/Canada, USA	1.0	242.1	24,27,35,36,38
St. Lawrence/Canada, USA	21.0	101.1	24,35,36,39,44
Susquehanna/USA	25.0	103.6	2,24,27,35,36
Tees/England	33.0	58.2	28,35,36,43
Thames/England	61.0	58.2	15,35,36,39
Tiber/Italy	46.0	84.8	26,36,38
Uruguay/S America	ND	ND	10,35,36,39
Vistula/Poland	47.0	55.9	24,27,35,36,38
Volga/USSR	13.0	36.8	2,24,27,35,36
Yangtze/China	10.0	116.8	11,13,34,35,36,45
Yukon/Canada	ND	ND	20,24,27,35,39
Zaire/Zaire	10.0	147.3	1,24,36
Zambezi/SE Africa	ND	ND	4,7,35
Average	9.8		
World Average	12.5		

Table 12.4. Sources for data listed in Tables 12.1-3. Reference numbers are the numbers listed in the "REF No." column in Tables 12.1-3. The column labeled "Collection" refers to data complied in the following sources:

* B.R. Davies and K.F. Walker, eds. (1986). *The Ecology of River Systems*.
§ E.T. Degens et al. (1982-85). *Transport of Carbon...* (3 parts). Scope/UNEP.
¶ B.A. Whitton, ed. (1984). *Ecology of European Rivers*.
\# Longhurst (1989). Acid Deposition.

Reference No.	Citation	Collection
1	Bailey 1986	*
2	Berner and Berner 1987 (Table 3.2)	
3	Bonetto 1986	*
4	Borchert and Kempe 1985	§3
5	Brunskill 1986	*
6	Cambray et al. 1986	*
7	Davies 1986	*
8	Day and Davies 1986	*
9	Descy and Empain 1984	¶
10	Di Persia and Neiff 1986	*
11	Edmond et al. 1985	
12	Friedrich and Müller 1984	¶
13	Galloway et al. 1987	
14	Gibbs 1972	
15	Goulding and Johnston 1989	#
16	Hart 1982	§1
17	Hennigan 1985	
18	Heuff and Horkan 1984	¶
19	Horkan 1984	¶
20	Kempe 1982	§1
21	Lee and Longhurst 1989	#
22	Liepolt 1972	
23	Limberg et al. 1986	

Table 12.4 (cont.)

Reference No.	Citation	Collection
24	Maybeck 1982	
25	Pantulu 1986	*
26	Pettine et al. 1985	§3
27	PRB 1989	
28	Raper et al. 1989	#
29	Richey et al. 1985	§3
30	Rosenberg and Barton 1986	*
31	Shehata and Bader 1985	
32	Skulberg and Lillehammer 1984	¶
33	Soliman 1982	§1
34	Szekielda and McGinnis 1985	§3
35	Times Atlas of the World 1973 (plate 6)	
36	Todd 1970 (Table 3-2, 1-8)	
37	Tofflemire and Hetling 1969	
38	UNEP 1987	
39	UNEP/WHO/UNESCO/WMO 1987	
40	Van Urk 1984	¶
41	Walker 1986	*
42	Welcomme 1986	*
43	Whitton and Crisp 1984	¶
44	World Resources Institute et al. 1988	
45	Worldmark Press 1976	

Table 12.5. Results of least squares regression analyses using the equation: $\log Y = \underline{m} \log X + \underline{b}$. Variables are river nitrate (μM), population density (inhabitants·km^{-2}), nitrate export ($\mu mol \cdot s^{-1} \cdot km^{-2}$), watershed area ($km^{-2}$), discharge ($m^3 \cdot s^{-1}$), nitrate deposition ($\mu mol \cdot s^{-1} \cdot km^{-2}$), and precipitation nitrate (μM).

	Y	X	n	$\underline{m}(\pm SE)$	$\underline{b}$	r^2	p
1	River NO_3^-	Pop. Dens.	42	0.56(.05)	0.67	0.76	<0.00001
2	NO_3^- Export	Pop. Dens.	42	0.64(.10)	1.51	0.53	<0.00001
3	River NO_3^-	W.S. Area	42	-0.19(.08)	2.49	0.12	0.02430
4	River NO_3^-	Discharge	42	-0.24(.09)	2.23	0.15	0.01274
5	NO_3^- Export	W.S. Area	42	-0.43(.10)	4.65	0.31	0.00013
6	NO_3^- Export	Discharge	42	-0.219.13)	3.06	0.06	0.12154
7	River NO_3^-	NO_3^- Deposition	37	0.87(.17)	-0.75	0.44	0.00001
8	Precip. NO_3^-	Pop. Dens.	37	0.34(.07)	0.65	0.42	0.00002

that, while accounting for 53% of the variation in export, is also highly significant at p<.00001 (Fig. 12.4; Table 12.5). Both relationships are striking because nitrate concentration and nitrate export are presumably affected by complex biotic, abiotic, and anthropogenic factors. Our data show that watershed characteristics such as area and discharge were, in general, poor predictors of either nitrate concentration or nitrate export (Table 12.5).

Humans might affect nitrate export by a number of processes, including sewage loading, industrial emissions to the atmosphere, agriculture, and deforestation. If our correlation between population and nitrogen is not spurious, one or more of these sources should be of sufficient magnitude to account for the observed pattern of nitrate export increasing with population. Therefore, we examined the potential nitrate contribution from sewage loading and atmospheric deposition. We compared the nitrate export for the 42 rivers to an estimate of per capita sewage release (Fig. 12.5). Assuming no retention of sewage N by the watershed (a reasonable assumption if sewage is discharged directly into the river), inputs from human sewage alone would be sufficient to account for the increase in

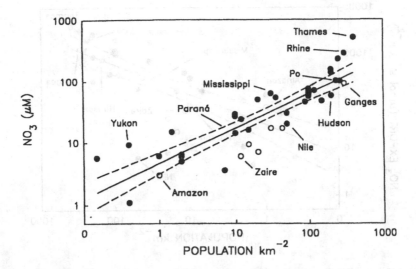

Figure 12.3. The average annual nitrate (NO_3^-) concentration in 42 rivers of the world versus the corresponding watershed human population density (pop. dens.). Nitrate is in μM and population density is inhabitants per km^2. Note log scale on both axes. Each point represents a different river; some rivers have been labeled for comparison. Open circles represent tropical rivers ($\leq 23.5°$ lat.); all others are indicated by closed circles. The solid line is the least squares regression: $\log[NO_3^-] = 0.56 \cdot \log[\text{pop dens}] + 0.67$ ($r^2 = 0.756$, $p < 0.00001$). Dashed lines are the 95% confidence interval for the regression. Data and sources are given in Tables 12.1-4.

nitrate export that is predicted from population increase (Fig. 12.5). Thus, human population density is strongly related to nitrate export, and at least one minimal consequence of population density, sewage, may be sufficient to explain the magnitude of the response we observe.

Atmospheric deposition may also be a significant source of nitrogen to rivers (Paerl 1985; Fisher and Oppenheimer 1991). There is a strong positive relationship between calculated atmospheric nitrate deposition and river nitrate, as well as between precipitation nitrate and population density (Table 12.5). We used precipitation nitrate as a surrogate for all forms of atmospheric N that might contribute to river nitrate; nitrate deposition may be greater than we estimated. Nevertheless, nitrate deposition can account for most of the export from sparsely populated watersheds. As population increases, however, the relative importance of nitrate deposition decreases with respect to sewage loading (Fig. 12.6). Watersheds with moderate to high human population will likely be dominated by sewage rather than precipitation inputs of N. Although we did not quantify the impact of additional human disturbances, previous work has shown cases where deforestation or agriculture contribute a significant percentage of the nitrate exported from watersheds (Likens et al. 1977; Kohl et al. 1971; Vollenweider 1968;

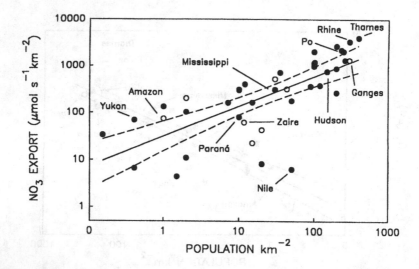

Figure 12.4.   Relationship between nitrate export from rivers and watershed human population density. Nitrate export has the units $\mu mol \cdot s^{-1} \cdot km^{-2}$. Format follows Fig. 12.3. The solid line is the least squares regression: $\log[NO_3^- \text{ EXPORT}] = 0.64 * \log[\text{pop dens}] + 1.51$ ($r^2 = 0.53$; $p < 0.00001$). Dashed lines are the 95% confidence interval for the regression. Data and sources are given in Tables 12.1-4.

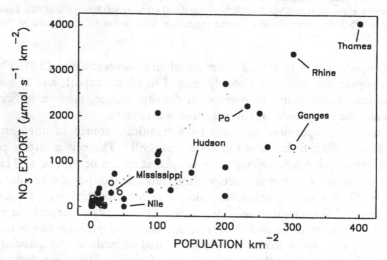

Figure 12.5.   The potential importance of human sewage to nitrate export in major rivers. Format follows Fig. 12.4 and nitrate export has the units $\mu mol \cdot s^{-1} \cdot km^{-2}$. The dotted lines represent an estimated range (4.13 to 12.4 μmol N$\cdot$ person$^{-1} \cdot s^{-1}$; Vollenweider 1968) of N loading to the river from direct sewage discharge. Note that sewage N loading is the same magnitude as nitrate export for virtually all the systems analyzed.

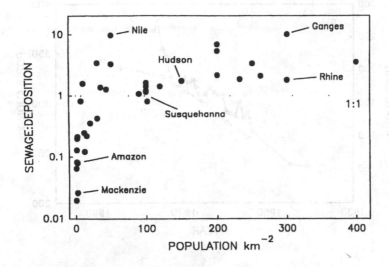

Figure 12.6.   Ratio of sewage loading to atmospheric nitrate deposition plotted against human population density in the river watersheds. The sewage N loading is calculated from the mean per capita loading in Vollenweider (1968) (9.9 μmol N· person^{-1}· s^{-1}) and the population density for each watershed. Nitrate deposition was converted to the same units as sewage loading (μmol· s^{-1}· km^{-2}). The y axis is log scale and the x axis is linear. Each symbol represents a different river system and some have been labeled for comparison. The line labeled 1:1 signifies equal sewage loading and deposition.

Haycock and Burt 1990). Accounting for these processes would only strengthen the conclusion that human activities dominate N export from land.

If nitrate export and concentration is correlated with human population in space, are these variables also linked in time? That is, as a watershed becomes increasingly populated over time, does its nitrate concentration or export of nitrate also increase? In addition to the extensive cross-system data discussed above, we were able to compile two time-series sets which documented changes in river nitrate and watershed human population (Fig. 12.7). For both the Mississippi River and the Rhine, increases in human population in the watershed have been accompanied by increases in river nitrate concentrations.

Implications

Our analyses suggest that projected population increases in the next few decades will lead to further increases in river nitrate. Based on Eq. 2 in Table 12.5, a doubling of population density (approximately 40 years) will

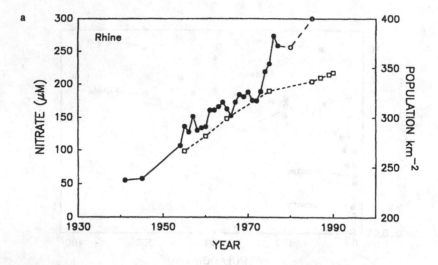

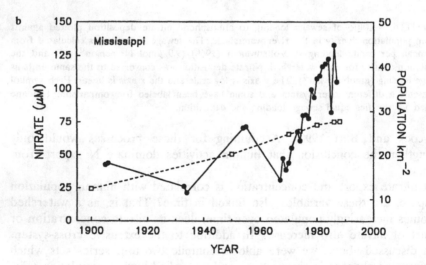

Figure 12.7.   Time series of annual average nitrate concentration (μM; filled and open circles) and human populations (numbers of people km^{-2}; squares) for two watershed systems, the Rhine (A) and the Mississippi (B). For the Rhine, nitrate data come from Bennekom (1981; filled circles) and from UNEP (1989; open circles). Human population for the Rhine watershed were calculated from data in CIA (1990); Kurian (1989); PRB 1976, 1989; UNEP (1989); and World Resources Institute (1988). For the Mississippi, nitrate data were modified from Turner and Rabalais (1991) and human population was calculated from data in Houghton Mifflin (1989); Kurian (1989); PRB (1989) and World Mark Press (1976).

result in about a 55% increase of river nitrate export and could accelerate eutrophication of the estuarine zone. Thus, coastal ecosystems around the world will continue to bear the impact of an ever increasing population.

An additional effect of increased human population will likely be an increase in the number of large rivers for which nitrate concentrations exceed the concentration recommended for drinking waters. The World Health Organization currently sees 10 ppm nitrate nitrogen (714 μM) as the safe limit for infant methemoglobinemia (Alexander 1977). Several rivers in our data set approach this limit already (Fig. 12.3); a doubling of population would likely push other important drinking supplies into this realm.

Although ecologists often try to study pristine ecosystems with as little human influence as possible, biogeochemists, especially those working at a global scale, have recognized for decades the human dominance of many chemical cycles. Our analysis suggests that at the scale of the world's larger watersheds, it would be very unwise to ignore the impact that humans have on the nitrogen cycle.

Acknowledgments

We thank L. Hedin, G. Likens, and D. Strayer for advice and comments. Financial support was provided by the U.S. Environmental Protection Agency, the Hudson River Foundation, the Andrew W. Mellon Foundation, and the Mary Flagler Cary Charitable Trust. This project is a contribution to the program of the Institute of Ecosystem Studies.

Recommended Readings

Berner, E.K. and R.A. Berner. (1987). *The Global Water Cycle*. Prentice Hall, Englewood Cliffs, New Jersey.

Likens, G.E., F.H. Bormann, N.M. Johnson, D.W. Fisher, and R.S. Pierce. (1970). Effects of forest cutting and herbicide treatment on nutrient budgets in the Hubbard Brook watershed-ecosystem. *Ecol. Monog.* 40:23-47.

Meybeck, M. (1982). Carbon, nitrogen, and phosphorus transport by world rivers. *Am. J. Sci.* 282:401-450.

Todd, D.K., ed. (1970). *The Water Encyclopedia*. Water Information Center, Port Washington, New York.

13
Humans: Capstone Strong Actors in the Past and Present Coastal Ecological Play

Juan C. Castilla

Most of the world's human population dwells on or near the coast of oceans and seas. Yet, the subtle effects of humans on coastal ecosystems has been little explored. This chapter will present key results from studies of humans as components of central Chilean coastal ecosystems, and point out the need for extending such studies to other areas throughout the world.

Prehistoric Evidence

The coastal zone has been a prominent ecotone attracting human populations, groups of hunters and food gatherers, since the Holocene (Bailey and Parkington 1988). Bailey and Parkington's book on the archaeology of prehistoric coastlines highlights the subject, focusing on European, South African, North American, Japanese, and Andean civilizations. Observations of diverse fishing and shellfish collecting patterns, and particularly the study of ancient shell mounds and middens, have shown the strong attraction exerted by coastal areas (including both rocky and sandy environments) both on nomads and sedentary non-farming populations. This attraction seems to have played a major socio-economical role along the South Pacific coast line of South America, and so theories related to the maritime foundations of Andean civilizations (Moseley and Felman 1988) have been proposed (but see Raymond 1981).

In Chile, two of the most ancient maritime archaeological coastal sites studied are the one reported by Llagosteras (1979) for northern Chile, dated about 10,000 B.P., and the recently discovered Curaumilla-1 site in central Chile (33°06'S, 71°44'W), dated about 8,500 B.P. (Ramírez et al.

1991; Jerardino et al. 1992). It must be kept in mind that Dillehay (1984) discovered in Monte Verde, southern Chile, human settlements dated about 13,000 B.P.

The marine site of Curaumilla-1, uncovered as a result of a joint project between ecologists and archaeologists (see Jerardino et al. 1992), seems particularly appropriate to the essence of this chapter. Indeed, the shellfish remains at the Curaumilla-1 site indicate beyond any question that, from the time of the ancient food gatherers' settlements to the present, the site has perisisted as a "rocky-shore environment." The unique record has allowed us to confront recent evidence of food gathering on rocky-shores in central Chile (Durán et al. 1987) with similar activities affecting the Curaumilla intertidal rocky-shores hundreds or thousands of years ago. We have reported cause-effect, direct and indirect evidence of the contemporary ecological impacts caused by human predation on these environments (Castilla and Durán 1985; Olivia and Castilla 1986; Castilla and Paine 1987), and could assume that, under similar human predation pressures, those impacts have taken place on rocky-shore Chilean ecosystems over extended periods of time. In fact, the shellfish species collected by ancient and modern coastal food-gatherers in central Chile are mainly the same (see Durán et al. 1987; Jerardino et al. 1992). Moreover, the same intertidal keystone molluscan predator, *Concholepas concholepas*, has been targeted by collectors for thousands of years.

Using a field experimental approach (see Castilla and Paine 1987), we have described the ecological effects of present human predation on these ecosystems. On the other hand, we have hypothesized (Jerardino et al. 1992) the likelihood of "historical ecological effects" accumulating through time in these communities. I believe that such a hypothesis can assist us in the critical understanding of the structure and dynamics of central Chilean rocky-shore communities. So far, the likelihood of these historical or subtle ecological effects has been totally ignored, not only in Chile, but also for most of the rocky-shore environments around the world.

Modern Evidence

What are the ecological effects caused by modern industrialized human communities on coastal areas? On the one hand, the industrialized era has brought technological developments resulting, at least, in: 1) the use of more efficient and sophisticated technologies, devices, and tools which have "negatively impacted" coastal ecosystem communities and habitats at an increasing rate (i.e., diving and fishing gear); and 2) the application of aquaculture technologies which, from an anthropogenic point of view, have "positively impacted" coastal areas.

On the other hand, the modern era has meant a significant increase in life expectancy and, overall, an exponential increase of human population from about 1 billion people in 1800 to about 6 billion people by the end of this century (Meadows et al. 1992). The above demographic changes have meant: 1) an always-increasing pressure on coastal areas due to the need to build new cities and extend coastal recreation areas; and 2) the establishment of large coastal industrial complexes. Indeed, according to Ray (1989) the coastal zone is by far the most populated and urbanized portion of the Earth.

Undoubtedly, the largest number of papers dealing with the coastal impacts are those addressing pollution issues (i.e., dredging, metal pollution, organic pollution, etc.; see Cole et al., Chapter 12, this volume) and those addressing coastal food production or services (i.e., resources over-exploitation, aquaculture, recreation, etc.). They cannot be summarized in this chapter, and fall into the category of obvious human impacts (McDonnell and Pickett, Chapter 1, this volume).

Nevertheless, there are a group of ecological effects caused by humans on coastal or nearshore ecosystems, in which the cause-effect relationship is well understood, either because the observational or experimental protocols or the hypothesis testing procedures have been duly performed, or because the "biological material" used has been adequately selected. Such is the case for: a) human alterations of rocky-shore ecosystems in California (i.e., Ghazanshahi et al. 1983); b) the negative effects of trampling on rocky-shore ecosystems (Liddle 1975) or on coral reefs (Woodland and Hooper 1976); c) the effects of intensive fishing pressure on coastal fish assemblages (i.e., Russ and Alcalá 1989); d) the "competition" for habitat between humans and birds, and the presumed extinction of some bird species (Hockey 1987); or e) the well-documented effects of human consumption on the populations and reproduction of rocky intertidal species (Branch 1975; Castilla and Durán 1985; Olivia and Castilla 1986; Ortega 1987; Moreno et al. 1986).

Furthermore, a few authors have focused on human effects on rocky-shore populations or species assemblages, using humans as the key factor causing disturbance or otherwise altering them, generating rippling processes which affect the functioning of entire communities or ecosystems (Estes and Palmisano 1974; Moreno et al. 1984; Castilla and Durán 1985; Siegfried and Hockey 1985; Hockey and Bosman 1986; Durán and Castilla 1989). Numerous insights regarding the functioning of these ecosystems and conservation and management practices have derived from these papers.

The rocky shore of Chile offers a unique opportunity to test human ecological effects. In fact, on Chilean rocky shores, food collectors (subsistence food-gatherers or "mariscadores de orilla") remove by hand several species of intertidal algae, shellfishes, and fishes. Durán et al. (1987) have reported on these activities and evaluated the intensity of human predation, the specific composition and size of prey, and estimated annual catches of the main prey species. The important point is that these "mariscadores" usually target a carnivorous gastropod mollusc species, *Concholepas concholepas*, unique in Chile and Perú and which, in turn, is a keystone predator in the rocky-shore ecosystem (Castilla and Durán 1985; Castilla and Paine 1987). Hence, Castilla and Durán (1985) have demonstrated the dramatic ecological effects caused by the exclusion of "mariscadores" from Las Cruces marine preserve (33°31'S, 71°38'W), in central Chile. Following a few years of the exclusion of the "mariscadores" at Las Cruces preserve, the population of *Concholepas concholepas* increased dramatically and caused profound changes in the middle rocky intertidal fringe. In fact, the mussel *Perumytilus purpuratus*, the competitive dominant, was practically "eliminated" by *Concholepas* and barnacles (*Jehlius cirratus* and *Chthamalus scabrosus*) which invaded those areas. Species diversity was strongly modified (Durán and Castilla 1989) and the "intertidal landscape" within the preserve switched to a new point entirely different from the previous one or the one outside the preserve. The structure and dynamics of the middle rocky intertidal fringe community was strongly affected by the exclusion of humans. Therefore, it can be concluded that, since rocky intertidal communities outside Las Cruces preserve are strongly affected by "mariscadores," their structure and dynamics are the direct result of predation by humans. In these communities, humans are a "very strong" interacting species. Most probably, the same can be said regarding central Chilean subtidal rocky communities where skin divers and professional "hookah-divers" also target mostly *Concholepas* populations.

The modifications—due to human predation—in the functioning of intertidal communities reported above, illustrate the most direct effects of human activities on such ecosystems. Nevertheless, there are several indirect, cascading, and subtle effects (see Castilla 1988; Godoy and Moreno 1989) through the trophic web. As an example, the *P. purpuratus* mussel beds have been described as one of the habitats in which many species of the intertidal key-hole limpets (*Fissurella* spp.) settle. Will the key-hole limpet populations of Las Cruces preserve their age structure through time as a result of the elimination of mussel beds? Or, are the "mariscadores" outside the Las Cruces preserve enhancing the population of key-hole limpets by preying on *Concholepas* and, in turn, enhancing the mussel beds?

The question of whether we can find examples in coastal communities of "very strong interacting species" other than mankind rather dramatically affecting community functioning, is pertinent to this chapter. The better documented examples refer to starfishes such as *Pisaster ochraceus*, *Heliaster helianthus*, and *Stichaster australis* (Paine et al. 1985); lobsters (Robles 1983; Barkai and McQuaid 1988); sea otters (Van Blaricom and Estes 1987); or sea urchins (Lawrence 1975). Perhaps "predatory sharks" might appeal to a larger audience as a "very strong interactor" in the ecological coastal play. In fact, van der Elst (1979) suggested that the exclusion of large predatory sharks by protective gill netting, in order to provide safe bathing on beaches from coastal areas of Natal, South Africa, has enhanced the abundance of small-sized sharks which escape gill netting. The populations of small sharks would have negatively impacted the angling sport-fish populations, thus affecting the total catch. The alternative hypothesis would be to explain the lowering of anglers' catches as a case of fish over-exploitation. The evidence suggests that Natal's nearshore small sharks seem to be more efficient than anglers in impacting—or devastating—coastal sport fish populations. Large sharks play a critical role in the functioning of nearshore fish populations and communities. Hence, the direct manipulation of large sharks by humans has cascaded into an indirect ecological impact affecting nearshore sport fish resources in Natal. Are there more indirect or subtle ecological effects on these coastal populations and communities?

The above examples demonstrate that humans do play a critical ecological role in coastal ecosystems around the world, since they cap the trophic webs and show such a number of peculiarities. Direct, indirect, historical, or subtle ecological effects due to mankind's activities or to other strong actors in such communities ought to be disentangled if we aspire to get closer to the real answer explaining the beauty of nature's functioning.

Acknowledgments

The financial support of FONDECYT Project No. 3503-89 to Dr. J.C. Castilla is acknowledged.

Recommended Readings

Bailey, G. and J. Parkington, eds. (1988). *The Archaeology of Prehistoric Coastlines*. Cambridge University Press, Cambridge.

Castilla, J.C. and R.T. Paine. (1987). Predation and community organization on eastern Pacific, temperate zone, rocky intertidal shores. *Rev. Chil. Hist. Nat.* 60:131-151.

Dillehay, T.D. (1984). A late Ice-Age settlement in southern Chile. *Sci. Am.* 251:100-109.

Durán, L.R. and J.C. Castilla. (1989). Variation and persistence of the middle rocky intertidal community of central Chile, with and without human harvesting. *Mar. Biol.* 103:555-562.

Godoy, C. and C.A. Moreno. (1989). Indirect effects of human exclusion from the rocky intertidal in southern Chile: a case of cross-linkage between herbivores. *Oikos* 54:101-106.

14
Modification of Nitrogen Cycling at the Regional Scale: The Subtle Effects of Atmospheric Deposition

John D. Aber

Introduction

There are two aspects to the problem of subtle human effects on ecosystems. The first is designing scientific studies so as to be able to detect subtle effects (see also Russell, Chapter 8, this volume). The second, and perhaps more difficult, is deciding whether such subtle changes are "bad" and require regulatory correction. In this chapter, I will first present an example of the subtle effects of nitrogen (N) deposition on forest ecosystem function over large areas, and then will attempt to deal with the question of environmental change and its perception by humans.

Nitrogen Deposition as an Environmental Problem

It is well established that the combustion of fossil fuels is causing increased deposition of nitrogen and sulfur (S) to large areas adjacent to heavily industrialized regions. It is also generally accepted that deposition rates decline with distance from concentrated source areas, with the "plume" of deposition created in the direction of the prevailing patterns of wind movement (to the east in temperate zones of the northern hemisphere, e.g., Zemba et al. 1988; Summers et al. 1986). Thus, deposition in the northeastern corner of the United States comes in part from within the region, and in part from the Ohio River and Great Lakes regions to the west.

As a result, there is a strong gradient in total N and S deposition from west to east across the northeast. This gradient is significant for both N and S (Fig. 14.1). While the effects of S have been dealt with at length during the lifetime of the National Acid Precipitation Assessment Program (NAPAP), the effects of N were largely ignored during that effort.

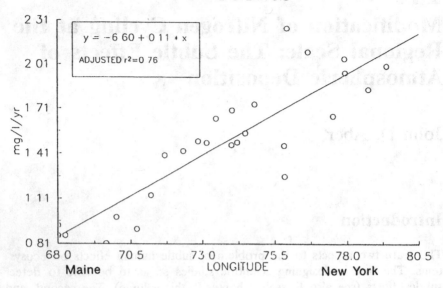

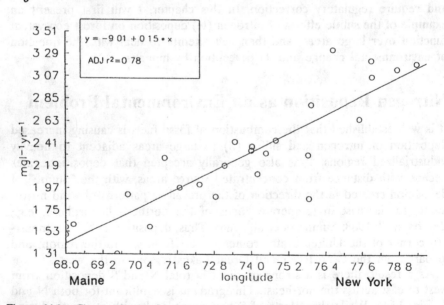

Figure 14.1. Changes in the concentration of nitrate (a) and sulfate (b) sulfate in wet deposition with longitude for the region from eastern New York State to eastern Maine (Ollinger et al. 1992, data from the National Atmospheric Deposition Program).

There is now an increasing awareness that N deposition has more complex, but perhaps more serious environmental consequences, than S. This awareness has become codified in a protocol signed between the United States and the European Community to establish "critical loads" for the deposition of both N and S to sensitive terrestrial ecosystems (Nilsson and Grennfelt 1988).

The realization that too much N can cause negative effects in terrestrial ecosystems required a fundamental change of viewpoint. In the 1970s and much of the 1980s there was concern that intensive harvesting of forests for wood products (short-rotations and whole-tree harvesting) would lead to net reductions in N stocks within forest soils, reductions in N availability to plants, and reduced productivity (Aber et al. 1982; Waide and Swank 1977; Boyle 1973; White 1974). In one sense, acid deposition has "solved" that problem for much of the northeastern U.S., and created the opposing problem: that availability of N in excess of biological demand will lead to elevated concentrations of nitrate in stream water, with the accompanying problems of soil and water acidification and increased aluminum mobility (Driscoll et al. 1987).

Nitrogen Deposition and Nitrogen Cycling in Spruce-Fir Forests

The spruce-fir (*Picea-Abies*) forests of the northeastern U.S. are in a particularly vulnerable position with regard to N deposition. Over most of New York and New England, these forests occur on mountaintops above 600-700 m in elevation. Atmospheric deposition of all elements is increased by several-fold over adjacent forests at lower elevations because of orographic wind patterns (which increase air movement through canopies), frequent immersion in clouds (which leads to direct deposition of cloud droplets on foliage), and very high leaf-surface area (which makes for a highly efficient air filtering system). In addition, a short growing season, frosts in all months, and very shallow and acidic soils all combine to reduce the potential for photosynthesis, tree growth, and N uptake.

The congruence of these factors causes the spruce-fir forest type to experience both the highest rates of N deposition in the region (Lovett and Kinsman 1990), and the lowest growth and N uptake potential. This suggests that this forest type should be the first in the region to show signs of "nitrogen saturation" or an excess of N availability over plant and microbial N demand (Aber et al. 1989).

The effects of N saturation are likely to be subtle rather than obvious. Changes in soil and plant chemistry, and in the chemistry of the water draining from these systems, are the first likely effects. None of these arevisible and all require sophisticated instrumentation to detect. Can we find evi-

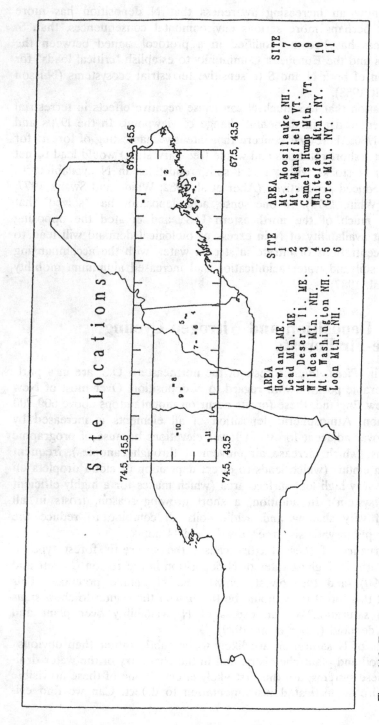

Figure 14.2 The northeastern United States with the locations of 11 study sites from which 161 samples of spruce-fir foliar and soils chemistry were collected (McNulty et al. 1991).

SITE   AREA
1   Mt. Moosilauke NH.
2   Mt. Mansfield VT.
3   Camels Hump Mtn. VT.
4   Whiteface Mtn. NY.
5   Gore Mtn. NY.

SITE   AREA
7
8
9
10
11

SITE   AREA
1   Howland ME.
2   Lead Mtn. ME.
3   Mt. Desert Il. ME.
4   Wildcat Mtn. NH.
5   Mt. Washington NH.
6   Loon Mtn. NH.

dence for regional changes in spruce-fir forests in response to N deposition?

The answer is yes. A survey of soil and foliar chemistry in 161 spruce-fir stands in 11 locations along the N deposition gradient (Fig. 14.2) shows some very consistent and significant trends. Those trends reveal a strong interaction between N deposition and the nitrogen and carbon cycling within this forest type. These results have been described in detail by McNulty et al. (1990, 1991), and will only be summarized here.

First, there is a highly significant relationship between current concentration of N in deposition across the region, and the concentration of N in, or the carbon to nitrogen ratio of, soil organic matter (O horizon, Fig. 14.3). While this has been offset to some degree by an apparent decline in total soil organic matter content, there still has been a significant increase (approximately 900 kg N ha^{-1}) in total N storage in this part of the system. If current high-elevation N deposition rates are on the order of 30 kg N ha$^{-1} \cdot yr^{-1}$ (Lovett and Kinsman 1990), then the difference in total N storage between the western and eastern extremes of the region is equal to 30 years of N deposition. Soils are the only compartment within these systems which can provide this magnitude of storage and retention of added N. Changes in N storage within the plant component are minor by comparison.

Increased N availability is reflected in the plant compartment by a large decline in the lignin content of foliage and a lesser increase in N concentration. These combine to cause large reductions in the lignin to nitrogen ratio in foliage (Fig. 14.4). Lignin is a complex, carbon-rich, polyphenolic compound which provides both structure and defense against herbivory. Changing lignin:nitrogen ratios in foliage reflect a decline in the ratio of the availability of carbon and nitrogen to plants (Waring et al. 1985).

The lignin:nitrogen ratio is also a critical determinant of the rate of decay of foliage litter (Melillo et al. 1982; Aber et al. 1990). Thus, a decline in the lignin content of foliage should mean an increase in the rate of needle litter decay. Increased litter decay may account, in part, for the reduced mass of the organic horizon in the heavily affected stands at the west end of the gradient. It is also seen in an increase in relative net N mineralization (release from soils by microbial decay, as estimated after a one-month laboratory incubation).

Finally, the increase in N content of soils, and the rate of release of N from organic matter by decomposition, result in the initiation of net nitrification, or the production of nitrate, in soils (Fig. 14.5).

Nitrification is a particularly critical process. In the absence of nitrification, a simplified view of the N cycle in forest ecosystems (Fig. 14.6a) shows the movement of N from the ammonium (NH_4^+) form to plants through plant uptake, to litter and soil through litter fall and decomposition, and back to ammonium by net mineralization. As a positively charged ion, ammonium tends to be retained in temperate zone forest soils through the process of cation exchange. Leaching losses below the rooting zone are

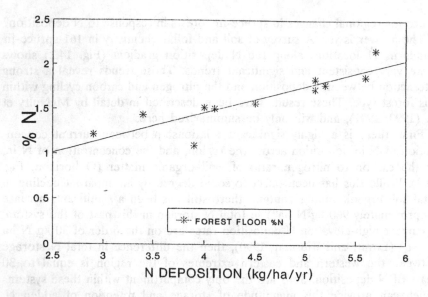

Figure 14.3. Concentration of nitrogen in the O horizon (forest floor) of spruce-fir stands across the N deposition gradient. (N deposition rates are for low-elevation stands. The gradient represents regional changes in deposition fields, not specific values for high-elevation sites. See McNulty et al. [1991] for further explanation).

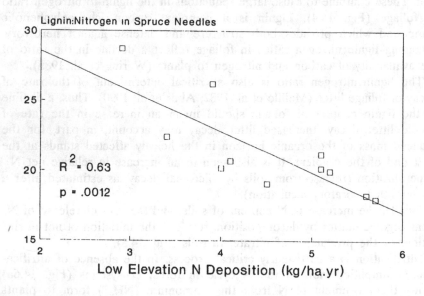

Figure 14.4. Changes in the ratio of lignin to nitrogen in red spruce (*Picea rubens*) foliage along the N deposition gradient (data from McNulty et al. 1991).

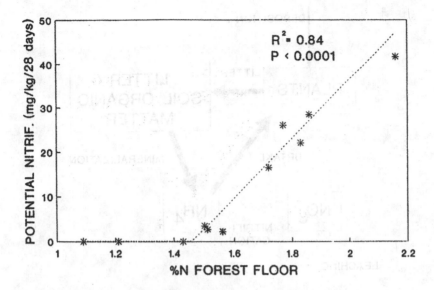

Figure 14.5. Potential net nitrification, as determined by a one-month laboratory incubation, as a function of N concentration in the organic horizon (forest floor) in spruce-fir forests across the northeastern U.S. (McNulty et al. 1991).

low and the system is efficient at retaining added N. As a system is enriched in N (Fig. 14.6b), the ammonium pool increases in size. This tends to induce the occurrence of nitrification, even in very acidic soils (McNulty et al. 1990; Novick et al. 1984; Smith et al. 1968). Nitrate (NO_3^-) is an anion with much greater mobility in soils than ammonium. Excess nitrate in soils will tend to leach out to streams and groundwater. During this process, the negative anion charge must be balanced by a positive charge so that first the nutrient cations calcium, magnesium, and potassium will be removed, followed by the acidifying and toxic cations, hydrogen and aluminum.

It is safe to say that, before the increases in N deposition resulting from fossil fuel combustion, nitrification was not a major process in the spruce-fir forests of the northeastern U.S. The shift from an ammonium-based to a partially nitrate-based nitrogen budget represents a fundamental, but subtle, shift in the biogeochemical dynamics of these forested ecosystems.

Is Nitrogen Deposition to Spruce-Fir Forests a Problem?

If we accept the trends described above, do they represent a problem which is serious enough to require corrective action? This question really has two parts: 1) do the changes *to date* require action; and 2) are these changes

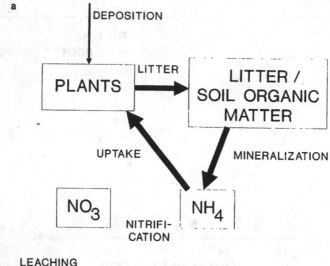

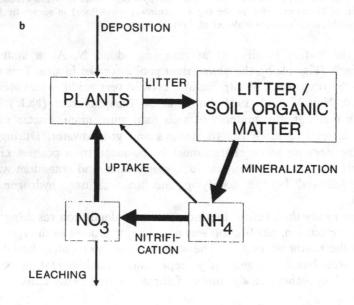

Figure 14.6. A generalized view of the nitrogen cycle in acid forest soils. Under conditions of low N availability (a), N cycles mainly as ammonium (NH_4^+). As N availability increases beyond the requirements of plants and microbes (b), nitrification becomes increasingly important, and nitrate leaching from the system begins.

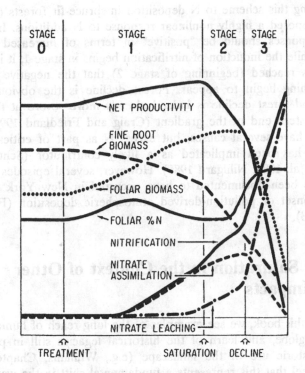

STAGE 0   STAGE 1   STAGE 2   STAGE 3

NET PRODUCTIVITY

FINE ROOT BIOMASS

FOLIAR BIOMASS

FOLIAR %N

NITRIFICATION

NITRATE ASSIMILATION

NITRATE LEACHING

TREATMENT   SATURATION   DECLINE

Figure 14.7. A diagram of hypothesized changes in N cycling and plant production in response to chronic additions of N to a nitrogen-limited forest ecosystem (Aber et al. 1989).

indicators of future changes which will be more significant (i.e., an "early warning system")? The first of these two questions can often be dealt within specific sites because the degree of change or stress is known. The second part deals with the unknown, and is more difficult.

The notions of current and future change can be combined conceptually using a continuum concept introduced first by Smith (1974) and Bormann (1982). They identified four stages in the response of an ecosystem to chronic or continuous pollutant inputs. We have adapted these to the process of N additions and saturation in spruce-fir forests (Aber et al. 1989; Fig. 14.7). The first stage (0) represents the pre-pollution condition. In stage 1, effects of pollution loading are very subtle, and, if the "pollutant" is an essential element for biological function, as with N, may result in actual enrichment of the system and higher rates of biological production. In stage 2, negative effects begin to occur, but they are still subtle and difficult to detect. Most of the responses described above would fall in stages 1 and 2. In stage 3, negative responses become highly visible or environmentally critical. It is important to understand that, according to this approach, the different stages can represent a continuous, delayed response to chronic pollutant inputs.

In adapting this scheme to N deposition in spruce-fir forests (Fig. 14.7), we have proposed a highly nonlinear response to N additions. Indeed, the stage 1 responses should be "positive" in terms of increased biological function. While the induction of nitrification begins in stage 1, it is not until saturation is reached (beginning of stage 2) that the negative effects of nitrate leaching begin to appear. Forest decline is the obvious stage 3 response, and forest decline is indeed already more severe at the heavily affected, western end of the gradient (Craig and Friedland 1991). Forest decline can have several causes, but excess N as part of critical nutrient imbalances, has been implicated as a likely contributor (Schulze 1989; Friedland et al. 1984; Nihlgard 1985). However, several episodes of spruce decline have been documented to have occurred in New York State well *before* the onset of pollution-derived atmospheric deposition (Foster and Reiners 1983).

Nitrogen Saturation in the Context of Other Human Impacts

Throughout this book, we see examples of the long reach of human habitation of the globe, and learn of the historical legacies still in place even from pre-historic use of the landscape (e.g., Williams, Chapter 3, this volume). I feel that this represents a fundamental shift in the way we view the natural world. The concept of "wilderness" as a place untouched by human activity is a cornerstone of much of our environmental policy. While useful as a policy signpost, this concept puts a wedge between humans and the rest of nature. It leaves us only the role of despoilers of nature, with any of our actions, by definition, degrading the initial pristine quality of the Earth. This approach requires no understanding of how the Earth and its ecosystems function. From the anthropological and archaeological evidence presented throughout this volume, true wilderness must represent a very small fraction of the Earth's land surface, and many of the places that we hold sacred for their wild characteristics have seen the effects of one-to-many waves of human occupation.

The alternative view that humans are part of the natural system opens a Pandora's box of difficult questions. The extreme argument is that, if we are indeed part of nature, then our actions are "natural" and the outcomes should not be considered unnatural or negative. This also requires no understanding of ecosystem function, and is more or less *carte blanche* with respect to environmental change.

The "middle ground" (Aber and Jordan 1985) substitutes, perhaps, conservation or stewardship for the polarizing dichotomy between preservation and destruction, and leaves open the possibility that human actions (such as ecological restoration, see Jordan, Chapter 21, this volume) can contribute to either the recovery of degraded systems, or even the aesthetic and physiological enhancement of natural landscapes. An example of aesthetic concern is the preservation of farmland in Europe (Grossman, Chapter 18, this volume), and of physiological enhancement, the initial phase of N-deposition (i.e., fertilization) in the northeastern United States. This is indeed a liberating world view in that humans can once again take on the role of active participant in the shaping of the global environment, hopefully with sensitivity and intelligence. This view requires, in fact, considerable understanding of how ecological systems function, and means a loss of absolutes within the environmental debate. Perhaps less palatably, it also calls for trade-offs in terms of environmental control, balancing environmental needs with other societal values.

How might this line of thought apply to the spruce-fir forest example presented above? The preservationist view might be that the "natural" spruce-fir forest is a system of low N availability and slow growth, and that even the addition of small amounts of N from fossil fuel combustion would alter the system to an unnatural state. The *carte blanche* approach would hold that since humans are part of nature, even complete forest decline in the spruce-fir zone resulting from human activity is just as natural as if the decline had been cased by another organism such as an insect or pathogen.

Stewardship of the spruce-fir forest may mean preserving the major functions of the ecosystem intact, without an attempt to hold the current species assemblages constant (Pickett et al. 1992). For example, early work by Siccama (1974) identified the spruce-fir zone as one particularly sensitive to perturbations. Partially as a result of this work, the State of Vermont placed restrictions on developing lands above about 660 meters (2,000 feet) elevation. Such a decision provides maximum protection with minimal real impact on other societal goals.

The question of altering the nation's energy policy in order to increase the potential for continuing the existence of spruce and fir on the mountaintops of New England is more difficult. It is important to recognize that these areas are not becoming devegetated as a result of forest die-back, but are being converted to increased occurrence of fir and several deciduous hardwood species. It is also important to realize that change is part of ecosystem function in general, and that cyclical die-backs of spruce, in particular, have been reported (Foster and Reiners 1983).

Here, perhaps, we need to be open to the notion that a human-created landscape, if that is what emerges, might not be less valued than what we perceive as the "pre-settlement" condition at these elevations in these mountains. Indeed, in this same region, efforts are made in many states and communities to preserve agricultural land from either development or reversion to forest as a contribution to the historical, aesthetic, and environmental well-being of the community.

Behind all of these arguments is the notion of change. The preservationist view is that all human-induced change is bad. The *carte blanche* view is that all human-induced change is fine. The stewardship view is that change is inevitable, and that, by working with the system and understanding its dynamics, we can make informed economic and aesthetic choices. The middle ground based on stewardship is, in fact, what we generally achieve in democratic countries, although this tends to come about by compromise between extreme positions, and hence, is generally not completely satisfying to any of the advocacy groups involved.

Recommended Reading

Aber, J.D. and W.R. Jordan. (1985). Restoration ecology: an environmental middle ground. *BioScience* 35:399.

Aber, J.D. and J.M. Melillo. (1991). *Terrestrial Ecosystems*. Saunders College Publishing, Philadelphia.

Aber, J.D., K.J. Nadelhoffer, P. Steudler, and J.M. Melillo. (1989). Nitrogen saturation in northern forest ecosystems. *BioScience* 39:378-386.

Bormann, F.H. (1982). The effects of air pollution on the New England landscape. *Ambio* 11:338-346.

Driscoll, C.T., C.P. Yatsko, and F.J. Unangst. (1987). Longitudinal and temporal trends in the water chemistry of the North Branch of the Moose River. *Biogeochemistry* 3:37-62.

Summers, P.W., V.C. Bowersox, and G.J. Stensland. (1986). The geographical distribution and temporal variations of acidic deposition in eastern North America. *Water Air Soil Pollut.* 31:523-535.

15

The Application of the Ecological Gradient Paradigm to the Study of Urban Effects

Mark J. McDonnell, Steward T.A. Pickett, and Richard V. Pouyat

Introduction

Urban areas are clearly highly polluted and ecologically altered (Stearns and Montag 1974; Graedel and Crutzen 1989; Berry 1990). Therefore, critical ecosystem processes and structures should be greatly affected in urban natural areas (Airola and Buchholz 1984; Cwikowa et al. 1984; Loeb 1987; White and McDonnell 1988; Rudnicky and McDonnell 1989). For a variety of conceptual and methodological reasons (see McDonnell and Pickett, Chapter 1, this volume; Egerton, Chapter 2, this volume), North American ecologists have historically neglected the study of ecological systems embedded in urban environments (Cairns 1987, 1988; Ludwig 1989). Ecologists have traditionally attempted to understand the relationship between pattern and process in a prodigious assortment of other natural and modified ecological systems. Urbanization, with its human modification of the landscape and associated environmental degradation, provides a "new" and unique pattern in which to extend the knowledge gained from other systems. The application of the well-established and successful "gradient paradigm" to the study of populations, communities, and ecosystems that occur along urban-rural land-use gradients provides a context for addressing both basic ecological questions and critical environmental issues. The purpose of this chapter is to: 1) develop in greater detail the concept of urban-rural environmental gradients (McDonnell and Pickett 1990); and 2) illustrate its application in the study of forests in the New York City metropolitan region.

The Gradient Paradigm

During the late 1950s and 1960s, Robert Whittaker changed the way North American ecologists view the landscape by championing the gradient concept (Whittaker 1967). This concept and the analytical tools used to demonstrate it became the foundation of the larger gradient paradigm. The gradient paradigm can be summarized as the view that graduated spatial environmental patterns govern the corresponding structure and function of ecological systems, be they populations, communities, or ecosystems (McDonnell and Pickett 1990). In the case of direct gradients, environmental variation is ordered in space. Over the past 30 years, ecologists have effectively studied natural environmental gradients in an attempt to elucidate the relationship between environmental variation and ecosystem structure and function (Siccama 1974; Austin 1985; Keddy 1989; Vitousek and Matson 1991). In exemplifying this body of work, we must recognize that there are two types of direct gradients that have been described in nature: simple gradients and complex gradients (Whittaker 1967; Ter Braak and Prentice 1988).

A simple gradient can be defined as an environmental series based on a single measured environmental factor, such as soil moisture, which is typically described with univariate statistics. A complex gradient, on the other hand, is an environmental series based on several factors, some of which may interact. The changes in temperature, moisture, and edaphic features that occur up a mountain side create a complex gradient of environmental conditions which ecologists commonly refer to as an elevation gradient (Siccama 1974). Similarly, a transect from subtidal to beach habitats reveals a direct, but complex gradient of exposure (Buss 1986; Roughgarden 1986). Such gradients are typically described using multivariate statistics.

Ecologists have developed a variety of gradient analysis techniques, but they fall primarily into two categories: direct and indirect analysis techniques (Whittaker 1973; Gauch 1982; Austin 1985). Direct gradient analysis is utilized when the underlying environmental factor is ordered on the ground such that it can be represented by an unbroken, straight line across the landscape (Fig. 15.1). Typically, the population, community, or ecosystem response parameter of interest is represented as a function of the environmental variable (Fig. 15.1; Ter Braak and Prentice 1988).

Indirect gradient analysis techniques, on the other hand, are utilized when underlying environmental factors are not initially obvious or not organized in a linear fashion across the landscape (Fig. 15.2). Indirect gradient analysis techniques involve measuring population, community, or ecosystem parameters first, and then ordinating these values to represent the underlying environmental gradient (Whittaker 1967; Ter Braak and Prentice 1988). The ordinated response variables serve as surrogate variables until

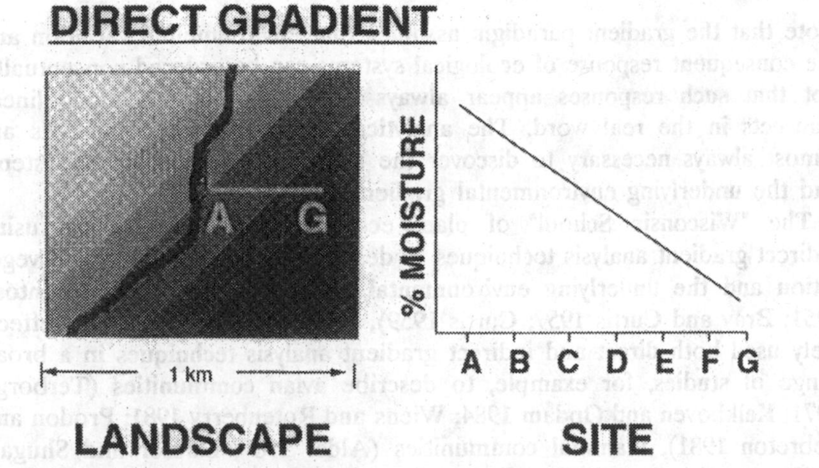

DIRECT GRADIENT

LANDSCAPE **SITE**

Figure 15.1. A diagrammatic representation of a simple moisture gradient along transect A-(
that is best defined using direct gradient analysis techniques.

INDIRECT GRADIENT

LANDSCAPE

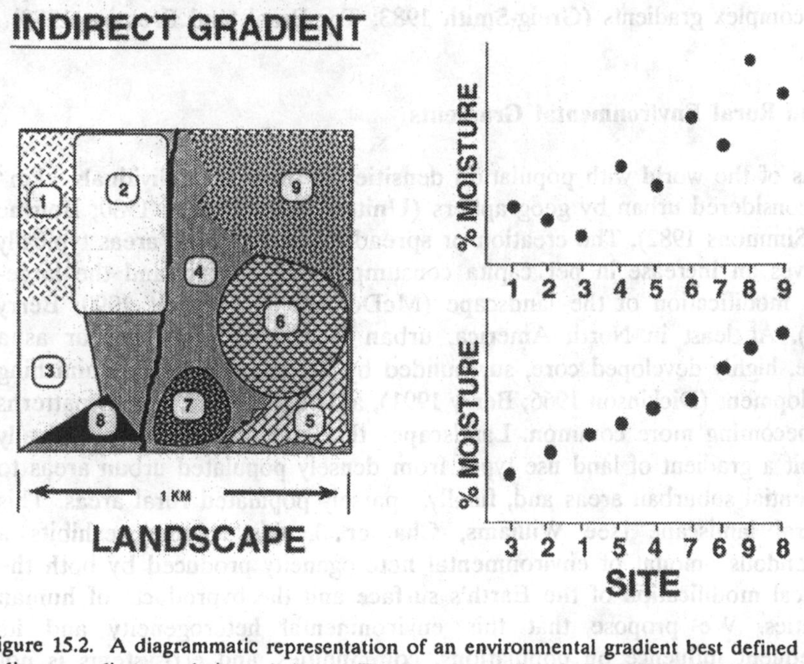

Figure 15.2. A diagrammatic representation of an environmental gradient best defined using
indirect gradient analysis techniques. The upper graph of site-moisture relationships is
unordered, while the lower has been ordinated.

the true nature of the underlying environmental gradient can be discovered.
Thus, when using indirect gradient analysis techniques, the characterization
of the underlying environmental gradient requires one to compare the
surrogate variables with actual environmental measurements, and ultimately,
initiate mechanistic studies to determine causal relationships (Keddy 1989).

Note that the gradient paradigm assumes that environmental variation and the consequent response of ecological systems can be ordered conceptually, not that such responses appear always as direct gradients along linear transects in the real word. The analytical steps of gradient analysis are almost always necessary to discover the response gradients in ecosystems and the underlying environmental gradients.

The "Wisconsin School" of plant ecologists were pioneers in using indirect gradient analysis techniques to describe the spatial pattern of vegetation and the underlying environmental gradients (Curtis and McIntosh 1951; Bray and Curtis 1957; Curtis 1959). Since then, scientists have effectively used both direct and indirect gradient analysis techniques in a broad range of studies, for example, to describe avian communities (Terborgh 1971; Kalkhoven and Opdam 1984; Wiens and Rotenberry 1981; Prodon and Lebreton 1981), mammal communities (Alder 1987; Dueser and Shugart 1978), aquatic communities (Ormerod and Edwards 1987; Furse et al. 1984), and to study ecosystem function (Minshall 1984). Direct and indirect gradient analysis techniques have been effectively used to study both simple and complex gradients (Greig-Smith 1983; Ter Braak and Prentice 1988).

Urban-Rural Environmental Gradients

Areas of the world with population densities of over 620 individuals $\cdot$ km^{-2} are considered urban by geographers (United States Census 1980; Bourne and Simmons 1982). The creation or spread of metropolitan areas typically involves an increase in per capita consumption of energy, and the large-scale modification of the landscape (McDonnell and Pickett 1990; Berry 1990). At least in North America, urban landscapes often appear as a dense, highly developed core, surrounded by irregular rings of diminishing development (Dickinson 1966; Berry 1991), although diffuse, patchy patterns are becoming more common. Landscapes that contain cities then typically exhibit a gradient of land use types from densely populated urban areas to residential suburban areas and, finally, sparsely populated rural areas. This cultural landscape (see Williams, Chapter 3, this volume) exhibits a tremendous amount of environmental heterogeneity produced by both the physical modification of the Earth's surface and the byproducts of human activities. We propose that this environmental heterogeneity and its subsequent influence on populations, communities, and ecosystems is not unlike the magnitude and complexity of environmental heterogeneity which occur in other places on earth. By examining the relationship between environmental variation along a transect running from urban to suburban to rural land-use types, and ecosystem structure and function on that transect, we can define an urban-rural environmental gradient. Due to the new combination of environmental factors, as well as the novel patterns of stress and disturbance present in urbanized landscapes, we hypothesize that

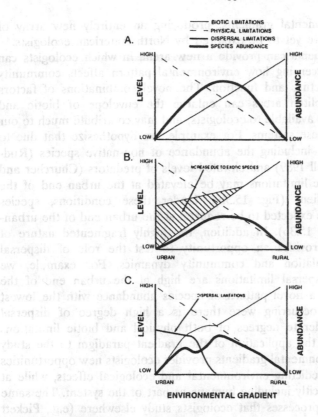

Figure 15.3. The relationship between species abundance and the levels of physical, biotic, and dispersal limitations along environmental gradients. (A) A typical environmental gradient would exhibit high species abundance at moderate levels of biotic and physical limitations. (B) An urban-rural environmental gradient could have especially high levels of biotic limitations (represented by the shaded area) due to the introduction of exotic species, consequently changing the species abundance curve. (C) Dispersal limitations present at the urban end of the gradient could be expected to further reduce species abundance at the urban end of the gradient.

the urban-rural environmental gradient is complex and nonlinear, and thus, can be elucidated using indirect gradient analysis techniques.

Characteristically, the distribution and abundance of species along environmental gradients is the result of the ability of organisms to cope with the range of physical and biotic conditions present (Whittaker 1967; Austin 1985). The greatest abundance of organisms at any point along a gradient of environmental conditions is commonly found in those areas with moderate levels of both physical and biotic limitations (Fig. 15.3a). Because urban-rural environmental gradients are new, the nature and pattern of both the physical and biotic limitations are virtually undescribed. In addition, human activities have altered the diversity and abundance of species present along such environmental gradients, producing an entirely new array of interactions that have yet to be explored by North American ecologists.

Urban-rural gradients can provide a new arena in which ecologists can test hypotheses concerning how environmental pattern affects community and ecosystem structure and function. The novel combinations of factors present in metropolitan areas can enlarge the envelope of biotic and ecosystem behaviors available to ecologists, and may contribute much to our understanding of these systems. For example, we hypothesize that due to a variety of factors, including the abundance of non-native species (Rudnicky and McDonnell 1989) and elevated levels of predators (Churcher and Lawton 1987), biotic limitations may be elevated at the urban end of the environmental gradient (Fig. 15.3b). Under these conditions, species abundance would be expected to be reduced at the urban end of the urban-rural gradient (Fig. 15.3b). In addition, the highly fragmented nature of urban landscapes provides an opportunity to test the role of dispersal limitations in population and community dynamics. For example, we hypothesize that dispersal limitations are high at the urban end of the gradient, producing a novel pattern of species abundance with the lowest species abundance occurring were there is a high degree of dispersal limitations and moderate degrees of both physical and biotic limitations (Fig. 15.3c). Thus, the application of the gradient paradigm to the study of urban-rural environmental gradients provides ecologists new opportunities for explaining or predicting environmental and ecological effects, while at the same time, explicitly including humans as part of the system. The same kinds of ecological processes that ecologists study elsewhere (e.g., Pickett and McDonnell 1989) would be expected to occur at the urban end of the urban-rural land-use gradient, but the relative contribution, or order of operation of the factors would likely differ.

In the next section we will describe the application of the urban-rural gradient paradigm to the study of forest ecosystems in the New York City metropolitan area, USA. This region encompasses both densely developed and rural areas. We will primarily address two questions: 1) Is there an ecosystem response signal from forests arranged along an urban to rural land-use gradient?; and 2) Is there an urban-rural environmental gradient of measurable environmental factors that correspond to the land-use gradient? We thus seek to determine whether the land-use gradient, as described by geographers, has functional ecological significance. The existence of a geographic or land-use gradient does not necessarily guarantee an ecological gradient. This is a matter requiring an empirical test.

The Study Area

Over the past five years, we have been studying the structure and function of forests along a 20 km wide by 140 km long belt transect located in

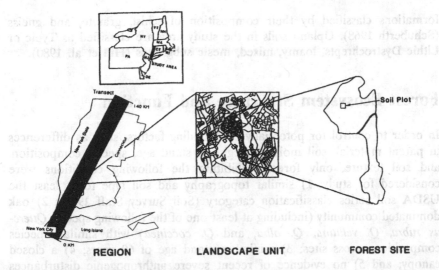

REGION   LANDSCAPE UNIT   FOREST SITE

Figure 15.4. The New York metropolitan region showing the location of the study transect, running from highly urbanized New York City to rural Litchfield, Connecticut. Central Park, which is located in the center of New York City, serves as a familiar 0 km point on the transect, although no forests within the park were examined in the study. Nine forest study sites (dark squares) have been established.

southeastern New York and western Connecticut that follows an urban-to-rural land-use gradient (Fig. 15.4). To avoid confounding factors, the transect is located entirely within the region having similar bedrock and soils as our urban forest stands. Indeed, the transect encompasses most of the appropriate lithology in the region. The transect starts in highly urbanized New York City and extends north to rural Litchfield County, CT (Fig. 15.4). Remnant forest stands still exist in New York City's park system, providing excellent study sites for investigating long-term human impacts on forest ecosystems in the city (Rudnicky and McDonnell 1989). Suburban sites are located in Westchester County, New York, and rural sites are found in Litchfield County, Connecticut (Thompson 1977; Regional Plan Association 1987).

The climate of the region is characterized by warm humid summers and cold winters with average annual air temperatures ranging from 12.5°C in New York City to 8.5°C in northwestern Connecticut (National Oceanic and Atmospheric Administration [NOAA] 1985). The temperature range, in part, reflects the well known urban "heat island". Precipitation in the area is quite uniform, ranging from an annual average of 108 cm in New York City to 103 cm in northwestern Connecticut (National Oceanic and Atmospheric Administration 1985). Precipitation is evenly distributed throughout the year for the entire study area.

The study area lies entirely within the Northeastern Upland Physiographic Province (Broughton et al. 1966). The bedrock consists of highly metamorphosed and dissected crystalline rocks which are separated into various formations classified by their composition of schist, granite, and gneiss (Schuberth 1968). Upland soils in the study area are classified as Typic or Lithic Dystrochrepts, loamy, mixed, mesic subgroups (Hill et al. 1980).

Forest Ecosystem Structure and Function

In order to control for potentially confounding factors, such as differences in parent material, soil moisture regimes, stand age, species composition, and soil texture, only forests exhibiting the following conditions were considered for study: 1) similar topography and soil type to at least the USDA soil series classification category (Soil Survey Staff 1975); 2) oak dominated community (including at least one of the following species *Quercus rubra*, *Q. velutina*, *Q. alba*, and *Q. coccinea*) with similar species composition across sites; 3) minimum stand age of 60 years; 4) a closed canopy; and 5) no evidence of recent severe anthropogenic disturbances such as soil excavation or tree cutting.

Nine forest study sites were established (Fig. 15.4; Pouyat and McDonnell 1991). They were stratified to have three forests within each of the urban, suburban, and rural land-use types that occur along the transect. Within each forest study site, three 20 m x 20 m plots were randomly established at least 30 m from the edge of the forest patch, giving a total of 27 plots along the transect (Fig. 15.4). The data reported here were collected within all or a subsample of these plots.

Abiotic Environment

A close examination of the 1989 temperature regime from NOAA/National Weather Service stations along the transect, indicates the southern end of the transect is approximately 2°C warmer than the other sites (Fig. 15.5). But there is little or no change in temperature within the 25 to 140 km section of the transect (Fig. 15.5). The relatively steep change in temperature in the first 25 km of the transect can be explained by the fact that Central Park, located in Manhattan, is the most urban station and is influenced by a strong heat island effect (Bornstein 1968). Although the NOAA/National Weather Service sites are the most reliable source of climate data, they only approximate the temperature regimes of the study forests. The southern-most forests we studied were over 10 km north of Central Park and, thus, may not exhibit quite as warm temperatures as the extremely urban Central Park site. More intensive micro-meteorological data collection is required in these and other urban natural areas.

Based on New York State data, air quality is, as expected, poor at the urban end of the gradient, and improves in rural areas, as illustrated by the

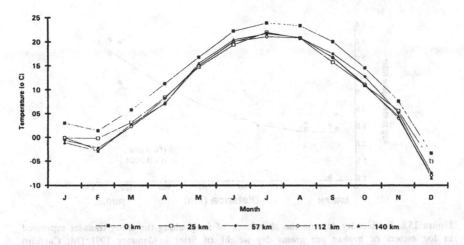

Figure 15.5. The mean average monthly temperature during 1989 at NOAA/National Weather Service sponsored stations along the study transect. Distances refer to locations along the study transect.

decline in particulate sulfate and total particulate concentration with increasing distance from the urban core (New York State Department of Environmental Conservation 1989; Graedel and Crutzen 1989). Unfortunately, because existing monitoring programs are focused primarily on human health, there are few data for sparsely populated rural areas of the study transect. One of the future goals of research along urban-rural gradients must be to better quantify air quality along the entire range of sites.

Characterization of forest soils along the study transect revealed elevated levels of lead, copper, and nickel in urban forests, declining to background levels in rural sites (Pouyat and McDonnell 1991). We also found soil hydrophobicity to be highest in forests at the urban end of the transect (White and McDonnell, unpublished data).

Biotic Response

Because the ecological study of urban-rural environmental gradients is new, we focused our studies on one particular process, litter dynamics, rather than attempting at the outset to build a complete community or ecosystem model. We focus on the dynamics of litter for several reasons. Litter: 1) affects plant community regeneration; 2) is a critical bottleneck in determining ecosystem nutrient flow, and, thus, availability of resources in a community; 3) is the principal initial site of incorporation of heavy metals in ecosystems; and 4) serves as both a resource and habitat for invertebrates and fungi. Why should litter be sensitive to the environmental changes be

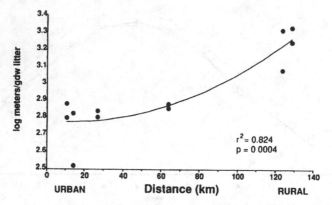

Figure 15.6. Preliminary data on the abundance of fungi along the study transect expressed as log meters of hyphae per grams dry weight of litter in January 1991 (M. Carreiro unpublished data).

tween urban areas and the surrounding countryside? There are functional relationships among the abiotic controls of litter dynamics (Meentenmeyer 1978; Day 1983; Facelli and Pickett 1990), the biotic influences on the litter pool (Garden and Davis 1988; Findlay and Jones 1990), and the resultant dynamics of the litter pool (Prescott and Parkinson 1985; Seastedt 1984). All of these factors and processes can vary along urban-rural environmental gradients.

We have specifically chosen to examine fungi and soil invertebrate abundances as examples of biotic responses to the environmental gradient, due to their sensitivity to pollution and their important role in litter dynamics. Total hyphal length and biomass of litter fungi increased toward the rural end of the gradient (Fig. 15.6; Carreiro unpublished data). We also found declining microarthropod abundance along the gradient, fewer nematodes in the middle range of the gradient, and the highest numbers in rural forests (Parmelee and Pouyat unpublished data).

Ecosystem Response

Our first working model (Model I) of how these forest ecosystems function is based on preliminary measurements of reduced soil fungi and microinvertebrates in urban forest soils, combined with information from the literature on the sensitivity of these organisms to pollution and anthropogenic stress (Tyler et al. 1989; Bengtsson and Tranvik 1989; Lee 1985; Freedman 1989; Baath 1989). This model predicts that forests at the urban end of the gradient would have reduced litter decomposition, nitrogen mineralization and nitrification rates. However, we found litter decomposition (Fig. 15.7) and potential N mineralization (Fig. 15.8) were higher in urban forest

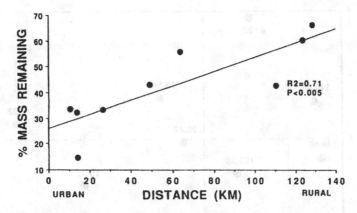

Figure 15.7. Litter decomposition measured as % mass lost of original mass in nylon mesh bags exposed in the field for 24 weeks from forests along the study transect (Pouyat 1992).

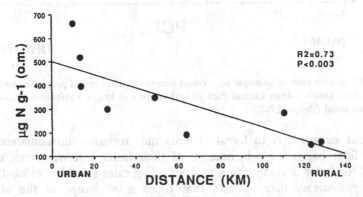

Figure 15.8. Net nitrogen mineralization in μg N per g dry wt soil, measured in laboratory incubations at 15°C at field capacity from forests along the study transect (Pouyat 1992).

stands than in rural forests (Pouyat 1992; Pouyat and McDonnell unpublished). N mineralization data are especially telling because the soils were incubated in controlled temperature and moisture conditions.

These results suggest an alternative model (Model II) that proposes the observed ecosystem response is produced by high levels of anthropogenically derived N deposition at the urban end of the gradient. High levels of exogenous N deposition could accelerate rates of litter decomposition and potential N mineralization by increasing available N (especially NH_4 for nitrification), and improve litter quality by increasing foliar N (Agren and Bosatta 1988; Aber et al. 1989). One of the most interesting and yet unstudied effects of elevated levels of N deposition on forest ecosystems, is the potential shift in the relative abundance of fungi and bacteria in the litter and soil (Baath et al. 1981; Soderstrom et al. 1983). The addition of

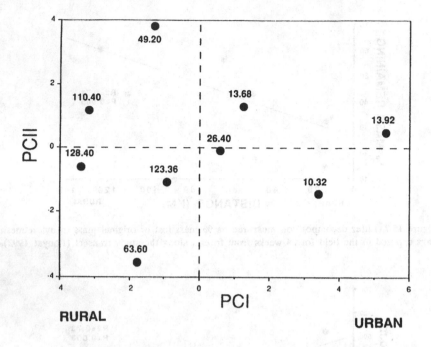

Figure 15.9. Scatter plots of principal component scores of soil chemical properties along the study transect. Distance from Central Park (0 km) is given in km as a reference location for each forest stand (Pouyat 1992).

exogenous inorganic N to forest systems may increase the abundance of soiland litter bacteria which, because of their faster turnover rates, could account for faster decomposition and N cycling rates (Hendrix et al. 1986).

Our preliminary data indicate that there is a change in the abiotic environment and a shift in the relative importance of the biotic controls of nutrient cycling along the urban-rural environmental gradient. To answer our first question presented above, we can say, yes, there is a clear ecosystem response signal along the urban-rural land-use gradient. The underlying causal mechanisms producing the signal as proposed by the alternative Models I and II have not yet been elucidated and are the focus of future research efforts.

Characterization of the Urban-Rural Environmental Gradient

Up to this point, the data have been presented as a function of distance from the urban core. The nonlinearity and the suggestion of thresholds in the biotic and ecosystem responses (Figs. 15.7 and 15.8) imply that the

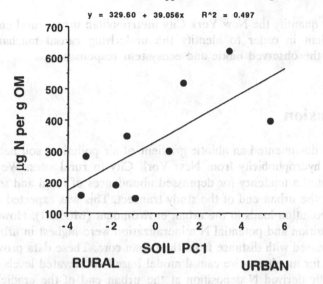

$y = 329.60 + 39.056x$ $R^2 = 0.497$

Figure 15.10. Net N mineralization presented as a function of the Principal Component Axis 1 which represents the urban-rural gradient (Pouyat and McDonnell, unpublished data).

actual underlying environmental gradient is complex and nonlinear on the landscape and, this, would best be defined using indirect multivariate gradient analysis techniques. In an attempt to reveal the true nature of the underlying gradient, we ordinated 20 forest soil variables using Principal Components Analysis (PCA) (Fig. 15.9; Pouyat 1992). The first four components of the PCA accounted for nearly 90% of the variation. The first Principal Component (PC1) accounted for over 40% of the variation. Positive loadings on PC1 correspond to high concentrations of heavy metals (especially lead, nickel, and copper) high salt concentration in the soil solution, high organic matter, high concentrations of base cations, and elevated soil acidity. As expected, the forest sites are not arranged on PC1 according to distance (Fig. 15.9).

These results confirm our hypothesis that the urban-rural environmental gradient is not linear across the landscape. The new arrangement of forest sites represents the first approximation of the urban-rural *environmental* gradient present in the New York City metropolitan area. The influence of this complex, indirect gradient on forest ecosystems can now be assessed by determining the relationship between the urban-rural environmental gradient and the biotic and ecosystem response variables. Using potential N mineralization, a key ecosystem process to test the relationship, we find that the urban-rural environmental gradient accounts for almost 50% of the observed variation (Fig. 15.10). Considering the potential confounding factors, such as differences in land-use history and microclimate effects, the strength of the response signal is quite remarkable. Future research efforts will incorporate additional environmental variables, including soil moisture and temperature, air pollutant levels, landscape structure, and land-use history, to better quantify the New York City metropolitan urban-rural environmental gradient in order to identify the underlying causal mechanisms that produce the observed biotic and ecosystem responses.

Conclusion

We have documented an abiotic gradient of air pollution, soil heavy metals, and soil hydrophobicity from New York City to rural sites. We have also documented a tendency for depressed abundances of fungi and soil invertebrates at the urban end of the study transect. This was expected due to the heavier pollution loads in the urban environment (Model I). However, litter decomposition and potential N mineralization were highest in urban forests, and decreased with distance from the urban core. These data provide strong evidence for an alternative causal model based on elevated levels of anthropogenically derived N deposition at the urban end of the gradient (Model II). Using indirect gradient analysis we were able to initially characterize the complex urban-rural environmental gradient that exists in the New York City metropolitan area. The existence of a quantifiable urban-rural environmental gradient and the counter-intuitive ecosystem responses we discovered serve to reinforce our belief that the study of urban-rural environmental gradients provide important opportunities for explaining or predicting environmental and ecological effects, while at the same time explicitly including humans as part of the ecological system.

Acknowledgments

We thank Drs. Margaret Carreiro, Kimberly Medley, Robert Parmelee, and Wayne Zipperer for help in developing the ideas presented here and for sharing unpublished data. Financial support was provided by the Lila Wallace-Readers' Digest Fund, the USDA Forest Service, and the Mary Flagler Cary Charitable Trust. This is a contribution to the program of the Institute of Ecosystem Studies.

Recommended Readings

Berry, J. L. (1991). Urbanization. In B. L. Turner II, W.C. Clark, R.W. Kates, J.F. Richards, J.T. Mathews, and W.B. Meyer, eds. *The Earth as Transformed by Human Action,* pp. 103-119. Cambridge University Press with Clark University, Cambridge.
McDonnell, M.J. and S.T.A. Pickett. (1990). The study of ecosystem structure and function along urban-rural gradients: an unexploited opportunity for ecology. *Ecology* 71:1231-1237.

Pouyat, R.V. and M.J. McDonnell. (1991). Heavy metal accumulation in forest soils along an urban-rural gradient in southern New York, U.S.A. *Water Air Soil Pollut.* 57-58:797-807.

Ter Braak, C.J.F. and I.C. Prentice. (1988). A theory of gradient analysis. *Adv. Ecol. Res.* 18:272-327.

Whittaker, R.H. (1967). Gradient analysis of vegetation. *Bio. Rev.* 49:207-264.

16
The Process of Plant Colonization in Small Settlements and Large Cities

Eduardo H. Rapoport

Introduction

Some authors consider cities as ecosystems of a singular type (Meier 1976; Halffter et al. 1977). Their climate generally differs from that of the surrounding areas (Landsberg 1981; Miess 1979) and their functioning differs from natural processes (Wolman 1965). Humans are accompanied by a rich urban flora and fauna, composed mainly of species introduced voluntarily, such as ornamental and utilitarian plants, mammals, birds, reptiles, fish, and other taxa, or involuntarily, such as weeds, pathogens, and animal pests. Other anthropophilous native vertebrates, invertebrates, and plants can be added to the list. The anthropization process involves a series of adaptations to peculiarities of rural and urban environments (Faliński 1968; Trojan 1982; Kowarik 1990).

In this chapter, I will focus upon the colonization of urban areas by plants based on relevés performed between 1976 and 1990. This chapter consists of an account on the floras, with no hypotheses to test or philosophical considerations to pose. Such an initial descriptive approach is justified because ecological studies of urban areas are still rare and preliminary (McDonnell et al., Chapter 15, this volume).

Exotic Elements in a Temporary Town

In 1973, construction of a dam on the Limay River was begun about 100 km east of Bariloche (41°40'S; 70°45'W) in the Patagonian xerophilous shrubland. The town of Villa Alicura was built for the workers and contained about 20 houses, some general-use buildings, warehouses, parking lots, and unpaved roads between the paved Highway 237 and the Limay River.

Table 16.1   The percentage of exotic plants at different locations in Villa Alicura, northern Patagonia.

Location	%
Town (urbanized zone) between 60 and 83%	74
Road contiguous to town (0-3 m wide strips)[a]	63
Highway 237 (Km. 1538) paved	57
Main road between warehouse and dam	50
Limay River shore in front of Alicura Town	48
Road from warehouse to dam	43
Road from town to highway	40
Road contiguous to the town (second strip 3-6 m)[a]	39
Warehouse parking lot (barely used)	38
Road from town to river	29
Provincial road No. 40 to Paso Flores	27
Road from town to river (abandoned section)	25
Provincial road No. 40 (abandoned section)	13
Areas surrounding the town (unaltered)	0

[a] Correspond to the same site.

Between February and March 1977, 24 species of native plants and 105 species introduced by people were recorded. Of the latter, 70 were cultivated in gardens and orchards, and 35 were weeds (Rapoport and Ezcurra 1977). Percentage of Exotic Species (PES) was calculated per hectare, except on roadsides where plots of 100 x 3 m were considered (Table 16.1). Values obtained varied considerably according to the type and intensity of human alteration. Based on the total colonizing flora in the village, instead of the average values per hectare, the PES reached 78.3. Three-fourths of these plants are of Eurasian origin.

In the surrounding semi-arid ecosystems, unaltered by soil removal, trampling, transportation, irrigation, or urbanization, there is no colonization by alien species. Rather, the exotic plants colonize just to the border between the altered and unaltered zone. Exotic invaders prosper, however, when irrigation or leakage water is present, and occasionally along wadis or arroyos, i.e., in areas disturbed by occasional floods. Watering, in these cases, is a disturbance factor breaking down the "resistance" of natural communities to invasion in this cold-temperate desert.

In other words, introduced plants in Alicura (three-fourths of Eurasian origin), though adapted to arid conditions and regional soil types, do not prosper in natural ecosystems except when the soil has been altered and there is a good provision of propagules coming from nearby areas. Natural ecosystems surrounding the town with only 10 dominant and subdominant

Table 16.2   Mean lateral colonization by weeds on the shoulders of the Bariloche-Challhuaco road, crossing a natural *Nothofagus pumilio* forest. Measurements in meters (n = 100 transects per sector) from the edge of the road. Distance between both sectors: 800 m. Sector length: 200 m.

Weed	First Sector mean	First Sector S.D.	Second Sector mean	Second Sector S.D.
Rumex acetosella	2.4	1.5	6.0	6.0
Muehlenbeckia hastulata[a]	2.6	1.6	1.0	—
Trifolium repens	1.4	0.8	1.8	0.5
Plantago lanceolata	0.9	0.4	0.6	0.5
Taraxacum officinale	0.7	0.7		
Matricaria matricarioides			0.7	0.1
Chenopodium album			0.6	0.2
Polygonum aviculare			0.5	0.1

[a] The only native colonizer.

species are, in some way, "resistant" to invasion.

Grouping the data by the type of alteration, it is possible to list habitats in order of increasing richness of exotic invaders: abandoned roads (5 species); roads in use (8); uninhabited buildings (9); roads close to houses (15); river banks (28); and houses (35).

The Flora of a Mountain Road

Roads crossing natural communities constitute good corridors for weed dispersal (Frenkel 1970, 1974). In Bariloche, exotic plants, with the exception of a few possibly introduced or cosmopolitan species (e.g., *Stellaria media, Galium aparine, Cerastium fontanum*) do not invade pristine habitats. To assess weed invasion, in January 1977 I studied 100 transects at 2 m intervals along 2 sectors of a dirt road which crossed a lenga (*Nothofagus pumilio*) forest, 10 km south of Bariloche. For each transect I measured the width of the lateral bands of colonizing vegetation across the shoulder of the road. Exotic invaders penetrate only between 0.7 and 6 m into the aboriginal forest (Table 16.2). A different situation is observed when exogenous factors such as timber extraction, fire, cultivation, or foraging by livestock occur in fields beside the road. In these cases, there is a massive invasion by exotic weeds measured in kilometers rather than in meters, where species composition may remain relatively constant for several years.

Table 16.3 List of exotic plant species around Refugio Neumeyer, Mt. Challhuaco, northwestern Patagonia.

	1977	1990
Agropyron repens	-	+
Carduus nutans	+	+
Cerastium fontanum	-	+
Chenopodium album	+	+
Dactylis glomerata	+	-
Galium aparine	+	+
Lolium multiflorum	-	+
Matricaria inodora	-	+
M. matricarioides	+	+
Plantago lanceolata	+	+
Poa annua	+	+
Polygonum aviculare	+	+
Prunus domestica	+	-
Rumex acetosella	+	+
R. longifolius	+	+
Stellaria media	+	+
Taraxacum officinale	+	+
Veronica serpyllifolia	-	+

Exotic Elements Near a Mountain Shelter

Nahuel Huapi National Park, in northwestern Patagonia, has practically no signs of being invaded by exotic plants in undisturbed areas. In sites under natural disturbance, some aliens such as *Rumex acetosella* may prosper on barren, stony mountain tops, or *Taraxacum officinale* on the borders of streams in mountain bogs, however, they are always in low abundance. Aliens are present along pathways connecting mountain shelters (Club Andino Bariloche) and around these buildings, i.e., on ground trampled by tourists. One of these isolated shelters (Neumeyer Shelter, Mt. Challhuaco) has been built in an old growth *Nothofagus pumilio* forest at the end of the above-mentioned road. This has been a summer tourist attraction since 1945. There is no crop or ornamental cultivation, practically no livestock, and no wood clearing, except around the campsite. By means of eight transects taken radially from the cabin, I measured the PES per square meter. In January 1977, the exotic elements on average did not extend more

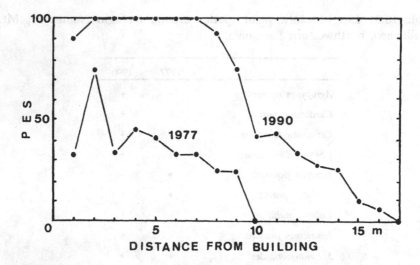

Figure 16.1. Dispersion of exotic plant species around a shelter in the Patagonian Andes, south of the city of Bariloche. Mean percentages were calculated on the basis of 1 m² samples along 8 transects radially centered on the building. Small, isolated patches of aliens also exist beyond the limits illustrated by the figure.

than 10 m away from the cabin (21 native plus 12 introduced species) (Fig. 16.1). Thirteen years later, the average distance colonized by exotics had increased to less than 17 m.

Between 1977 and 1990, there were two extinctions and five additions to the local flora around the shelter (Table 16.3). The PES, in a 50 m radius circle around the building, increased from 36.4 (1977) to 44.7 (1990). This represents a rather low increment, considering the number of tourists visiting the area.

Exotic Elements on the Outskirts of the City of Bariloche

The City of Bariloche (population 45,000 in 1977 and 81,000 in 1991) is adjacent to the Nahuel Huapi National Park, Rio Negro Province, Argentina. The situation on the border of the city is quite different from the case mentioned previously. Several processes of environmental alteration occur simultaneously or successively: road construction, utility wiring and fencing, lumbering, bulldozing, house building, trampling, soil removal, rubble accumulation, introduction of organic soil for gardening, soil covering by turf, planting of ornamentals, fertilizing, and watering, etc. In Fig. 16.2, the final results of these activities are illustrated as the "instantaneous" situation for April 1977. The area was divided into ninety-one 5 x 5 m

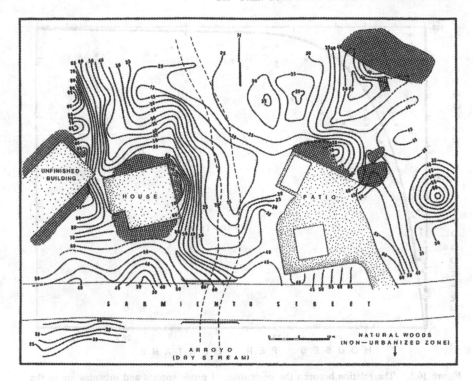

Figure 16.2.   Idealization of the spatial distribution of exotic species around houses on the border of the city of Bariloche, Argentina. Isolines represent percentages with respect of the total number of plant species. Cross-hatched areas correspond to heavily trampled soil without vegetation. See Fig. 16.4 for site location. Rodrigo Dalziel and E.H. Rapoport (unpublished).

squares, in which the native and introduced, cultivated and "spontaneous" (non-cultivated) plant species were considered. The map was drawn as an ideal model under the hypothesis that PES values were operative only in the center of each quadrat, and that gradual gradients exist between contiguous quadrats. Considering this, a map of intermediate isolines was drawn. Strictly cultivated species are restricted to locations immediately around the buildings; non-cultivated weeds and escaped cultivars cover most of the land. The mean PES was 41.3 (S.D. 18.9), ranging from 9.5% to 100% per quadrat. Note that the highest values appear around houses, and the lowest along a stream which flows only during the rainy season. Across the street toward the south (Fig. 16.2, bottom), the north slope of Mt. Otto begins, covered by a forest of Patagonian cypress (*Austrocedrus chilensis*) as the dominant species. The PES decreases to 4% near the mountain top where there is a new increase due to the presence of a road and a tourist resort. Aliens decrease again, down to 0% in some small patches on the southern slope in a forest with about 18 understory species. This constitutes

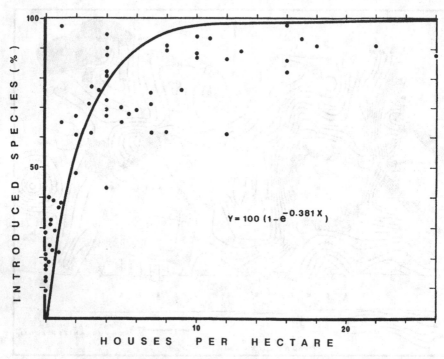

Figure 16.3.   The relation between the percentage of exotic species and urbanization in the city of Bariloche, Argentina, under the assumption of an asymptotic value of 100%. At equal housing densities, samples taken in low income neighborhoods tend to be below the prediction curve, while high income neighborhoods generally show higher values of PES. Optimization by Powell's method (minimum squares) gives a predictive curve described as $Y = 81.1$ $(1-e^{-0.833X})$ valid for suburban, residential areas.

a rather harsh environment (1,350 m altitude) because of its lower temperature, associated with longer periods without sunlight, and thick accumulation of snow in winter. In December 1989 a second survey showed similar results, with the addition of one species, *Galium aparine*, probably alien, in very low frequencies.

Although urbanization in Bariloche is a complex process, it is relatively simple to predict its final effects on the natural flora (Fig. 16.3). The curve describes the floristic changes that may occur in a district evolving into a densely built commercial area, under the hypothesis of an asymptotic value of PES = 100. In fact, in 1977, several blocks in the downtown area showed a PES of 100. However, the increment of PES with the increase of buildings in residential areas is better described by the equation which appears in the legend to Fig. 16.3, with an asymptote of 81.1. The curve reaches its maximum asymptotic value very rapidly. This indicates that there

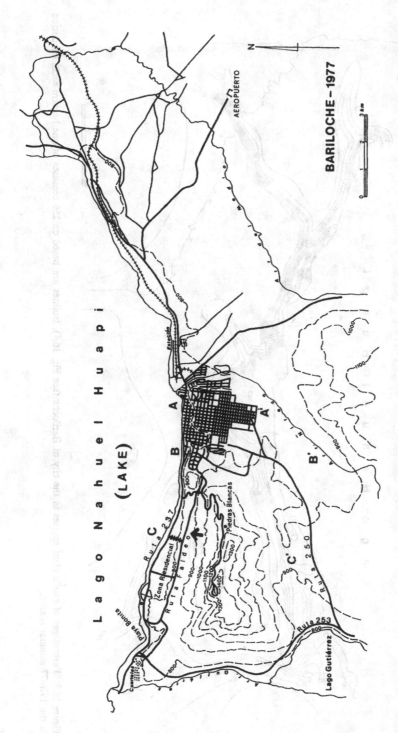

Figure 16.4. The city of Bariloche and surroundings. Suburban areas extend 30 km to the west and 10 km to the east of the city center. Transects are indicated by letters. The arrow points out the site where the survey illustrated in Fig. 16.2 was performed.

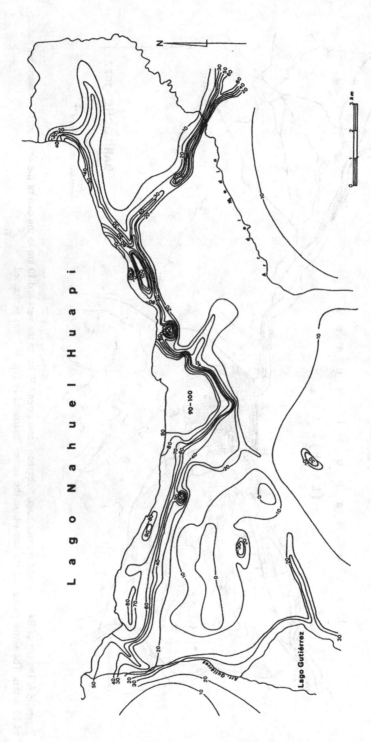

Figure 16.5. Percentages of exotic plant species in the city of Bariloche (see Fig. 16.4). Isolines are based on 156 censuses of one hectare each, conducted in the 1976-77 summer season.

is not only an alteration of natural communities by human intervention but also sociological and psychological factors. People generally do not prefer native Patagonian plants, no matter how beautiful they are, and they replace them with exotic ornamental species (Rapoport 1979). Escapes from gardens and agriculture constitute 45% of the exotic element among the 228 invaders of northwestern Patagonia (Rapoport 1989).

Exotic Elements in the City of Bariloche

In order to obtain an assessment of the exotic component of its flora (cultivated and adventive species per hectare), transects crossing the city of Bariloche from the lake (Fig. 16.4) toward the suburbs were surveyed (41 relevés of one hectare each). Another set of 115 sample plots (one hectare each) in selected sites of the suburbs was included. Each plot was inspected for approximately one hour and the recorded species were listed. In this way, the most abundant species were represented in the surveys and a comparison among the plots could be performed. Areas with a pronounced slope, e.g., roadsides, were surveyed in 30 x 300 m or even narrower strips. In these cases, surveyed areas were smaller than 1 ha, but results probably did not excessively misrepresent the situation because the proportion between native and exotic species was the relevant item. This first survey was performed during the 1976-1977 summer season. Outside the urbanized area, the main factors altering the natural vegetation are agriculture, livestock grazing, deforestation, and wild fires. Wild cattle, wild dogs, and some introduced animals, such as the European hog, European hare, European red deer, brown rat, and others, abound in the surrounding indigenous vegetation. In the downtown area, the exotic component ("spontaneous" and cultivated species) of the flora averages 91% (Fig. 16.4 and 16.5). One hectare plots consisting of 100% foreign plant species are not uncommon downtown. The values for the strictly weedy aliens is 80.6% (S.D. 14.8) for 59 censuses performed in urbanized plots. High values also occur along roads and around the railway station and airport. These values rapidly fall at the outskirts of the city, in places with a low number of houses per hectare. Regardless, values between 10 and 20 PES appeared throughout in the surroundings: up to ca. 15 km to the east and 30 km to the west, even in areas where there are no houses. In Fig. 16.5, some places with zero values can be observed, surprisingly quite near the city. This phenomenon occurs on the southern slopes of Mt. Otto, a place mentioned before as having a rigorous microclimate.

In the summers of 1987-1987 and 1989-1990, I completed 51 new censuses in plots studied in 1976-1977. From this sample, there were 34 cases (67%) in which PES increased (between 5 and 356%, with an average of 105.6% and S.D. = ±87.4). There were 15 cases in which values did not change, and only 2 cases showed a decrease in PES (from 33 to 28% and from 10 to 3%). In the latter cases, however, no weed species became totally extinct. This fact may be due to variations in germination caused by different environmental conditions during the period between the surveys, or to recolonization by some native plants. The total number of alien weed species in Bariloche was 139 in 1972; it increased to 156 by 1976, and to 228 in 1989 (Dimitri 1972; Rapoport 1989). Percentage isolines in Fig. 16.5 correspond to summer observations, but, of course, they vary according to the season. For example, a 3 km inventory along the Bariloche-Llao Llao paved route, made in January 1977, recorded exotic species as 85% of the total 78 non-cultivated species, against 52% of the 43 recorded species at the end of the winter (September 1977). This apparent reduction is due to the large number of annual species which overwinter as seeds, as well as to the presence of life-forms less conspicuous in the dormant season, such as hemicryptophytes and cryptophytes.

The Flora of Mexico City

Based upon 100 one-hectare plots, where roadside and vacant lot plant species were annotated (Rapoport et al. 1983), and 400 surveyed houses (Díaz-Betancourt et al. 1987), an overview was obtained on the cultivated and non-cultivated urban flora in Mexico City. Surveys were performed at random in an area of 26 x 38 km.

Roadsides and Vacant Lots

In the first survey (Rapoport et al. 1983), 564 species were recorded, of which 70% were native and 30% exotic. Among native species, 364 were non-cultivated, 10 cultivated, and 21 cultivated and non-cultivated. Among the exotic species, 90 were cultivated, 53 non-cultivated, and 26 cultivated and non-cultivated.

As there is a distinct rainfall gradient from 400 mm·y^{-1} in the northeast to 1300 mm·y^{-1} in the southwest, Mexico City shows some floristic differences per district, influenced also by urbanization type (residential, industrial, or commercial) and by income level. There is a positive linear correlation between annual rainfall and the number of non-cultivated plants per hectare on roadsides and vacant lots (r = 0.530, p<.001). The same relation holds for cultivated plants, though with a lower correlation coefficient (r = 0.193), p<.05) and lower regression slope. This is an indication that, in this city, humans are still not independent from rains with respect to the number of cultivated species they support. When measuring the percentage of area urbanized (covered by buildings and roads) as an independent variable, and the number of cultivated species per hectare as a dependent variable, the correlation was positive ($r = 0.352$), $p < .001$), but it was negative with respect to the number of non-cultivated species ($r = -0.573$, $p < .001$).

Air pollution by suspended particles (measured between 0 and 500 $\mu g \cdot m^{-3}$) was negatively related to the number of cultivated species per ha ($r = -0.287$, $p < .01$), with a low regression line slope ($b = -0.021$), and it was also negative with respect to non-cultivated plants ($r = -0.325$, $p < .001$), though with a larger slope value ($b = -0.058$). In the case of pollution by SO_2 (measured between 0.01 and 0.06 ppm), there was also a negative correlation with non-cultivated species richness per hectare ($r = -0.485$, $p < .001$); however, there was no significant correlation with cultivated species. This is due to the fact that when roadside cultivated plants die from pollution, most people replace them by others. Among cultivated plants, 15% were native to the Central Valley of Mexico while 38% included all New World species. The Paleoarctic Region contributed about 42%. On the contrary, among non-cultivated plants, 80% were native; the New World contributed 85%, and the Paleoarctic Region added only 13% of weeds and escapes. A more precise analysis of the data, adding 107 species by including those reported by other authors, indicated 67% of the species as native and 33% as exotic, including cultivated and non-cultivated ones. Of the 174 exotic cultivated species, 37% were aggressive invaders.

Private Gardens and Indoor Plants

Frequently, when calculating urban green areas, private garden, balcony, indoor, flat roof, and courtyard plants are not considered. However, this flora can significantly contribute to the city's greenness and embellishment. This aspect of urban ecology was studied by Díaz-Betancourt et al. (1987) in Mexico City. In a 400-house survey, 750 species were recorded, of which 89% were cultivated and 11% were non-cultivated. Among cultivated species, 30% were native and 70% were exotic, i.e., exactly the opposite to what was found with the street and vacant lot flora mentioned above. Among strictly cultivated plants, 40% came from the Paleoarctic Region, a percentage equal to that of native species. Social status affects the flora of houses and neighborhoods. First, medium income families have smaller garden areas than high or low income families (Table 16.4). This is because, in general, they inhabit urbanized areas with high population densities where land cost is, in relative and absolute values, much higher than land in suburban areas where low income people live. In spite of the

Table 16.4 Garden and pot plants in Mexico City (Díaz-Betancourt et al. 1987). Values indicate the mean and (standard deviation).

	High income	Medium income	Low income
Number of houses surveyed	83	146	171
Garden area (m²)	295.6 (1269.2	36.0 (62.0)	73.3 (99.2)
Courtyard area (m²)	11.3 (18.0)	19.3 (20.9)	60.7 (170.9)
Houses with garden (%)	85	30	20
Houses with yard (%)	10	50	70
Houses with garden and yard (%)	5	20	10
Number spp. per house (cultivated)	26.6 (19.0)	25.3 (20.2)	22.6 (14.9)
Number spp. per house (non-cultivated)	1.3 (3.2)	1.4 (1.9)	2.6 (3.3)
Number flowerpots/house (outdoors)	16.6 (28.1)	26.6 (41.9)	29.9 (43.7)
Number flowerpots/house (indoors)	6.0 (15.1)	3.4 (7.8)	0.6 (2.0)
Houses with outdoor flowerpots (%)	72	86	88
House with indoor flowerpots	53	42	15

differences in areas dedicated to gardening and in garden styles, the three social classes show few differences in relation to the total number of cultivated species. Second, the number of outdoor flowerpots increases from wealthy to low income families, whereas the opposite phenomenon occurs in the case of indoor flowerpots. Poor people have ten times fewer indoor flowerpots then rich people because they are more crowded in smaller homes with smaller living rooms. Also, they avoid indoor plants because they believe that they may consume substantial amounts of oxygen during the night.

Gardens and Balconies in Other Cities

Quite possibly, plant species composition and richness variations in urban gardens are higher than those in natural communities. This depends not only upon aesthetic preferences and time and money to be invested by each individual, but also upon very complex psycho-social factors. In Mexico City, Díaz-Betancourt et al. (1987) found that species richness not only depends upon water availability and garden area, but also upon the garden age (this is valid only for people with high and medium incomes).

For example, in the city of Buenos Aires, I surveyed a suburban (Quilmes) garden nine times between November 1973 and October 1985. The front garden (10 m^2) had, on average, 8.9 ±29 cultivated plants and 12.9 ±3.9 weeds. The back garden (90 m^2) showed 37.8 ±9.7 cultivated plants and 23.6 ±8 weeds. There was also a variable number of outdoor pots (between 18 and 55) which had an average of 23.3 ±55 cultivated plants and between 2 and 4 weeds. As some species were planted in the front and back gardens and in pots, the species total for that household showed oscillations between 44 and 61 cultivated species (average 55.1 ±14.3) and between 17 and 36 weeds (average 27.0 ±6.2).

In downtown Buenos Aires, with tall, high population density apartment buildings, there is a high proportion of buildings with balconies. Between June 1977 and September 1986, I surveyed a balconied apartment (Barrio Palermo, 8th floor) five times with the following results: 4-19 indoor pots with 4-16 cultivated plants and no weeds; 37 to 51 outdoor pots (on the balcony) containing 30-61 cultivated species and 9-13 weeds. Without considering repetitions, the total number of cultivated species fluctuated between 45 and 65 species. That is to say, in spite of a smaller available area, an apartment may have the same species richness as a well-tended suburban garden.

Occasional semi-randomized surveys[1], performed in different downtown areas in which I counted how many balconies had pots with plants and how many were empty, gave the following tally: London, England (June 1976), 141 balconies with and 353 without plants; Buenos Aires, Argentina (September 1976), 518 balconies with and 222 without plants; Caracas, Venezuela (June 1977), 260 balconies with and 151 without plants; Mexico City (November 1978), 327 balconies with and 560 without plants; Gdańsk, Kraków, Radom, Warsaw, and Grojce, Poland (July 1976), 205 with and 102 without plants. A t-test shows that there are no significant differences between Polish cities and Buenos Aires, but London had significantly fewer cultivated balconies ($p < .05$). Of course, not all city districts showed similar tendencies. For example, in the strictly office-building district in downtown Buenos Aires, the proportion of cultivated balconies was significantly lower in comparison to other districts: according to a survey I performed in 1977, 272 balconies without and 99 with plants. A contingency table and X^2-test indicate that the proportion of balconies with and without plants differ significantly for an $\alpha = 0.05$. As an indirect inference when analyzing the X^2 values, the cities of London and Mexico strongly deviate from other

---

[1] Selecting the first house or apartment sighted in each block, along different bus routes crossing the city.

cities by their lower proportion of cultivated balconies. In contrast, Buenos Aires deviates by a higher proportion of cultivated balconies. In residential areas, 70% of the balconies were vegetated, while in non-residential areas, only 27% were vegetated.

Table 16.5 shows results obtained in occasional surveys without a statistical test but which can give an idea of the cultivated and weed flora in some New and Old World cities. From this table, we can infer it is likely that the urban balcony flora can be as rich in cultivated and weed species as the garden flora, although the former is concentrated in small areas. From the point of view of space, balconies are better used than gardens. It is surprising that, in spite of the fact that it might be easier to control weeds in pots than in a garden, the number of species and the ratio of weeds to cultivated plants in pots do not differ greatly from those in gardens. This could be due to the fact that people living in apartments have a greater tolerance toward invading plants than those people who have gardens.

In general, gardens with similar areas tend to have a similar number of cultivated plants and weeds, and similar variances as those observed in natural communities. When the number of surveyed gardens increases, the cumulative number of cultivated species and weeds becomes higher, following an exponential curve typical of species/area curves. In a floristic survey I performed in 42 front gardens in Northumberland Place, London W2 (June 1976), the increase in species (Y) is linearized with the logarithm of the area ($\log X$) as follows:

$$Y = -28.2 + 64.1 \log X, \ r = 0.992 \qquad (16.1)$$

Discussion and Conclusion

In the urbanization process, a replacement of the local flora and fauna by a cosmopolitan (or at least Eurasian) biota can be observed (Pearson 1976; Nuorteva 1971; Pisarski and Czechowski 1978; Rapoport, 1979, 1991; Bornkamm 1982; Sukopp and Hejny 1990). In the first stages of urbanization, plant and animal species richness generally decreases, but with the passing of time, a new biota starts to organize, composed of some native as well as exotic anthropized species (Guthrie 1974). In certain taxa, species diversity as well as their abundances and biomasses can increase and even

Table 16.5 Number of cultivated plants and weeds in gardens and balconies. Standard deviation in parentheses (Rapoport unpubl.).

Site	Mean No. cultivated plant species	Mean No. weed spp.	Mean No. of pots per balcony	Mean area (m2) per garden[a]	Weeds/cultivated plant ratio	Sample size
GARDENS						
London (1976)	14.1 (7.9)	2.5 (2.7)	-	11.9 (25.7)	0.19 (0.24)	65
Gdansk (1976)	25.1 (13.3)	2.7 (1.8)	-	55.3 (86.0)	0.16 (0.19)	7
Warsaw (1976)	7.2 (7.6)	2.6 (2.1)	-	40.8 (16.1)	0.71 (0.97)	9
Szczecin (1976)	14.5 (8.5)	5.8 (2.1)	-	17.4 (6.2)	0.52 (0.32)	8
Poland (6 cities, 1976[b])	13.1 (13.3)	8.3 (10.8)	-	115.4 (259.9)	0.63 (0.68)	34
Mexico City (1980)	15.6 (7.1)	3.8 (3.8)	-	35.8 (54.4)	0.23 (0.21)	19
Buenos Aires (1976)	16.7 (17.8)	8.8 (10.4)	-	53.4 (133.3)	0.79 (0.79)	21
BALCONIES						
Mexico City (1980)	12.2 (9.3)	1.9 (1.9)	23.0 (25.2)		0.16 (0.18)	31
London (1976)	5.7 (5.6)	c	c		c	27
Buenos Aires (1977-86)	36.4 (7.4)	9.3 (2.9)	38.6 (10.9)		0.25 (0.07)	7

[a] Front and back gardens.
[b] Warsaw + Gdansk + Szczecin + Brok + Ploty + Kraków.
[c] Not registered. Cultivated species range between 1 and 20.

exceed the values normally existing in surrounding areas (Owen 1971; Vale and Vale 1976; Luniak 1980). In the case of plants, species richness may decrease with the increasing isolation of the village but increase with village size (Hanski 1982).

From the above observations, we may conclude that non-native plants in northwestern Patagonia generally do not colonize undisturbed areas. Urban growth constitutes a complex disturbance which increases permanently. Exotic plants are favored by this situation, as well as some native species which are transformed into anthrophilous (apophytes) species, i.e., they are adapted to cohabit with the human being.

The City of Bariloche in northwestern Patagonia has a 100-year history and was built in an area without a permanent agricultural tradition. On the other hand, Mexico City has endured millennia of interaction between humans and nature. Both cities show notable differences in their floras. Exotic weeds predominate in Bariloche, while native anthropophytes predominate in Mexico City. Another difference between both cities is that in Bariloche, the total number of non-cultivated species per hectare in roads and vacant lots (natives and exotics) does not differ significantly between urban (18 to 57 species) and outlying suburban (31 to 49 species) areas, except in the downtown area where species richness obviously decreases due to the absence of vacant lots. In Mexico City, a linear decrease can be observed from the suburbs (30 to 80 species·ha^{-1}) toward downtown Tlatelolco (3-10 species species·ha^{-1}). This marks significant differences with what was observed in Hertfordshire, England, where the number of non-cultivated species increases linearly with increasing urbanization in exotic as well as in native species, in a 2 x 2 km grid census (Rapoport 1977, based on Dony 1967). Two factors determine this feature. First, English gardeners use exotic as well as native ornamental and utilitarian plants, and second, urbanized areas act as artificial refuges for species normally eradicated by long-term farming and agricultural practices. These results emphasize the fact that the process of urbanization is not always deleterious, thus opening an important field of research for ecologists and planners searching for new ways of interaction between humans and nature.

Acknowledgments

I am indebted to R. Dalziel, M. Esteban, E. Ezcurra, D. Grigera, C.R. Marino, and A. Salibián for their collaboration in the first botanical survey of Bariloche City. My thanks to M. Gross and A. Ruggiero for their help in processing part of the data.

Recommended Reading

Bornkamm, R., ed. (1982). *Urban Ecology*. Blackwell, Oxford.
Frenkel, R.E. (1970). *Ruderal Vegetation Along Some California Roadsides*, 2nd ed. 1977. University of California Press, Berkeley.

Rapoport, E.H. (1991). Tropical versus temperate weeds: a glance into the present and future. In: P.S. Ramakrishnan, ed. *Ecology of Biological Invasions in the Tropics*, pp. 41-52. SCOPE Workshop, International Scientific Publications, New Delhi.

Sukopp, H. and S. Hejny, eds. (1990). *Urban Ecology. Plants and Plant Communities in Urban Environments*. SPB Academic Publishing bv, The Hague.

Trojan, P., ed. (1982). General problems of synanthropization. *Memorab. Zool.* (Warsaw) 37:1-147.

17
Ecological Implications of Landscape Fragmentation

Robert H. Gardner, Robert V. O'Neill, and
Monica G. Turner

Introduction

Understanding the effects of extensive and rapid changes in land-use patterns on the distribution and abundance of organisms is an important ecological issue (Burgess and Sharpe 1981; Forman and Godron 1986; Risser et al. 1984; Wiens et al. 1985; Naiman et al. 1988; Lubchenco et al. 1991). Although many physical and biotic factors, such as topography, disturbance, and resource gradients, are associated with habitat fragmentation, the direct effects of human activities (O'Neill et al. 1988a), on the size, shape, and the spatial arrangement of habitat types are particularly important (Krummel et al. 1987; Odum and Turner 1990). Sudden changes in broad-scale patterns have been predicted (Gardner et al. 1987) from gradual change in land-use, producing patterns which may be very sensitive to future disturbances (Franklin and Forman 1987).

The effects of patch size, isolation, and regional abundance on forest bird communities were assessed by van Dorp and Opdam (1987) in 235 small (0.1-39 ha) deciduous woodlots. Woodlot size was the best single predictor of species number and probability of occurrence. While regional levels of abundance affected the probability of occurrence of avifauna in all woodlots, interior species were affected most by the degree of woodlot isolation. Bolger et al. (1991) compared the species-area relationship in unfragmented chaparral habitat with that in urban chaparral fragments and found rapid population extinction of resident bird species as the habitat was fragmented. Because more abundant species persist longer in fragmented landscapes, species remaining in small fragments do not reflect the original diversity of biota in the unfragmented landscape (Bolger et al. 1991).

Recent theoretical and empirical studies indicate that a multi-scaled, hierarchical approach may be needed to relate landscape pattern and process. Systematic examination of a number of different landscapes (O'Neill et al. 1991a, 1991b) has shown that landscape patterns shift as the grain and extent of the data set change. On all of the landscapes examined, the small-scale patterns of biota (e.g., distribution of individual plants) are superimposed on the broader-scale patterns of abiotic variables (e.g., soils, topography, climate, etc.). Because there is a multitude of factors which affect biota, and because these factors are distributed across a hierarchy of scales, the nature of community persistence must be analyzed and interpreted at more than one scale (Rahel 1990).

As human effects accelerate, a general approach is needed to characterize spatially heterogeneous systems and understand what makes different species sensitive to landscape fragmentation (Lubchenco et al. 1991). This chapter examines the effect of scale-dependent changes in spatial patterns by: 1) generating a series of simple random and hierarchically structured maps; 2) comparing the results with patterns of forest habitat from digitized landscape data; and 3) simulating two consumer populations with different life history characteristics. The effect of spatial patterns on population abundance (number of sites and spatial extent of habitat utilized) were compared to determine how abundance of biota can be predicted from scale-dependent changes in landscape patterns.

Methods

Generating Simple Random Maps

Two-dimensional random maps were formed by creating arrays with M columns and M rows and randomly setting the M^2 elements of the array to 1 with a probability of P, or to 0 with a probability of $(1.0 - P)$. Array elements set to 1 represent suitable habitat, while elements set to 0 represent sites that are unsuitable. Figure 17.1a shows an example of a random map size of $M = 64$ and P of 0.64, with the dark pixels representing areas of suitable habitat. Large random maps will have an average of PM^2 suitable sites, S, and $1 - PM^2$ unsuitable sites. The patterns of simple random maps were characterized by generated arrays with M fixed at 216 (46,656 total sites) and P ranging from 0.3 to 0.8 (Table 17.1).

A cluster or habitat patch is defined as a group of suitable sites which have at least one common edge along the vertical or horizontal directions of an array but not along the diagonals (i.e., each suitable site has four possible neighbors). The number, size, and shape of clusters are known to change as a function of P, with rapid changes occurring near the critical probability, p_c, when the largest cluster extends, or percolates, from one

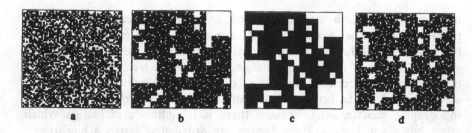

Figure 17.1. Four randomly generated maps with 64 rows and columns (4,096 individual sites). All four maps have a total fraction of suitable sites, P, of 0.64. The dark sites represent locations of suitable habitat: (a) a simple random map; (c-d) three-level hierarchically random maps with different level-dependent probabilities of site suitability. The probabilities are: (b) $p_1 = p_2 = p_3 = 0.861$; (c) $p_1 = p_2 = 0.8$ and $p_3 = 1.0$; (d) $p_1 = 1.0, p_2 = p_3 = 0.8$.

edge of the grid to the other. The value of p_c for extremely large maps has been determined to be 0.5928 (Stauffer 1985) when exactly 50% of the observed maps have clusters that extend from one edge to the other. Although percolation on an individual map is a discrete phenomena, the probability of percolation can be estimated by sampling many maps at the same value of P. Analysis of arrays generated by percolation theory has provided a means of applying these methods to ecological systems (see Gardner et al. 1992).

Generating Hierarchically Structured Maps

The recursive algorithm for generating hierarchical structured random maps is derived from methods of fractal geometry referred to as curdling (Mandelbrot 1983), since the process transforms an initially uniform distribution of material into many small clumps of high density (Feder 1988). Our adaptation of this algorithm requires the specification of the number, L, of successively finer scales within the map; the number of units, m_i, within each level, I (i = 1...L); and the fraction of units, p_i, that are randomly designated as units that contain suitable habitat. Units at the finest level, L, are sites whose values have been set to 1. The recursive procedure for generating a three-level hierarchical map ($L = 3$) is summarized in Fig. 17.2. A matrix of (m_1 x m_1) elements is created with the elements randomly set to 1 with a probability of p_1. Each element that is set to 1 is subdivided into an (m_2 x m_2) matrix and these (m_2)^2 elements are randomly set to 1 with a probability of p_2. The process is repeated a third time by subdividing those elements set to 1 at level 2 into an (m_3 x m_3) matrix and randomly setting to 1 the elements at this finest resolution with a probability of p_3. Because sampling is performed without replacement, the algorithm produces a random hierarchical map with a total

Table 17.1.   Characteristics of two populations on simple random maps.[a]

		Map[b]			SAW[c]		DAP[c]	
P	ΣC	LC	ΣE	ξ^2	V_s	ξ^2	V_s	ξ^2
0.3	6030.0	30.4	39300.0	5.7	3.0	1.3	4.2	0.8
0.4	5040.0	70.0	45000.0	17.1	2.4	1.0	2.0	0.5
0.5	3140.0	307.0	46900.0	117.0	8.6	4.8	20.0	3.5
0.54	2340.0	927.0	46700.0	498.0	25.8	17.6	51.6	5.1
0.56	1940.0	2480.0	46300.0	1990.0	17.0	13.3	239.0	16.1
0.58	1580.0	5460.0	45800.0	4250.0	51.2	29.7	625.0	28.1
0.6	1250.0	15700.0	45100.0	5530.0	65.0	68.5	204.0	8.6
0.62	994.0	24900.0	44300.0	3580.0	27.6	17.9	2500.0	29.6
0.64	780.0	27600.0	43400.0	2970.0	60.8	53.9	6270.0	70.3
0.66	614.0	29300.0	42300.0	2650.0	1060.0	1030.0	8740.0	131.0
0.68	470.0	30700.0	4100.0	2430.0	1660.0	1410.0	8290.0	168.0
0.7	365.0	31900.0	39700.0	2270.0	5070.0	3410.0	13000.0	556.0
0.72	282.0	33100.0	38000.0	2130.0	9470.0	4230.0	16700.0	688.0
0.74	208.0	34200.0	36300.0	2010.0	14200.0	3760.0	17700.0	17990.0
0.76	147.0	35300.0	34400.0	1910.0	14500.0	2910.0	18000.0	28070.0
0.78	100.0	36200.0	32600.0	1820.0	22400.0	3390.0	15500.0	3410.0
0.8	79.0	37200.0	30500.0	1740.0	27300.0	2870.0	21600.0	4880.0

[a] The results are means of 5 replicate maps with dimension, M, of 216 by 216 (M^2 = 46,656 sites per map).

[b] The means characteristics of the maps are P, the proportion of suitable sites; ΣC, the total number of clusters; LC, the number of sites in the largest cluster; ΣE, the total amount of edge (the total number of non-habitat sites adjacent to each habitat site); and ξ^2, the correlation length (see text for details).

[c] SAW and DAP are results of simulations of a Self-Avoiding Walker and the Dispersal of an Annual Plant, respectively (see text for details of simulation methods). V_s is the number of sites visited during the simulation by SAW and the number of sites populated by DAP. ξ^2 is the correlation length of visited or populated sites for SAW and DAP, respectively.

Generating Hierarchically Structured Random Maps

A three-level hierarchical map with 64 rows and 64 columns (4,096 individual sites, s) is iteratively generated by:

Step 1: Forming a $m_1 \times m_1$ matrix with $(m_1)^2$ sites, then randomly setting individual sites to 1 with probability p_1 and to 0 with probability $(1.0 - p_1)$;

Step 2: Each site set to 1 in Step 1 is subdivided into $m_2 \times m_2$ sites. The finer-grained sites are randomly set to 1 with probability p_2, and to 0 with probability $(1.0 - p_2)$; and

Step 3: The procedure is iterated a final time by further subdividing those sites set to 1 in Step 2 into $m_3 \times m_3$ sites. These finest-grained sites are then set to 1 with probability p_3, and to 0 with probability $(1.0 - p_3)$.

The overall dimension of the hierarchically structure map, m, will be equal to $m_1 \times m_2 \times m_3$, and the overall probability, p, will be equal to $p_1 \times p_2 \times p_3$. Three hierarchically structure maps are illustrated in Figure 17.1.

Figure 17.2. Recursive procedure for generating a three-level hierarchically random map.

number of sites, M^2, equal to $(m_1 \times m_2 \times m_3)^2$, a total fraction of suitable sites, P, equal to $(p_1 \times p_2 \times p_3)$, and the total number of suitable habitat sites, S, will be $(m_1 \times m_2 \times m_3)^2 (p_1 \times p_2 \times p_3)$.

Figure 17.1 compares a simple random map (Fig. 17.1a) with three hierarchically random maps (Fig. 17.1b-d). All maps in Figure 17.1 have M = 64 (4,096 individual sites) and P = 0.64. In addition, the values of L and m_i for the three hierarchical maps has been fixed at 3 and 4, respectively. Figure 17.1b shows a self-similar hierarchical map (i.e., the pattern is the same at each level) with p_i = 0.861 (0.861^3 = 0.64). The fact that identical values of the m_i and p_i produce a self-similar map is confirmed by observing that exactly two empty "units" are found at each hierarchical level. Figure 17.1c shows a hierarchical map with m_i and m_2 = 0.8, while m_3 = 1.0 (i.e., all sites are suitable at the finest scale if the units at higher levels are also suitable). Figure 17.1d shows a hierarchical map with the m_1 = 1.0 and m_2 and m_3 = 0.8.

If we fix L = 3 and require that the m_i are identical, it is possible to generate a number of hierarchical maps to compare with the random maps. We chose to generate 7 maps with P = 0.64, M = 216, and L = 3. The 7 maps are produced by limiting the permutations of the p_i's to combinations of (1.0, 1.0, 0.64), (1.0, 0.8, 0.8), or (0.86, 0.86, 0.86).

The LUDA Maps

The U.S. Geological Survey (USGS) digital land-use and land-cover data base (LUDA) provides a spatial description of landscapes across the United

States (Fegeas et al. 1983). The data originate from National Aeronautics and Space Administration (U2/RB-57 high-altitude aerial photo coverage taken in the 1970s, and from these photos, land-cover types were delineated into 1 of 37 Level II (Fegeas et al. 1983) land-cover categories. The minimum polygon (patch) size of forested land in the LUDA data base is 16 ha, based on the provision that the minimum width of a polygon must be greater than 400 m. Because the original USGS data divides the 1:25,000 quadrangles into 24 separate sections, a special computer program was written to reformat the arc and node topological elements of the polygons and remove section boundaries that arbitrarily dissect polygons along those boundaries.

Seven quadrangles were selected from the LUDA data to form a transect from southern Georgia to northeastern Tennessee. Each quadrangle was sub-sampled by selecting 5 maps, each with M = 216, from the northeast, southeast, northwest, southwest, and central portion of each quadrangle. The forested regions of each map were analyzed and compared with the random and hierarchical maps. The percent forested habitat of the 35 LUDA maps ranged from a P of 0.27 to 0.94, and the total number of habitat patches ranged from 9 to 279.

Method of Map Analysis

A computer program was written to analyze the spatial patterns of simple random, hierarchically random, and actual LUDA landscapes. This program calculates for each map: S, the number of suitable habitat sites; P, the total fraction of map sites that are suitable ($P = S/M^2$); $\sum C$, the total number of clusters or patches of suitable habitat; LC, the size of the largest cluster; and $\sum E$, the total amount of edge (the total number of non-habitat sites adjacent to each habitat site). In addition to these metrics, the mean squared radius of patches was estimated as a measure of habitat dispersion. The mean squared radius, R_s^2, also called the "radius of gyration" (Stauffer 1985), is calculated by enumerating the number, s, and position, r, of individual sites within a given cluster:

$$2R_s^2 = \sum_{i,j} \frac{|r_i - r_j|^2}{s^2}$$

(17.1)

The information provided for R_s^2 for each cluster can be used to estimate the correlation or connectivity length, ξ, of the entire map. Because ξ is the average distance of two sites belonging to the same cluster, the average value of $2R_s^2$ for all clusters is weighted by the size and frequency distribution of the clusters to estimate the squared correlation length for the map:

$$\xi^2 = \frac{2\sum_s R_s^2 s^2 N_3}{\sum s^2 N_s} \qquad (17.2)$$

where N_s is the number clusters of size s per site.

Simulating Two Types of Consumer Populations

The generation of simple random and hierarchically random maps provides a spatial description of habitat resources for simulating the patterns of growth and dispersion of organisms. We performed simulations by using a set of rules which describe the patterns of habitat utilization and spread of different consumer populations, and analyzed the final patterns of abundance at the conclusion of the simulations. The population-specific rules for habitat utilization and spread consisted of: 1) the probability, i, of spread from a populated site to adjacent suitable sites; 2) the probability, e, per unit time and per site for local extirpation of the population (the expected residence time of a population at a site will be $1/e$); and 3) the probability, h, that the local site will be unsuitable for recolonization after a local population has been extirpated. If $h = 0.0$, the sites will be unaffected by colonization, if $h = 1.0$, then resources are permanently depleted and the site cannot be recolonized. Values of h between 0.0 and 1.0 indicate that colonization diminishes (but does not deplete) local resources, lowering the probability of reestablishment.

Although these rules are probabilistic, certain combinations of parameters will produce deterministic results. For instance, when i is small and e and h are set to 0.0, the population will slowly spread into all sites within a cluster. When i, e, and h are equal to 1.0, then the population spreads very rapidly throughout the cluster, destroying sites as it moves, much like a forest fire. Therefore, we chose two sets of parameters that produced different spatial dynamics. The first consumer populations (SAW) had values of i, e, and h of 0.8, 1.0, and 1.0, respectively. The high value of i causes rapid spread to adjacent suitable sites, but site colonization lasts for only one time step ($e = 1.0$) and sites cannot be recolonized ($h = 1.0$). The landscape dynamics of SAW are much like a self-avoiding walk and similar to a herd depleting resources as it moves through a landscape. The second population (DAP) had values of i, e, and h of 0.8, 0.25, and 0.25, respectively. DAP is similar to a weedy, annual plant that moves quickly to adjacent sites ($i = 0.8$), has a moderately low probability of local extinction

(e = 0.25), and may frequently recolonize previously occupied sites (h = 0.25).

Simulations were begun by defining the parameters of i, e, and h, initializing a single map site near the center of the map at random, recording the spread of the population through time, and analyzing the patterns of abundance at the end of 200 time steps. Because populations were initialized at a single site, and because the rule for growth is restricted to adjacent sites, the upper limit of population size is set by the size of the cluster that is randomly selected by the initialization process. Five iterations were performed for each map and population type.

The Map Patterns

The total number of clusters, $\sum C$, the size of the largest cluster, LC, the total amount of edge, $\sum E$, and the correlation length, ξ, were calculated as measures of pattern for 5 replicate simple random maps with values of P ranging from 0.3 to 0.8 and M (the number of rows and columns in the map) equal to 216 (Table 17.1). The effect of the critical threshold, p_c = 0.5928, is evident in the sudden changes in $\sum C$ and LC near that value of 4 PC; LC increased nearly three-fold from 5,460 at P = 0.58 to 15,700 at P = 0.6, and the number of clusters declined rapidly from 1,580 at P = 0.58 to 994 at P = 0.62. The effect was even more dramatic if expressed as a percent of available habitat. At P = 0.56, only 10% of available habitat was found in the largest cluster (% = $LC/(M^2P)$), while at P = 0.6, 56% of available habitat is present in one large cluster. The total amount of edge, $\sum E$, reached a maximum at 0.5, and declined gradually as P increased.

Changes in ξ^2 with increasing P were more complex (Table 17.1; Fig. 17.3a). For values of P below 0.5, clusters were small, consequently the correlation length was also small. The aggregation of clusters that occurred above the critical threshold caused ξ^2 to increase rapidly, reaching a maximum at P = 0.6. Values of ξ^2 then declined as the cluster became dense and was confined by the boundary of the map, resulting in a decrease in the average distance between sites.

Differences in P among maps caused a corresponding change in the number and size distribution of clusters on the map (Stauffer 1985). We controlled for this effect by generating a set of hierarchically random maps with P fixed at 0.64 (Table 17.2). Although the total fraction of available habitat was the same among maps, the differences in pattern were large. The total number of clusters, $\sum C$, showed a nearly 300-fold change from a low of 2.6 for the hierarchical map with p_is of (0.64, 1.0, 1.0), to a high of 779 for the simple random map. Differences in the amount of edge, $\sum E$,

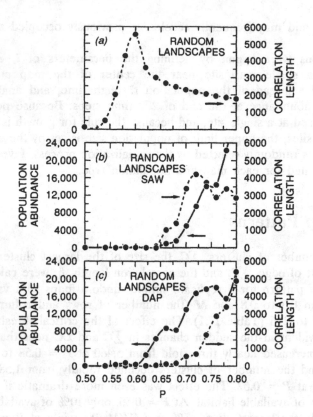

Figure 17.3. The correlation length (dashed line) and landscape abundance (solid line) for: (a) habitat patterns of simple random maps and the simulation of two organisms types; (b) SAW; and (c) DAP. See text for details of simulation procedures.

were also large. The simple random map and the hierarchical map with p_is of (1.0, 1.0, and 0.64) had the greatest amount of edge, while the hierarchical map with p_is of (1.0, 1.0, and 0.64) showed nearly 30-fold fewer edges. The size of the largest clusters, LC, did not differ greatly among maps (ranging from 23100 to 30200), although smaller LCs were observed when $p_1 < 1.0$ and larger LCs occurred when $p_3 < 1.0$. Because the correlation length was affected most by the size of the largest cluster, the values of ξ^2 did not differ greatly among maps (Table 17.2). When p_1 and p_2 equal 1.0 and $p_3 = 0.64$, the pattern of the hierarchical map was nearly identical to the simple random map.

The landscape pattern of forest patches extracted from the LUDA quadrangles (Table 17.3 and Fig. 17.4) showed greater variation than the precisely defined simple random and hierarchically random maps. Total number of clusters peaked near 0.5. The LC increased with increasing P(Fig. 17.4b), but declined gradually at p_c. The amount of total edge was

Table 17.2. Characteristics of random and hierarchical maps when P = 0.64.[a]

Random Map

		Map[b]			SAW[c]		DAP[c]	
P	ΣC	LC	ΣE	ξ^2	V_s	ξ^2	V_s	ξ^2
0.64	779.0	27500.0	43400.0	2970.0	60.8	53.9	6270.0	70.3

Hierarchical Maps

						Map[b]		SAW[c]		DAP[c]	
P_1	P_2	P_3	ΣC	LC	ΣE	ξ^2	V_s	ξ^2	V_s	ξ^2
0.64	1.0	1.0	2.6	23100.0	1530.0	3640.0	16500.0	4100.0	10100.0	3480.0
1.0	0.64	1.0	27.4	24500.0	7670.0	3450.0	7360.0	3500.0	11900.0	6070.0
1.0	1.0	0.64	662.0	28000.0	44500.0	2890.0	149.0	142.0	6330.0	60.3
1.0	0.8	0.8	60.8	30200.0	27800.0	2520.0	14200.0	4450.0	10400.0	1580.0
0.8	0.8	1.0	3.4	29100.0	4910.0	2660.0	22800.0	3710.0	17200.0	5040.0
0.8	1.0	0.8	43.0	29200.0	25100.0	2710.0	22000.0	3730.0	14100.0	5930.0
0.86	0.86	0.86	14.0	29700.0	20100.0	2570.0	15600.0	4100.0	15400.0	5560.0

[a] The results are means of 5 replicate maps with dimension, M, of 216 x 216 (M^2 = 46,656 sites per map). The hierarchical maps are generated with $m_1 = m_2 = m_3 = 6$ (6^3 = 216). P, the fraction of suitable sites, is equal to 0.64 for all maps. For hierarchical maps, the fraction of suitable sites is calculated as $P = p_1 \times p_2 \times p_3$.

[b] The means characteristic of the maps are: P, the proportion of suitable sites; ΣC, the total number of clusters; LC, the number of sites in the largest cluster; Σ, the total amount of edge (the total number of non-habitat sites adjacent to each habitat site); and ξ^2, the correlation length (see text for details).

[c] SAW and DAP are results of simulations of a Self-Avoiding Walker and the Dispersal of an Annual Plant, respectively (see text for details of simulation methods). V_s is the number

Table 17.3.   Samples from LUDA maps.[a]

Quadrangle	Sample	P	ΣC	LC	ΣC	ξ^2
Valdosta	a	0.57	221.0	16975.0	17372.0	3934.0
	b	0.42	279.0	7363.0	14238.0	4325.0
	c	0.66	53.0	27361.0	11444.0	2931.0
	d	0.59	237.0	15454.0	15876.0	5144.0
	e	0.52	258.0	12175.0	16294.0	8940.0
	mean	0.412	135.0	5871.0	11163.0	3661.0
Waycross	a	0.27	190.0	1493.0	11185.0	1690.0
	b	0.37	173.0	3028.0	12617.0	2478.0
	c	0.58	61.0	12543.0	10184.0	5510.0
	d	0.47	93.0	7664.0	11516.0	4416.0
	e	0.37	159.0	4625.0	11986.0	4213.0
	mean	0.412	135.2	5870.0	11629.0	3661.0
Macon	a	0.49	121.0	8909.0	11317.0	4904.0
	b	0.66	41.0	29357.0	8846.0	2680.0
	c	0.72	25.0	22611.0	8908.0	3399.0
	d	0.49	115.0	11592.0	14432.0	4285.0
	e	0.68	57.0	27480.0	9422.0	2914.0
	mean	0.608	71.8	19990.0	10584.0	3636.0
Athens	a	0.59	73.0	25444.0	12648.0	3450.0
	b	0.81	19.0	37000.0	7240.0	1650.0
	c	0.76	24.0	33935.0	8870.0	1960.0
	d	0.90	9.0	41630.0	5590.0	1400.0
	e	0.76	36.0	34700.0	8770.0	1930.0
	mean	0.764	32.2	34541.0	8623.0	2078.0
Greenville	a	0.69	94.0	29067.0	13528.0	2653.0
	b	0.88	17.0	40945.0	5888.0	1379.0
	c	0.75	40.0	33776.0	8902.0	1983.0

Table 17.3 (con't)

Quadrangle	Sample	P	ΣC	LC	ΣC	ξ^2
	d	0.38	261.0	3942.0	14706.0	2053.0
	e	0.76	64.0	33947.0	10624.0	1978.0
	mean	0.692	95.2	28335.0	10730.0	2009.0
Knoxville	a	0.92	18.0	42525.0	4678.0	1322.0
	b	0.65	159.0	24736.0	9418.0	3357.0
	c	0.81	49.0	36123.0	6622.0	1733.0
	d	0.78	48.0	34795.0	7956.0	1936.0
	e	0.94	14.0	42716.0	3768.0	1252.0
	mean	0.82	57.6	36179.0	6488.0	1920.0
Johnson City	a	0.53	134.0	10717.0	13764.0	5972.0
	b	0.88	26.0	40308.0	6202.0	1427.0
	c	0.76	30.0	34502.0	7634.0	1921.0
	d	0.54	120.0	15200.0	10817.0	4787.0
	e	0.66	61.0	14286.0	10876.0	5627.0
	mean	0.674	74.2	23001.0	9858.0	3946.0

[a]Five maps with dimension, M, of 216 by 216 (M^2 = 46,656 sites per map) were extracted from each LUDA quadrangle. The characteristics of the maps are: P, the proportion of suitable sites; ΣC, the total number of clusters; LC, the number of sites in the largest cluster; ΣE, the total amount of edge (the total number of non-habitat sites adjacent to each habitat site); and ξ^2, the correlation length (see text for details).

LUDA

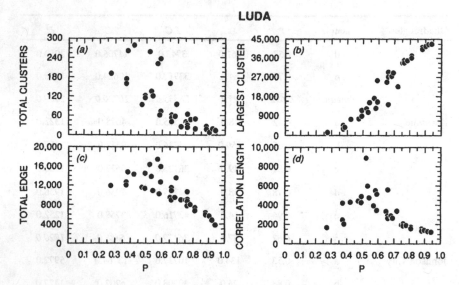

Figure 17.4. Patterns of forested areas for 35 LUDA maps plotted as a function of the percent of the map, *P*, that is forested: (a) total number of forest clusters; (b) the size of the largest cluster; (c) total amount of edge; and (d) the correlation length of forest patches.

of sites visited during the simulation by SAW, and the number of sites populated by DAP. ζ^2 is the correlation length of visited or populated sites for SAW and DAP, respectively.highest near 0.6, and declined with higher values of *P*. The pattern of change of the correlation length (Fig. 17.4d) was similar to that for random maps (Fig. 17.3a), although there was a great deal of variation among the LUDA maps. A comparison of the LUDA maps with *P* near 0.64 with the appropriate simple random and hierarchically random maps (Table 17.2) showed that the cumulative frequency distribution of cluster sizes of LUDA maps were significantly different from random, but quite similar to a hierarchical map with p_i, p_2, p_3 of 1.0, 0.64, 1.0, respectively.

Interaction Between Pattern and Process

The effects of changes in available habitat (e.g., changes in *P* for simple random maps) and the scale of habitat availability (e.g., hierarchical maps) are summarized in Tables 17.1 and 17.2, respectively. Previous results have shown probability of a site being altered by a population, *h*, has little effect on landscape abundance (the number of sites successfully colonized during a simulation), V_s, when *i* = 1.0, but dramatically decreases population abundance when *i* <1.0. The reason is that alterations of local sites (*h*

>0.0) effectively "disconnects" large patches of habitat, preventing the population from spreading across the landscape (Gardner et al. 1992).

For simple random maps, the landscape abundance, V_2, of SAW showed a 17-fold increase between P = 0.64 and 0.66 (Fig. 17.3b). The fact that this increase occurred well above the critical threshold (p_c = 0.5928) is an indication of the interaction effect of parameters P and i. Although V_s increased continually with increasing P, the correlation length showed a peak at P = 0.72, indicating that the sites occupied by SAW became more densely packed above P = 0.72, but fractured into many sub-populations below P = 0.72. This pattern was confirmed by the ratio of V_s to LC, which was 0.5 at P = 0.68, 0.29 at P = 0.72, and 0.42 at P = 0.76.

The DAP simulations showed a different pattern of landscape abundance on simple random maps. DAP was more abundant at smaller P, and rapidly increased in abundance at P = 0.62 (Fig. 17.3c). V_s and ξ^2 increased continuously with P, with a maximum increase for ξ^2 occurring between a P of 0.7 and 0.74.

The irregularity of the lines in Figure 17.3b and 17.3c for V_s and ξ^2 is due to the random initialization of populations at the beginning of each simulation. Although each point on the lines in Fig. 17.3b and 17.3c represents the mean of 5 iterations, if a population were randomly placed on a small cluster at the beginning of a simulation, the resulting growth would be dramatically different than if the population was initially placed on the large, percolating cluster. Hence, large variance between iterations will occur if random initialization results in 1 or more populations being placed on a small cluster. Because cluster size increases rapidly above the critical threshold, the probability of randomly initializing a small cluster diminishes as P increases.

Peak values of the correlation length occur when the population growth has just managed to reach the farthest extent possible. Randomly generated maps will produce habitat clusters that are larger than the map when P >0.5928 (i.e., infinite clusters; see Stauffer 1985; Feder 1988), and the border of the map will truncate the pattern. This effect is clearly illustrated in Fig. 17.3a, where the correlation length increases as P increases until the critical threshold is reached. Above this threshold, the extent of population growth is limited (i.e., truncated by finite map boundaries) and the cluster begins to fill in rather than expand. The growth of DAP showed a similar pattern (Fig. 17.3b), with a peak value for the correlation length occurring at P = 0.72. The continued increase in landscape abundance of DAP (solid line, Fig. 17.3b) indicates that above P = 0.72, growth is truncated by the map border and the population growth results in densely packed clusters. The effect of cluster truncation is a lower average mean squared distance between sites. Conversely, the landscape abundance of SAW (solid line, Fig. 17.3c) continues to increase as P increases, with no peak in the correlation length evident (dashed line, Fig. 17.3c). The different responses of these two populations indicates that the spatial extent of the map is adequate for characterizing DAP until $P = 0.72$ and for SAW until at least $P = 0.8$.

Table 17.2 shows the effect of hierarchically structured maps on the patterns of abundance and correlation length for SAW. The abundance, V_s, of SAW ranged from 60.8 to 22,800 sites, while DAP ranged from 6,270 to 17,200. Both SAW and DAP showed smallest abundances on the simple random map. It is important to note that this result is not related to the size of LC because there is little difference in LC among the maps (Table 17.2). What does differ among maps is the scale at which the habitat pattern changes. When sites were fragmented at the finest scale ($p_3 = 0.64$), the lowest values of V_s occurred for SAW and DAP. Population abundances increased as fine-scale fragmentation diminished (e.g., p_1, p_2, p_3 = 1.0, 0.8, 0.8), with V_s highest when (p_1, p_2, p_3) was equal to (0.8, 0.8, 1.0) for both SAW and DAP.

Discussion

The unique response of individual species to fragmentation must be determined if patterns of abundance and extinction in response to human use of landscapes are to be predicted. Even simple differences among consumers show that the relationship between pattern (the amount and spatial arrangement of suitable habitat) and process (the growth and persistence of populations at landscape scales) can be complex. Previous results (Gardner et al. 1987; Gardner et al. 1992) demonstrated that the gradual loss of habitat (i.e., decreasing values of P, Fig. 17.3) produces critical thresholds in habitat connectivity that dramatically affect population abundance. Above the critical threshold ($p_c = 0.5928$) growth is a monotonic function of population-specific attributes. For instance, the simulated expansion of SAW above p_c is less than that of DAP because SAW has a high probability of resource depletion (h - 0.25 for DAP and 1.0 for SAW), preventing reestablishment of populations in previously inhabited sites. The effect of resource depletion becomes less important as habitat connectivity increases ($P = 0.8$, Table 17.1). Below the critical threshold, differences between species are small (Table 17.1) because the characteristically small size of habitat clusters limits population expansion.

The hierarchical maps allow the response of the SAW and DAP populations to scale-dependent changes to be systematically evaluated. The algorithm for generating hierarchically random maps (Fig. 17.2) can produce a variety of patterns while controlling the overall fraction of available habitat, P (Fig. 17.1). Hierarchically structured maps with $P = 0.64$ do not differ greatly in the size of the largest cluster, but show a 5-fold difference in the amount of edge and a 300-fold difference in the number of habitat clusters (Table 17.2).

The range of simulation results of SAW and DAP (Table 17.2) shows that population abundance is not predicted by the size of the largest cluster, as results from simple random maps might indicate, but rather by the scale at which habitat fragmentation occurs. Fragmentation at the finest scale (i.e., map with p_1, p_2, p_3 =1.0, 1.0, and 0.8, Table 17.2) produces the lowest levels of abundance for the SAW and DAP populations. Although differences between the two populations do exist, it is not surprising that the patterns are similar, because the neighborhood rule and rate of expansion (i = 0.8) are the same for both populations.

The correlation length, ξ^2, is a measure of the average, taken across the entire map, of the distances over which habitats or populations are connected. Highest values of ξ^2 will occur at critical thresholds (Fig. 17.3) when a single cluster extends across the map. Because clusters at the critical threshold are large but very irregular (Stauffer 1985; Milne 1990), the average distance between sizes will be maximal. Values of ξ^2 decline above p_c as cluster size is truncated by the map boundary and sites become densely packed (Fig. 17.3a). Very low values of ξ^2 occur below the critical threshold because the map is composed of numerous, small clusters.

Peak values in the correlation length of population abundance (i.e., the correlation length calculated for those habitat sites which have been successfully colonized by the simulated population) provide a useful metric for determining when critical thresholds between pattern and process occur. Maximum values of ξ^2 for SAW occur on random maps at P = 0.72 (Table 17.1), while values of ξ^2 for DAP continue to increase with increasing P. The upper bound for the correlation length of habitat (Table 17.1) is 5530 at P = 0.6. The correlation length of population abundance for SAW and DAP approach the maximum possible value at P = 0.72 and P = 0.8. Values of ξ^2 for populations that are greater than the corresponding habitat values (Tables 17.2 and 17.3) indicate that the populations are more dispersed than habitat.

The discontinuities produced by thresholds in habitat connectivity have important implications for landscape studies. However, several important points must be kept in mind when interpreting these results: 1) thresholds in connectivity are dependent on the neighborhood rule (i.e., spread is limited to adjacent sites) used to define connectivity among sites. Because these rules may vary among species, and because different species are likely to respond uniquely to changes in landscape pattern, no single threshold will characterize the interaction between a variety of species and the landscape pattern. Populations capable of spreading to non-adjacent sites (i.e., long-range dispersal of seed) will show thresholds at values of P below that defined from nearest-neighbor rules of habitat connectivity; 2) the parameters used to simulate SAW and DAP were quite similar, yet slight differences caused substantial shifts in the observed thresholds. Therefore, differences among biota are likely to show broad ranges of thresholds, implying that as landscapes are progressively fragmented, successive shifts in species composition can be expected; and 3) because all landscapes are finite, observed changes are not likely to be as sudden as those produced from theoretical studies of infinitely large systems. However, the identification of critical thresholds provides a quantification of the zone within which small changes in habitat connectivity will result in disproportionate changes in population abundances.

Fragmentation of natural landscapes is probably never random. The forested habitats analyzed from actual landscapes (i.e., the LUDA scenes) were quite different from random maps, but remarkably similar to hierarchically structured maps. However, there was still considerable variation among LUDA maps even when differences in P among maps were controlled. This range of patterns is not surprising because multiple factors, such as topography, climate, urban development, and other human land-uses, all contribute to producing these patterns. It is also noteworthy that although critical thresholds in habitat connectivity exist in real landscapes, the values of P at which connectivity occurs does not always occur at 0.5928, as it does for infinitely large random maps. Scale-dependent patterns within a landscape will influence the observed value of this threshold. Thus, our analyses of landscape pattern suggest that real landscapes exhibit scaled patterns that reflect a variety of causal factors.

The correlation length calculated for the LUDA maps provides a useful measure of habitat fragmentation (Table 17.3). For example, an examination of the four LUDA maps (Valdosta c, Macon b, Greenville a, and Johnson City e) which have a similar value of P near 0.66, shows that the pattern for Johnson City is quite different from the others. The largest cluster on the Johnson City map is considerably smaller while the correlation length is greater, indicating that, on average, the forested habits are more fragmented than the other landscapes at a similar value of P. Based on our simulations of SAW and DAP, we expect these differences in habitat patterns to produce large differences in the abundance of the biota.

The complex patterns of species response to the amount (Table 17.1) and scale (Table 17.2) of landscape fragmentation illustrate the importance of understanding relationships between pattern and process at landscape scales. It is clear that the data necessary to test the results of simple spatial models must be gathered at appropriate scales. An understanding of both the grain and extent of species interaction with the heterogeneity of landscapes is necessary to identify key processes and develop general principles for landscape management, the preservation of biodiversity, and habitat conservation.

Summary

The effects of scale-dependent changes in landscapes are an important aspect of human interactions with ecosystems. Therefore, we investigated changes in habitat patterns by simulating two consumer populations on random and hierarchically random landscapes. The first consumer depleted resources as it randomly walked across the landscape, while the second consumer spread outward to suitable sites from an initial point. Simple model landscapes were generated by randomly selecting the habitat suitability of each site at a given probability, while hierarchically random maps were generated by a nested series of rules which changed the probabilities of site suitability with spatial scale. These two map types were compared to digitized land-use data.

Three effects were observed: 1) the reduction of available habitat always reduced population abundance and spread; 2) small changes in landscape pattern produced sudden changes in abundance which could be predicted from the interaction between landscape characteristics and population-specific life history parameters; and 3) the hierarchically structured landscapes affected abundances most when habitat fragmentation coincided with the scales at which the consumer populations utilized spatial resources.

The relationships among life history characteristics, landscape pattern, and the scale of disturbance must be jointly considered when predicting the broad-scale change in population abundance which may result from human-induced fragmentation of the landscape.

Acknowledgments

Special thanks are due to Sidey Timmons for making the analysis of LUDA data possible. Frequent comments and helpful suggestions for improvements in the manuscript by Roy Plotnick, Bill Hargrove, and Richard Flamm are appreciated.

This research was funded by the Ecological Research Division, Office of Health and Environmental Research, U.S. Department of Energy, under Contract No. DE-AC05-84OR21400 with Martin Marietta Energy Systems, Inc.

Publication No. 3956, Environmental Sciences Division, Oak Ridge National Laboratory.

Recommended Reading

Burgess, R.L. and D.M. Sharpe, eds. (1981). *Forest Island Dynamics in Man-Dominated Landscapes*. Springer-Verlag, New York.

Lubchenco, J., A.M. Olson, L.B. Brubaker, S.R. Carpenter, M.M. Holland, S.P. Hubbell, S.A. Levin, J.A. MacMahon, P.A. Matson, J.M. Melillo, H.A. Mooney, C.H. Peterson, H.R. Pulliam, L.A. Real, P.J. Regal, and P.G. Risser. (1991). The sustainable biosphere initiative: an ecological research agenda. *Ecology* 72:371-412.

Pickett, S.T.A. and P.S. White, eds. (1985). *The Ecology of Natural Disturbance and Patch Dynamics*. Academic Press, Inc., New York.

Roughgarden, J., R.M. May, and S.A. Levin, eds. (1989). *Perspectives in Ecological Theory*. Princeton University Press, Princeton, New Jersey.

Turner, M.G. and R.H. Gardner, eds. (1990). *Quantitative Methods in Landscape Ecology: The Analysis and Interpretation of Landscape Heterogeneity*. Springer-Verlag, New York.

Section III   Implications for Ecosystem Management and Construction

The chapters in Section III continue to expand our appreciation of the commonness and long ecological history of human populations and subtle effects on ecosystems. However, we have grouped these chapters together to emphasize the practical advantages of recognizing humans as components of ecosystems. Goal 1 is to highlight the policy implications and insights from integrating humans and ecosystems. Several of the chapters in Sections I and II have had a similar message. A point is that management or conservation strategies that neglect the various, but often inconspicuous, roles of humans in ecosystems are flawed. Thus, the chapters collected in Section III have a second goal, to stimulate the establishment of new ways to set standards for management. Given that the traditional concept of what is "natural" developed under the assumption that humans are not a part of ecological systems, the old standards must be rethought. Finally, the section will illustrate new strategies for understanding ecosystems and providing for their management by active restoration, or by the construction of new ecosystems.

Section III begins with studies of existing landscapes. Grossmann (Chapter 18) presents a multifaceted modeling approach that considers the reciprocal relationship of humans and landscapes. The resultant model is effective for ecosystem management and policy, and explicitly exposes subtle human effects. The important issue of biodiversity is defined functionally by Jutro (Chapter 19), who emphasizes its links with inconspicuous or distant human actions. Policy makers need to act on current predictions, and to compensate for novel rates of processes and constraints affecting biodiversity. Chapter 20 (Wagner and Kay) provides a practical example, the question of the health or sustainability of the Yellowstone ecosystem. The assumption that pre-Columbian and non-industrial populations had no effect on the landscape is the basis for policies now found to be wanting. The actual histories and human roles, both conspicuous and subtle, must be considered in designing management strategies.

The final two chapters take radical approaches to the problem of humans as components of ecosystems. More familiar to ecologists is the need to restore damaged ecosystems. Jordan (Chapter 21) takes an experimental approach to restoration, one that requires direct, subtle, and historical effects of humans as its sine qua non. The need to construct entirely new, sustainable ecosystems for closed systems on Earth and space travel provides an exciting opportunity for a closely monitored ecological experiment in the form of Biosphere 2 (Nelson et al., Chapter 22). Subtle and unexpected effects must be measured, and in some cases, immediately compensated for, in order to protect the health of the human occupants and the novel system itself.

18
Integration of Social and Ecological Factors: Dynamic Area Models of Subtle Human Influences on Ecosystems

Wolf-Dieter Grossmann

Subtle Human Influences on Ecosystems: A New Area of Investigation

Extensive, obvious, and direct influences of humans on ecosystems are well investigated. Methods are available to deal with these, although much more work is needed. Subtle influences are now becoming an important topic in ecosystems studies. "Subtle," according to McDonnell and Pickett (Chapter 1, this volume), denotes human influences on ecosystems or interactions of humans with ecosystems that are not obvious, conspicuous, or direct, but are lagged or influenced by past developments or consequences at a distance in time and space, or any combination of these. The word "subtle" does not imply that the influences are negligible (see Russell, Chapter 8, this volume). Indeed they could be extensive.

"Subtle" is a relative notion. The impact of DDT was seen in the early 1950s as a subtle effect or was disputed, if recognized at all. One reason was that instruments for measuring very small concentrations of substances were not available by that time. Now, effects by chlorohydrocarbons (CHCs) on the food web are regarded as large; very low concentrations of CHCs can easily be detected and measured. Nor was the enrichment of some CHCs in the foodweb known at that time. The enrichment of CHCs in foodwebs is indirect, an influence at a distance, and is subtle; but it may have large and extensive effects.

Subtle influences pose novel problems to research and management compared to direct, obvious effects (Russell, Chapter 8, this volume). They are more difficult to perceive, investigate, explain, and prevent than direct effects. They will most likely also be far more numerous than direct influences and might be far more important. New problems such as these need effective new tools.

Problems in Dealing With Subtle Influences

Subtle human influences on ecosystems, including those inhabited by humans, are difficult to perceive due to a variety of reasons:

1)   Subtle effects, by their very nature, are easily hidden in a broad spectrum of continuous and erratic developments and changes (Strayer et al. 1986). Also, systems can often cope for a long time with a subtle influence before an effect becomes obvious;

2)   Subtle influences may further be hidden because human perception tends to adapt to slow changes. Forest die-back, for example, seems to have begun in the 1970s but was only perceived in the beginning of the 1980s (Grossmann 1991d);

3)   Terms of reference usually are not available against which unusual or unexpected or erratic developments could become obvious;

4)   In subtle human influences on ecosystems, two components with different characteristics interact: the human (or social) system and the ecosystem. If the human system influences the ecosystem and if this, in turn, influences the human system, then the two might even form one system. Therefore research on subtle human influences may need systemic studies concentrating on *complex interdependencies and their dynamics*; and

5)   Subtle human influences usually pose multifaceted problems due to the interaction of social systems and ecosystems, two very different systems with different facets. Methods in ecology usually are not multifaceted.

A Multifaceted Method to Deal With Subtle Human Influences

Ecosystems have three qualitatively different "aspects" or "facets." In the multifaceted method, which was applied in the research reported in this chapter, each of the different ecosystem facets is analyzed with appropriate methods, and the results are synthesized. The method has been described in more detail elsewhere (Grossmann 1983; Grossmann 1991a; Grossmann and Watt 1992). The term "multifaceted" was used by Zeigler (1979); see also Allen and Starr (1982) for a description of hierarchical analysis.

Highest or Strategic Layer

On the highest layer, reasons for actual or potential *structural* changes are analyzed. Systems tend to change structurally if they are open, as are ecosystems and human systems. The most important structural change is the emergence of a system or its disapppearance. Other important structural changes are evolution or autopoiesis, or the addition or acquisition, and loss or removal of parts or of connections.

This analysis is done with strategic criteria, so called "strateria," such as diversities within the system (e.g., of species, within species, of ecosystems, or of pollutants), its structural type (closely connected or composed of comparatively independent sub-units), its inherent variability (e.g., over time, adaptation to seasons, migration patterns), or availability of resources for and reserves in the system. Usually there exist relationships between strateria. Examples are: 1) the well-discussed relationship between diversity and "stability" which is a relationship between variety, the necessary effort of maintenance ("costs" for variety), and potential changes; and 2) general rules, such as those that exert caution with all changes that are both extended and difficult to reverse. If humans had always followed this rule, most probably no civilization would have come into existence. However, if this rule had been applied reasonably, the present carbon dioxide problem might be considerably lower, but the release of N_2O might have happened as it did (because this was not perceived initially) while the release of CFCs might not have happened at all. Such criteria are difficult to quantify. Often it seems far more appropriate to express them qualitatively than quantitatively.

The strategic insights derived from the strateria are used to write scenarios on the past and future development of the system structure and about its environment. An example is the analysis of forest damage that also takes into account political actions to decrease emissions of pollutants. The life-history of a system can be evaluated using such criteria.

The strategic insights add a new dimension to classical scientific analyses dealing with the aspect of partial unpredictability of systems. For this, reasonable notions have to be generated as to how the system might survive and develop even if some influences on the system are not at all known or are subtle and, therefore, difficult to assess. How robust or viable is the system? How easily will it adapt, or might it disappear? Which changes seem likely to increase the viability of the system?

One more insight helps in predicting likely structural changes of a system. Although ecosystems as well as human systems are to a smaller or larger extent *unpredictable*, they tend to be *reliable*. This difference between unpredictability and unreliability is important for most inhabitants. Insect-eating birds need food very regularly several times per day. The species of prey, that is the details of the feeding, is often not important. People need a lot of resources, too. For human systems, many types of external, even international, aid tend to be given if life-support from the immediate environment becomes unreliable. In both cases, the source or type of resource may be unpredictable, but the supply is reliable.

Middle Layer or Layer of Complex Dynamics and Complex Interrelationships

On the middle layer, the past, present, and potential future structures, as depicted and defined by the strategic analysis, are described and evaluated. One form of evaluation is to determine the dynamics of the system. A structure usually generates or *implies* dynamics. Different structures (determined with the strategic analyses) tend to exhibit different dynamics. The results are time-series, dependent on different structures.

Adequate methods for analyzing the middle layer are those that allow researchers to model the structure and discover system dynamics. These goals can be accomplished with many methods, for example flow diagrams, formal feedback-structure models, multi-loop feedback models, or dynamic linear optimization models. Each method has different advantages and disadvantages. Another method for determining dynamics is guesses from experts about the developments they expect to happen.

Lowest Layer or Layer of Details

On the lowest layer, the manifold details within particular structures and the elements that comprise a system are described and analyzed. Many details of ecosystems can be described along gradients (Vitousek and Mattson 1991; McDonnell et al., Chapter 15, this volume); many developments are area-related. Most human influences on ecosystems have a spatial dimension. Therefore, spatial variables are an important means of describing and analyzing details of systems. Area is an important interface between the human being and ecosystems.

For this level, all precise descriptions and evaluations of details, e.g., data banks, statistics, or induction, are appropriate. Geographical Information Systems (GIS) are adequate to deal with the spatial dimension of ecosystems. They allow combinations of different thematic maps and performance of area-related balancing (Schaller 1987; Ashdown and Schaller 1990); see below.

Combining Layers in the Multifaceted Method

Ecosystems exhibit all three facets described above. For example, if diversity is important on the highest layer, then the interactions between different species or sites are investigated on the intermediate layer, and details concerning species and sites are analyzed on the lowest layer. When all three facets interact, it is necessary to synthesize the results from the analyses on all three layers. This leads to the use of multifaceted methods which are more capable of dealing with complex problems than mono-dimensional methods. One important cause of past failures in analyses was the overextension of methods by addressing problems beyond their scope. For research on subtle human influences, each method is used within its own sphere of applicability; the results are synthesized. The most important source of failure is improper management of data. It has been traditionally important in science to collect data without further intentions or only vague ideas of how to process them. Insights might come later. For instance, without Tycho Brahe's incredible collection of data on planets, Kepler would not have been able to discover his famous laws. However, many data collections end up as "data grave yards." The data are poorly documented and not properly archived. Nobody will ever use such data. Given today's advanced methods, a data collection should be embedded into a project with very detailed plans about how to evaluate the data. Without this planning, the data may be useless and the project actually fail. But evaluation often needs close cooperation between scientists and a reward system for both the data collector and the data evaluator.

Dealing with Subtle Human Influences Using Dynamic Models

Dynamic models are adequate for the following characteristics of subtle influences:

1) *Indirect influences* are passed on through the structure of the system. A model can depict structures and, hence, reveal potential indirect effects of subtle influences and their pathways;

2) *Lagged effects*. Many modeling methods allow portrayal of delays between influence and effect, e.g., due to chained state variables or information delays (Odum 1982; Forrester 1968);

3) *Historic effects* can be mirrored by selecting initial values for variables that correspond to historic values due to past influences, e.g., lower initial availability of nutrients to take into account litter raking in the past;

4) *Biological legacies (cf.: biotic components, persistent changes, debris)* are often known from historical records. In such cases, their effects can be partially assessed with models which rely on state variables, if the legacies are put into corresponding initial values of those state variables (e.g., debris). Some persistent changes are changes of the

mode of behavior. Here, the appropriate methods come from synergetics (Haken 1978) or from branching theory in mathematics (see in particular for chaos theory, Lorenz 1963; May 1974; or as a summary, Gleick 1988). Some dynamic models can be switched between different modes of behavior, e.g., Thom's (1975) cusp model allows for two modes with identical data. The results of this analysis are useful for selecting the appropriate model; and for such a model; and

5)   *Echoes of the past.* Here, again, models can be used which show long-term developments. Such models help to locate echoes that would otherwise easily disappear in the "noise" caused by temporarily dominant variables.

In addition, models can deal with effects and influences that vary over time and with synergistic or antagonistic relationships between human-ecosystem influences.

Prediction is one of the most effective approaches for dealing with subtle human influences on ecosystems. Because dynamic models depict dynamic behavior, they allow predictions. Predictions are often very helpful for discovering unexpected effects or supporting perceptions of developments, which without model dynamics as a point of reference, would go undetected. However, predicting in complex systems is difficult due to practical limitations from complexity and, in many situations, it is impossible due to the inherent capability of almost all nonlinear systems to exhibit unpredictable behavior. But is actually an advantage of model predictions because predictions allow us to find differences between expectations and reality. Hence, it is important to use models for predictions so that terms of reference will become available to help us more rapidly perceive developments in nature, particularly subtle ones.

Dealing with Subtle Human Influences Using Spatial Variables and GIS

Geographical Information Systems (GIS) allows effective and detailed processing of spatial variables. One of the advantages of GIS over non-spatial data banks is the ability to depict most results in the form of maps. Maps, or patterns, are an easily read form of information which allow the inclusion of collaborators who otherwise would not contribute. For example, the spatial distribution of the frequency or likelihood of fog can be synthetically derived based on elevation, a time series of temperature, and a few other variables. The result can be depicted as a map and discussed with local people (e.g., farmers, foresters) who know the area well and can tell where fog is frequent or rare, and thus, where the map is correct and where the map is wrong. The local population could not react in the same way to numbers. This inclusion of local experts in the work allows a new level of data quality. Also, people can often detect features in patterns which would remain hidden in numbers.

Three basic uses of GIS are particularly applicable to subtle human effects:

1) Area-related balancing. GIS combines the maps with a data bank that stores descriptions of each polygon or raster element (a raster is a special form of a polygon) in a map. With this information, an area-related balancing can be performed; for example, each polygon can be numerically evaluated for the likelihood of a specific development. As an example, flood-prone areas can be calculated and depicted based on elevation to the right and left of a river;

2) Overlay of maps ("polygon overlay"). With GIS, different maps can be overlaid as if they were drawn on transparencies and put on top of each other. In the resulting map, the polygons from the different maps intersect so that the resulting polygons are more numerous and smaller. Within each one of the new polygons, the values of the variables from the source maps are about the same, e.g., the same tree species, same soil type, same orographical features, etc. The GIS builds a new data bank for the resulting overlay map, where, for each of the new polygons, the combined information is available. Again, area-related balancing can be done with this information. using assessment procedures. One example is the assessment of suitability of ecological conditions within an area for a species. The suitability depends on factors such as soil type, water supply in the soil, microclimate, orography, and sometimes existing vegetation. The result of the assessment of the whole map is a synthetic map of potential vegetation of the area; and

3) Base Assessment Maps. A Base Assessment Map (BAM) is the spatial basis for the generation of dynamic maps, as described below. It depicts the spatial distribution of an assessment with respect to a purpose, a risk, a suitability for a type of land use or type of land cover, or the likelihood of a development. More specifically, it helps determine the suitability of an area for infrastructural construction or agricultural use, or the risk of ecosystems being damaged by pollutants or tourism. Very often, an overlay map is used for this assessment.

Spatial variables processed with GIS can help deal with the following characteristics of subtle influences:

1) *Spatially indirect influences* originate from one area and are passed on through transport processes. A BAM can be constructed to show, e.g., likely flows of water along gradients. Such a map looks like a

synthetic river. This BAM must now be supplied with data from different water regimes (dynamics). The resulting maps can be compared with data from reality.

2)   *Lagged effects.* Transport processes take time, and in extreme cases, even years (e.g., dispersal of chemicals in the soil or in groundwater flows). Transports can be modeled in the same way as the spatially indirect effects, above.

3)   *Historic effects* can be exposed by comparing different ecosystems that are similar with respect to the spatial ecological conditions of the sites, but which differ in their history and, therefore, have acquired dissimilar structure and dynamics.

4)   *Biological legacies* (as biotic components, persistent changes, debris) usually have a spatial extension and are distributed along spatial gradients.

5)   *Unexpected action at a distance* can in retrospect be spatially modeled, using transport processes or other suitable mechanisms.

Dynamic Maps: A Method of Connecting Dynamic Models and GIS

Detailed geographical data tend to be static because updating is time consuming and expensive. In contrast, complex dynamic feedback models tend to be poor in detail because detailed complex models tend to become opaque and capable of the most unintelligible and crazy behavior (Lee 1973). Therefore, it is reasonable to process spatial variables with a GIS and to evaluate patterns of change with dynamic models, and then combine the results of the two processes (Grossmann et al. 1983).

Dynamics from models, estimates, and historical records can be combined with the "Base Assessment Maps" (BAM) mentioned above. The result is a dynamic that is valid for extensive areas. For example, the amount of forest damage due to pollution is calculated with a dynamic model. The result is *locally modified*, with the specific situation depicted in a map that assesses the local risk of damage. This local risk depends on spatially distributed factors such as concentrations of pollutants, species and age of trees, soil properties, orography (exposure, elevation, slope), and prevailing wind directions (Grossmann and Schaller 1986). These factors are used to generate a BAM which assesses the risk for different forest areas to pollution. If the model predicts that a certain amount of area will be damaged in a specific year, the polygons with the highest risk are depicted as the first ones affected by damage.

The dynamic model depicts one facet of a complex system, the complex structures, and their inherent dynamics. It does not handle details because interconnected complex models should not be burdened with many data.

The base map deals with the facet of the manifold details in their spatial distribution. It does not depict the dynamics. The combination of these two methods produces dynamic maps which depict the *spatial* and *temporal* aspects of complex processes. As dynamic maps result from a combination of different methods which deal with different facets of a complex problem, they are an example of a multifaceted method. For example, the method depicted development of erosion in an area (Grossmann et al. 1983) and implemented the results of strategic analyses (Grossmann 1991a). Dynamic maps are very suitable for analyzing subtle human influences on ecosystems.

Three Examples of Subtle Influences Studied With Multifaceted Methods

Proposed 1992 Olympic Winter Games in Berchtesgaden

The Olympic Winter Games would have changed the Berchtesgaden region in Germany in many respects, due to construction, increase in traffic, changes in the regional economy, population shifts, and changes in the mixture of tourist categories (demanding, standard category, one-day visitors) in both winter and summer (Haber 1985; Grossmann and Clemens-Schwartz 1985).

The income from tourism in Berchtesgaden has lagged behind expectations since the 1980s. Eighty percent of the economy of Berchtesgaden is dependent on tourism. The Winter Games were intended to attract more tourists through the construction of new ski slopes and other infrastructure. Ecologists were afraid that the Winter Games would destroy ecologically valuable parts of the region. A study of the potential effects was undertaken and the result was very unexpected: it turned out that the Winter Games would neither directly do much ecological damage nor improve the economic situation of the region. However, indirect economic and ecological effects from increased car traffic would be considerable. These conclusions were based on: 1) a comprehensive array of spatial data stored in one of the largest GIS available for research at that time; and 2) use of a dynamic model depicting relations among different types of tourists, infrastructure, natural attractiveness of the area, regional economy, etc.

The GIS was used for area related balancing in order to answer the following questions: 1) where are the ecologically most valuable areas located? 2) where are the most likely sites of tourist activities, including illegal paths through wilderness areas, and new ski-slopes (based on the orography)? and 3) where would the new infrastructure be built? The model demonstrated that more tourists and even more one-day visitors would be attracted by the Games, but due to overcrowding, the mix of tourists would shift toward the standard category and one-day visitors, both of which are less inclined to spend money.

As a reaction to these results, a "Revitalization Policy" was developed to overcome the predicted continued downturn: a higher standard of the tourist offerings would be needed to increase income from tourism. More wealthier tourists would increase the overall income even if the overall number of visitors were to decline considerably, but wealthier tourists would come only if the overall attractiveness of the region were increased. This could be achieved by decreasing the number of cars driven to the area and by revitalizing the *inner city* of Berchtesgaden and other local cities. The resulting decrease of traffic, pollution, and one-day visitors would be ecologically and economically beneficial. The ecological system and the economic system would both benefit from the same policy. Most ecological effects would be indirect: 1) decreasing car traffic would mean fewer animals, e.g., insects, killed by cars and fewer trees affected by pollution such as ozone from volatile organic compounds (which is a human effect at a distance because ozone is destroyed by nitrogen oxides if concentrations are too high close to roads); 2) decreasing tourists would cause less disturbance of animals; and 3) reducing the amount of roads in the region, instead of the normal ongoing expansion of the road network, would cause areas not separated by roads to become larger. For an ecological assessment of the importance of intersections of areas, see O'Neill et al. (1988b) and Gardner et al. (Chapter 17, this volume).

This revitalization policy was not adopted, but the Winter Games also did not happen in Berchtesgaden. However, contrary to our predictions, Berchtesgaden is now booming with tourists. The sudden blooming of algae in the Mediterranean, the Baltic, and the North Sea caused a dramatic shift in tourism into mountainous areas that still seem ecologically unspoiled. This development is an example of an action at a distance. The proposed revitalization policy was based on viability analysis of the region (using strateria). Its results would have remained valid under the new circumstances; the decrease of accessibility of the area, as proposed in the revitalization policy, would have allowed the same increase in earnings but with a lower number of visitors and less environmental degradation and damage.

Agricultural Decline Resulting in Ecological Problems

Agriculture has not always been the great devastator of ecosystems (Richerson, Chapter 11, this volume). The German cultural landscapes of the nineteenth century have been those with the largest species diversity during the last thousand years. Hence, in some regions, a decline of traditional agriculture is threatening ecosystems.

In many alpine regions, agricultural soils are poor and the climate is not favorable for agriculture. Hence, farmers tend to abandon farming and earn their living in different ways, such as increasing their offerings of "Holiday on the Farm" (Kerner and Spandau 1990). Mountainous pastures, created by humans and rich in species which can live only there, are no longer grazed by cattle; ecological succession begins. These developments were examined in another case study in the Berchtesgaden area. The types of succession were carefully evaluated using area related balancing. These pastures allow tourists to look down into the valleys from their mountain trails, whereas forests, which are the next successional state on most pastures, would block the view. With the loss of cultivated landscape, the attractiveness of the region also declines. This attractiveness is a precondition for tourism to flourish. The succession would cause a considerable decrease in species richness nearly everywhere; most pastures would become spruce-dominated forests, which already are abundant in that area.

The study showed that withdrawal from farming would be the worst option for everybody, whereas investments into "organic farming," with marketing of the products to demanding tourists, would increase the income of the farmers. Again, economy and ecology would simultaneously suffer from the first policy of abandoning farming, and benefit from the second policy of shifting to organic farming.

As in the case study reported above, the human and the ecological systems act as a closely connected system, not two separated systems where man influences the ecosystem and that is all there is. This close connection seems to be the rule, not the exception. Therefore, it could well be one of the great challenges of ecosystems science, in particular in the new international programs on "Global Change," to address this interconnectedness and to develop methods to adequately deal with this combination of social and ecological systems. Sometimes "esoteric" links may be sufficient to connect two such systems. Here the main linkage from the ecosystem back to the regional economy is the *attractiveness* of the landscape which mainly affects tourism and income. Such links are truly subtle. (The beauty of the landscape was carefully evaluated by a sophisticated application of photographs [Nohl and Neumann 1986]).

This study also had an unexpected result: the demand for "organic products" would so much surpass possible local production that resulting intensification of agriculture could endanger sensitive species. Demanding tourists would reject this intensification of agriculture. Hence, a careful surveillance of agricultural practices by unions of local farmers would become necessary. The dynamic maps showed where intense farming would be best suited and where it, hence, could occur in which year. These maps support monitoring of farming practices.

Useful Wrong Predictions on Ecosystems Development

This is a case study where the predictions were partially wrong but nevertheless important. One of the lessons of the International Biological

Programme for future large ecological programs was, in Jeffers' 1981 words: "Adoption, where practical, of predictive modeling using simulation, optimization, and other models."

Predictions were made in 1986 in this case study on progressive forest damage, that in 1990 turned out to be wrong. The cause was three different errors which, opposite to common expectations, did not cancel each other out, but instead amplified each other. To some extent, the alarm triggered by the prediction helped to prevent the prediction from coming true.

The method of dynamic maps (DM) was applied in 1986 to predict the location and severity of forest damage and its temporal development. Simultaneously, the state of the forest was evaluated using terrestrial observation (TER) and false-color infrared photography (FC-IR). In 1989, the terrestrial observation and the false-color infrared photography were repeated. This allowed two types of comparisons:

1)   *vertical comparison with respect to time*
        DM 1986 with DM 1989
        TER 1986 with TER 1989
        FC-IR 1986 with FC-IR 1989

2)   *horizontal comparison with respect to time*
        DM 1986 with TER 1986 and DM 1989 with TER 1989
        DM 1986 with FC-IR 1986 and DM 1989 with FC-IR 1989
        TER 1986 with FC-IR 1986 and TER 1989 with FC-IR 1989.

DM 1986 was more similar to TER 1986 and to FC-IR 1986 than was TER 1986 to FC-IR 1986. The calculated map was, so to speak, in the middle between the two observations. However, the prediction of forest damage by DM 1989 was, in most polygons, too high by one-half point on a scale of 5 (for more details see Grossmann [1991b-d]). The prediction was that damage would increase, whereas, in reality, the damage decreased. According to the prediction, not much healthy forest would remain in year 2008. This now seems to be wrong. In order of importance, the reasons for errors were as follows:

1)   Forest damage which already occured in the 1970s, was only observed in the 1980s. Hence, the inventories of forest health from the early 1980s most likely showed a much more dramatic development than actually occurred. The calibration of the model based on these observations (2% damage in 1981, 8% in 1982, 34% in 1983, 50% in 1984) was, accordingly, wrong. To achieve such an incredible increase of damage, the model used a so-called exponential collapse mechanism. But this selection of model structure was based on incorrect data;

2)   The development of damage by secondary pollutants, the underlying primary pollutants, SO_2, NO_x, and N deposition and the acidifying stress (Ulrich 1984, 1987) is dependent on climate. Future climate was assumed to be the same as in the period 1981-1985. Solar irradiation and temperature in the summers from 1981 to 1985 were above average, and from 1986 to 1989, below average; and

3)   Several effective policies to curb pollution were implemented in Austria. The catalyzer for car exhaust became obligatory in 1986, and for motorcycles in 1989. Emissions of SO_2 were cut to a quarter of the 1980 rate. The predicted doom of forests helped to push the adoption of these laws which restricted emissions.

The model was based mainly on damage done by secondary pollutants (or "products" as Altshuller (1983) named them, in particular, ozone, hydrogen peroxide, Peroxyacetylnitrate, nitric acid, formaldehyde, and organic acids). This was not the most popular hypothesis in 1986, but the importance of ozone for forest damage in Austria or Germany has now been generally confirmed. The hypothesis used in the dynamic model is still not mainstream in the emphasis put on secondary pollutants other than from ozone. The predicted spatial distribution was remarkably good.

After evaluation in 1989, the model was improved to include climate and the reductions of emissions achieved since the early 1980s. The already low impact attributed to SO_2 and NO_x was further decreased. The calibration was based on vastly improved data. But this version of the model now indicated that forest damage had already occurred in the 1970s. Data seemed to confirm this result but were too poor for a satisfactory confirmation. The development of forest health in the 1980s is nicely reproduced by the model, both the deterioration from 1980 to 1984, and the recovery from 1985 to 1988, as well as the severe damage in 1990. Only a few time series are used in the model: irradiation, temperature, and emissions of volatile organic compounds (VOCs), SO_2, NO_x and carbon monoxide. The results of the project seem to be quite satisfactory:

1)   First results had, by 1985, allowed a program to be devised to decrease VOCs (Grossmann and Grossmann 1985). The decision to install the automobile exhaust catalyzers was also based on these results;

2)   A term of reference was provided against which the actual development of forests could be compared;

3)   An early version of the model from 1984 was capable of demonstrating a very rapid recovery of forest health (Grossmann et al. 1984) which nobody believed, but which happened in the years 1985-1989; and

4)   A model which needed only very few data had become available and seemed capable of making good predictions, which will most proba-

bly be rendered wrong by new, at present unknown events or developments. The model might, however, help to detect these events at an earlier date.

The politicians were not upset about the incorrect prediction because it helped them to pass new laws. Now, such a dramatic prediction is no longer tenable; further legislation to curb VOCs and nitrogen oxides is promised but difficult to enact.

The Multifaceted Method and Subtle Influences

This chapter has shown that multifaceted models are a productive way to deal with subtle effects. The monitoring and prediction provided by such models improves our capability to deal with legacies of the past, echoes of the past, and rare events. Furthermore, the depiction of connectivity, that is, the system's structure in the models and the spatial neighborhoods on the base maps, provides better insights into possible pathways of indirect effects.

The Multifaceted Method takes care of the five problems listed in the opening section.

1)   Subtle effects become more easily detectable if a term of reference is provided;
2)   The method allows depiction of complex interdependencies and their dynamics;
3)   The method is multifaceted; and
4-5) The method provides a term of reference to support human perception.

Complexity is nearly inevitable if the human and the ecological systems are regarded, more properly, as one system. The multifaceted method decouples complexity, because different facets of a complex problem are treated with different, but adequate tools.

The method can compute space into time and vice versa, because some areas are subject to succession or to development of different phases. Dynamic maps allow depiction of how areas might look in the future or have looked so that comparisons of different areas at different times are possible. The method also supports monitoring because it provides a very obvious term of reference.

The method can provide predictions which, even in case of failure, are most helpful for seeing where data and hypotheses are good or bad, *or* where unexpected developments occur. A correct prediction in the study of subtle human effects is far less useful than one deviating from the actual future trajectory of the system, in contrast to the paradigmatic use of physics as the epitome of scientific prediction. The practical use of models of human-ecosystem interaction is most successful if the models yield an improved environment for people and ecosystems.

Because they provide predictions, the dynamic maps provide an early warning of environmental degradation. Because a series of dynamic maps supports environmental monitoring, any deviation between the predicted and the actual becomes rapidly very obvious. This is a basic requirement for the use of predictions.

Summary

Issues and methods for assessing subtle human influences on ecosystems have been presented. One suitable new method, integrating social and ecological factors by combining Geographical Information Systems and dynamic models, has been applied in about 30 case studies (Adisoemarto and Brunig 1978; Vester and von Hesler 1980: Haber et al. 1984; Haber 1985; Grossmann 1983, 1990). The method was used as the base for several large, integrated projects and as early as 1983, dynamic maps were produced by combining dynamic models and base assessment maps. Experiences from a number of studies are reported. The most important result may be that the approach of "including human influences" into ecosystem's studies often falls short by one step. The human systems usually, in turn, are influenced by reactions of the ecosystems which previously have been subject to human influences. In our case studies, we usually had to extend the "disturbance and impact" philosophy of the human system acting on an ecosystem by an integrated approach of dealing with *one interconnected socio-ecological system*.

Acknowledgments

This paper has indirectly benefitted very much from work done together with K.E.F. Watt (Grossmann and Watt 1992).

Recommended Readings

Ashdown, M. and J. Schaller. (1990). *Geographic Information Systems and their Application in MAB-Projects, Ecosystem Research and Environmental Monitoring.* (MAB Mitteilungen Nr. 34). Deutsches Nationalkomitee, Bonn. (German and English).

Gleick, J. (1988). *Chaos.* Bantam Books, New York.

Grossmann, W.D. (1991a). Model- and Strategy-Driven Geographical Maps for Ecological Research and Management. In: P.G. Risser and J. Mellilo, eds. *Long Term Ecological Research: An International Perspective*, pp. 241-256. SCOPE 47. John Wiley and Sons, Chichester, U.K.

Grossmann, W.D. and K.E.F. Watt. (1992). Viability and sustainability of civilizations, corporations, institutions, and ecological systems. *Syst. Res.* 1:3-41.

Jeffers, J.N.R. (1981). *Preparatory statement for the meeting of the "Reconvened Expert Panel on Systems Analysis" of UNESCO's Man and Biosphere Programme MAB.* Institute of Terrestrial Ecology, Grange-over-Sands. (This is a very good statement. I will send it to anyone who wants to read it.)

Appendix: Hierarchy of Scientific Explanation

The hierarchical ordering of systems (Fig. 18.1), see e.g., Bertalanffy (1969) mirrors the hierarchy of scientific explanation.

When scientists deal with a scientific problem, they select the focal or adequate layer of work. If they have to *explain* a phenomenon, they move one layer further downward, derive the explanation, and escape upward. A too ambitious explanation could cause a scientist from one layer deeper to say: "You didn't really explain this. You must also take into account this and this." Thus our scientist may end up doing particle physics. It is a convention in science that something is accepted as an explanation if it is based on facts from one layer downwards. This is partially justified due to the astonishing independence of higher layer phenomena from lower layer phenomena for most systems, named "nearly decoupled" by Simon (1973). If scientists want to generalize a result, they move up by one layer, leave details behind, move in parallel to the other area, and go down one layer. They must also avoid staying longer on the higher layer, otherwise they could be accused by their colleagues of "holistic thinking." Thus the layers act as attractors.

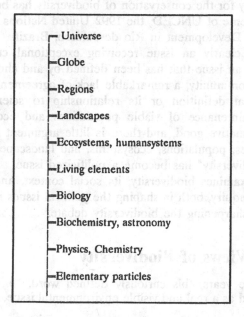

```
┌─ Universe
├─Globe
├─Regions
├─Landscapes
├─Ecosystems, humansystems
├─Living elements
├─Biology
├─Biochemistry, astronomy
├─Physics, Chemistry
└─Elementary particles
```

Figure 18.1. If a scientist does integrated research, this means horizontal movements. But the scientist is expert only in his area, not in adjacent scientific areas where he/she is an amateur and faces the risk of being accused of superficial work. Thus, the scientific areas tend to be absorbing from within, and well protected against outside intrusion. They prevent their prisoners from leaving their respective occupational fields. This hierarchy, with its phenomenal success in simple systems, is now discouraging for integrated research on complex phenomena.

19

Human Influences on Ecosystems: Dealing With Biodiversity

Peter R. Jutro[1]

Introduction

Loss of biodiversity is regularly identified, together with global climate change and stratospheric ozone depletion, as one of the greatest environmental risks facing mankind as the result of human activity. Organizations have mobilized to address it or protect it. Legislation has been proposed to preserve biodiversity and organize the government's approach to it. An international treaty for the conservation of biodiversity has been negotiated. It was a major theme of UNCED, the 1992 United Nations Conference on Environment and Development in Rio de Janiero, Brazil.

Biodiversity is clearly an issue receiving exceptional current political attention. But for an issue that has been defined by and chosen as a cause by the scientific community, a remarkable lack of agreement exists regarding its operational definition or its relationship to scientific research. Although the maintenance of viable populations and ecosystems is an unambiguous normative good, and there is little argument that it is these gene pools, species, populations, ecosystems, and landscapes that we wish to preserve, "biodiversity" has become a politicized issue.

This chapter examines biodiversity, its social context, and the potential role of scientific inquiry, both in shaping the political issues inherent in biodiversity, and in sharpening the biodiversity debate.

Traditional Views of Biodiversity

Over the last five years, this curiously defined word, "biodiversity" (Fig. 19.1), has emerged as a real and visible environmental issue. Although there

[1]The views represented in this chapter are those of the author and do not necessarily represent those of the United States Environmental Protection Agency.

♦ "The variety and variability among living organisms and the ecological complexes in which they occur."
Office of Technology Assessment (1987)

♦ "The variety of life and its processes."
U S. Forest Service (1990)

♦ "The full range of variety and variability within and among living organisms and the ecol ogical complexes in which they occur; encompasses ecosystems or community diversity, species diversity and genetic diversity."
Pending legislation (U.S. Congress 1991)

♦ "The degree of nature's variety, including both the number and frequency of ecosystems, species, or genes in a given assemblage."
McNeely, et al (1990)

♦ "The variety of life and its processes. It includes the variety of living organisms, the genetic differences among them, and the communities and ecosystems in which they occur."
Keystone Dialogue (1991)

♦ "The variety of life on all levels of organization, represented by the number and relative frequencies of items (genes, organisms and ecosystems).
EPA (1990)

♦ "Those environmental goals that go beyond human health concerns."
Environmental Law Institute (Fischman 1991)

♦ "The wealth of life on earth including the millions of plants, animals, and microorganism as well as the genetic information they contain and the ecosystems that they create."
AID (1989)

♦ "The variability among living organisms from all sources including, inter alia, terrestrial, marine and other aquatic ecosystems and the ecological complexes of which they are part; this includes diversity within species, between species and of ecosystems."
UN Convention on Biological Diversity (1992)

♦ A "variety or multiformity, the condition of being different in character or quality."
R. Patrick (1983)

♦ "The totality of genes, species, and ecosystems in a region."
Global Biodiversity Strategy (1992)

♦ "The variety of genes, genotypes and genepools and their relationships with the environment at molecular, population, species and ecosystem levels."
FAO (1990)

♦ "The variety and variability of all animals, plants and microorganisms on earth, ... can be considered at three levels -- genetic diversity (variability within species), species diversity, and habitat diversity."
Overseas Development Administration (1991)

♦ "I suggest a fourth category ... functional diversity -- the variety of different responses to environmental change, especially the diverse space and time scales with which organisms react to each other and the environment."
J Steele (1991)

Figure 19.1. Definitions of biodiversity.

have been differences over its precise definition, for the most part, biodiversity stems from a conservationist rather than a scientific root—from a recognition that, as a planet, we are losing enormous amounts of biological capital stock, and that it is happening all over the globe for many of the same reasons—because of people who are forced to pursue strategies or undertake actions that are logical, perhaps almost necessary, in the short run, but are undertaken without the vision, power, or incentives to cope with their inevitable biological consequences.

As a result of press attention and the emergence of an extensive popular literature, most reasonably well read people are aware of the problems of tropical deforestation and land conversion, the trade in endangered species, and the implications of agricultural chemical practices. One could go on to develop a formidable list of causes of pauperization of ecosystems and habitats. Simply, these are the activities that, for the most part, are envisioned when the word biodiversity is used in current parlance.

There is concern in the scientific community about the lack of empirical data for many current estimates of biodiversity loss, but if these estimates are correct, the magnitude of current and impending losses of the world's biota is staggering. Human activities appear to be responsible for a species loss rate that is the most extreme in the last 65 million years (Wilson 1985). Even allowing for disagreement on absolute numbers and rates, the general problem seems fairly straightforward. Through human activity, we are transforming land at a dramatic rate.

The challenge is to focus this loss of biodiversity as a social question: to find ways to grasp the meaning of this loss not only in terms of reduction and simplification of gene pools, species, ecosystems, and landscapes, but of the implications of these losses in terms of the opportunities forgone by their absence from the armamentarium of ingredients available for both natural evolution and future human use (see Hall, Chapter 5, this volume).

Climate and Biodiversity

Whereas an extensive literature has evolved on the subject of traditionally recognized direct threats to biodiversity, this is not yet the case for the potential impact of climate change on biodiversity. This may turn out to be the most problematic stressor of biodiversity (Jutro 1991). Should predicted climate change occur, the future ranges of species will be dramatically different than today's. A few examples indicate the potential disruptions. Tree species in the eastern United States, for example, would have to shift northward by hundreds of kilometers to adjust to changing climate regimes. Beech trees, which now grow in the east from Florida to Southern Canada would have to shift northward to a new range from New England to Hudson's Bay. Sugar maples would leave most of the eastern United States. Hemlock and birch would have to undergo similar radical changes (Davis 1989). But could they? When temperatures rose 3-5 degrees centigrade at the end of the pleistocene glaciation, beech forests moved less than 30 km per century—well under a tenth of the predicted rate of change.

In general, it is likely that the composition of communities will change. Species may move in the same direction, but they do not do so at the same rates. As a result, shifting forest ranges would also affect animals. Many species depend on complex relationships between soils and several other species of animals and plants. In addition, the theoretical notion of northward migration may be illusory. Soils, for example, in the predicted new range, may not support the immigrating species. Some animals might respond by moving up in elevation rather than north in latitude, but they will be successful only if appropriate flora move up as well. But the tops of mountains are smaller than the bottoms, and populations will, of necessity, decrease. Current high elevation species may have nowhere to go.

It has been theorized that with a 3°C change, the Great Basin, USA, will lose 44% of its mammals, 23% of the butterflies, and perhaps a fifth of the birds (Murphy 1991). Many other scientists have performed similar evaluations for different parts of the world, and predicted trends are similar (Schneider 1991).

Plants and animals evolve, as do the communities in which they live. Population and range constantly change. Paleoecological studies suggest that it is reasonable to expect a reassortment of species, communities, and habitats as climate changes (Davis 1989). Success in adapting to change will be dependent on species ability to move with the changing climate by dispersing colonists. In North America, these will have to move north or up. If all of a species' habitat becomes unsuitable, or barriers to migration are present, extinction may result.

Are species dispersal abilities adequate? Considering the biology of the species and the projected rates of change, it is unlikely that most species could disperse adequately without human help. In addition, the problem is complicated by obstacles or physical barriers to migration. Managed agricultural lands, cities, artificial lakes, roads, parks, golf courses—any number of anthropogenic land use changes threaten the ability of species to react successfully to environmental change. We may well see a twenty- first century in which most surviving species are found only in small patches of their original (or novel, but suitable) habitats, isolated by larger expanses of human-managed landscapes (e.g., the cultural landscapes described by Williams, Chapter 3, this volume).

Such shifts are not mere abstractions. We have preserved national parks and wildlife refuges around the world because of unique characteristics, more often than not, various plants, animals, and communities that we chose to treasure as a people. If the biota of parks or refuges are forced to migrate, they face a habitat which, even if available, offers none of the protection that they would continue to enjoy under a stable climate (Peters and Darling 1985). If these analyses are correct, we may well be doomed to lose many of our most cherished places.

Attention to this issue is relatively recent (Peters and Lovejoy 1992). Until lately, most of our concern with biodiversity has been focused on the tropics: the incredibly lush, rich, and diverse rainforests, the so-called megadiversity areas, where orders of magnitudes more species appear to exist than anywhere else on Earth. There, the problem is primarily one of loss of habitat, and these losses, in turn, can be attributed to a variety of social and economic factors.

Now we realize that climate change will have a similar and perhaps even greater effect on biodiversity than habitat loss, perhaps an even greater effect. And as a result, we see the threat of dramatic changes in the composition of natural communities in the temperate and northern latitudes resulting from climate change, as well as from other problems traditionally confronted by land managers and conservationists.

Emerging Biodiversity Concerns

We must also realize that although tropical systems have served as the touchstone for concern over loss of biodiversity, and global climate change represents a potential major worldwide perturbation, concern over the diminution of biological diversity among biotas other than tropical rain-forests is increasing. For example, a substantial literature on marine biological diversity is emerging, see the special issue of *BioScience* in 1991 (vol. 41 no. 7), and the Keystone Center dialogue on *Biological Diversity on Federal Lands* (Keystone 1991), which focused on government controlled land in the United States. Similarly, much early colloquy in the United Nations Environment Programme (UNEP) Ad Hoc Working Group preparing for negotiating a biodiversity treaty, ensured that biomes other than tropical forest were not ignored (UNEP 1990).

Approaching Large-Scale Environmental Problems

The notion of "global change" as an integrating conceptual device has emerged over the past several years (Committee on Global Change 1988). Although the concept was initially developed in order to incorporate a range of related issues—including climate change, ozone depletion, perturbation of biogeochemical cycles, land transformation, and changes in biological diversity—in reality, U.S. governmental activities conducted under its aegis have largely confined themselves to the atmospheric issues of climate change and ozone depletion. The broader scope may be more appropriate (Likens 1989).

These global change issues are highly complex, and of them, only the issue of ozone depletion appears to have a reasonably discrete, definable, and implementable technical solution. In the case of the global greenhouse, the fortuitous, straightforward, technical solution does not present itself. What scientists have been able to do so far, and it is a great deal, is identify the problem, estimate the possible physical and biological consequences and their probabilities, and project the potential rates of occurrence under different scenarios. Social scientists have worked with the natural scientists to develop those scenarios depending upon assumptions about technological developments and the ability of different societies to make changes.

For the narrower issue of biodiversity, much of the same is true. But whereas scientists have proposed research programs, and have met with some considerable success in assuring funding for programs to deal with the greenhouse effect, it has been difficult to make the same claim for the proposals that have to date been made in response to the biodiversity problem. The reasons for this are worth exploring.

One is the very success of the U.S. global change program as a coherent entrepreneurial enterprise (Committee on Earth Sciences 1990). Originally, the program was conceived of broadly as global change. As the concept translated itself into governmental investment in programs, however, the bulk of the emphasis turned to researchable scientific questions having to do with atmospheric chemistry and physics. These questions, in turn, led to a series of other questions regarding the development, use, value, and interpretation of the General Circulation Models (GCMs) upon which most climatic prediction is based.

Although there was no doubt that fundamental public interest in the climate issue resulted from a concern about the implications of climate change, to the extent that such concern is measured by research investment, the implication questions have but slowly begun to achieve stature commensurate with issues of causation. What about these biological implication questions? Are these global change issues? Are they climate issues? Are they biodiversity issues? Does it matter? These are not mere abstract questions, for research program results can be very sensitive to initial concepts and approaches. Since different organizations and agencies have responsibilities for different issues, how an issue is defined may determine what success we have in dealing with it.

In the context of planning global change research, *questions of ecological effects* were generally given low priority until a sophistication in analyzing the global climate question evolved. As scientists stressed the importance of biota in material and energy feedbacks, support for these areas grew. But the actual areas of investigation were limited. They first included questions relating to biogenic atmospheric constituents. Subsequently, other carbon-cycle issues, paleoecology, and more recently, deforestation as a carbon source/sink and albedo issue garnered support. But we have neither yet seen an appreciation, reflected in substantial financial support, of the fundamental question of ecological change, nor of the broader aspects of global change as originally conceived.

Approaching Biodiversity

Having made these observations about the evolution of investment in global change research, we must recognize that despite initial hopes of large-scale integration of global change research planning, biodiversity emerged as a separate and distinct issue. It is, therefore, appropriate to focus on developments in biodiversity apart and aside from those of climate change.

Biodiversity has been approached somewhat independently as both an international and a domestic issue. Internationally, biodiversity has emerged as something of a traditional North-South issue. As a broad generalization, developed nations argue for the worldwide maintenance of habitat and the preservation of species as a common concern of mankind, whereas the developing countries argue that exploitive practices of developed countries have *led* to degradation throughout the world, and that unencumbered financial and technical assistance is needed to ameliorate the situation and help them develop in an environmentally benign fashion.

Internationally, the biodiversity arena is quite active. Negotiations under the auspices of UNEP for an international treaty on the conservation of biological diversity were completed in May 1992. Negotiators from approximately 100 U.N. member countries, developed this treaty for signature at the United Nations Conference on Environment and Development (UNCED) in Brazil in June 1992 (UNEP 1991).

In turn, UNCED selected biodiversity as one of its three major themes, together with climate change and sustainable development (UNGA 1989). In the case of both the treaty negotiations and preparations for UNCED, non-governmental organizations were heavily involved in development of the issues. The World Resources Institute (WRI), the International Union for the Conservation of Nature and Natural Resources (IUCN), and UNEP jointly sponsored a series of regional consultations in support of their joint biodiversity activities (IUCN et al. 1990). Similarly, in an effort to better inform itself in preparation for UNCED, the U.S. Council on Environmental Quality (CEQ) held a series of public roundtables around the country to solicit input for the development of a formal U.S. National Report to UNCED.

Domestically, many U.S. governmental agencies have been involved in attempting to develop an approach to biological diversity. The Keystone dialogue on biological diversity on federal lands represents one effort of governmental officials, conservation groups, and commodity interests to achieve consensus on a range of biodiversity issues (Keystone Center 1991).

For the past several years, biodiversity legislation had been introduced in the U.S. House of Representatives. Hearings were held, but in each Congress, the proposed legislation was not enacted. In 1991, biodiversity legislation was introduced in both the Environment Subcommittee of the House Committee on Science, Space, and Technology (USHR 1991a), and in the Fish and Wildlife Conservation and the Environment Subcommittee of the Committee on Merchant Marine and Fisheries (USHR 1991b). Companion legislation was introduced and hearings held by the Senate Committee on Environment and Public Works (USS 1991). The likelihood of enactment remains small, but Congressional attention to the issue is growing. Although the Administration has expressed formal opposition to the enactment of biodiversity legislation as currently formulated, governmental entities have been exploring the possibilities of undertaking some of the activities envisioned by proposed legislation independent of passage. As of this writing, planning for the formation of a U.S. National Biodiversity Center is underway.

While biodiversity has become an active issue, it is hard to identify tangible progress, as measured in preservation, directly related to the activity currently underway. The greatest progress to date has been the work of the non-governmental organizations (NGOs) which have actually managed to save habitats in various parts of the world. Organizations such as The Nature Conservancy (TNC 1991), Conservation International (CI 1990), and the World Wildlife Fund (WWF 1989) have actually protected land or species, and others are trying to broaden the effort. Similarly, the U.S. Department of Agriculture, the International Board on Plant Genetic Resources and its affiliates, as well as museums, zoos, and botanical gardens have worked hard to assure the *ex situ* preservation of living exemplars.

But what appears to be the underlying political issue, the social force that keeps the issue of biodiversity alive, is the *in situ* preservation of species, which most frequently translates into preservation of habitat. If it is this conservation issue that is driving, and if such preservation is largely a social, economic, and political problem, then what role can or should science play?

Science and Biodiversity

In general terms, one can hypothesize a functional taxonomy of roles appropriate to an issue of this sort. Its elements would include:

1) Science can help define a problem;
2) Science can help identify questions susceptible to scientific analysis and define their boundaries;
3) Science can suggest or contribute to, or occasionally provide, workable solutions;
4) The results of scientific research can motivate political activity;
5) Science may contribute to mediation or change of underlying causes and problems; and
6) Scientific research results can help set priorities.

Scientists have recognized the above roles and have begun to define their role in biodiversity and related issues. Several individuals and groups have attempted to define appropriate research directions:

- The Committee on the Application of Ecological Theory to Environmental Problems of the National Research Council addressed the general problem of the relationship of ecological knowledge to environmental problem solving (Committee on the Applications of Ecological Theory to Environmental Problems 1986);

- The Science Advisory Board (SAB) of the U.S. Environmental Protection Agency (EPA) suggested an imbalance in EPAs investments relative to risks. SAB rated habitat alteration, species extinction, and biodiversity loss as among the greatest threats to the planet's environment (Science Advisory Board 1990);

- The Ecological Society of America analyzed the status of ecological research and addressed specific biodiversity and habitat related issues that could be better understood or better dealt with through further research (Lubchenco et al. 1991);

- The World Resources Institute listed a number of scientific activities necessary to enhance the foundation for decision-making about biodiversity (Reid and Miller 1989);

- Solbrig (1991), in preparation for a IUBS-SCOPE-UNESCO sponsored workshop on biodiversity research, proposed a program of scientific research in biodiversity, and suggested a number of testable hypotheses related to changes in biodiversity resulting from human activity;

- Soulé and Kohm (1989) proposed a group of research priorities for conservation biology; and

- The President's Council on Environmental Quality devoted a well annotated major chapter to the problem of biodiversity and ecosystems (CEQ 1991). In addition to defining the problem in considerable detail, it characterized the need for further information and the particular need for scientific information about the distribution and abundance of species and communities.

As might be expected, each of these analyses has its biases, strengths, and weaknesses. In their own way, for instance, each begs fundamental questions such as, what is biodiversity? What is its societal or ecological importance? How can we realistically manage for many millions of species? What is the relationship between biodiversity, current human needs, and environmental health? Similarly, some of these analyses provide little indication beyond sweeping generalities of how support for the research they propose will help solve the problem toward which they are said to be directed.

These questions become particularly important for an area such as biodiversity, in which it is difficult to describe problems both simply and coherently because of their scope and complexity. We recognize that, as we deal with large problems, understanding them involves elements of many different disciplines (e.g., Grossmann, Chapter 18, this volume). Our response is generally to disaggregate the larger problem into smaller, more manageable units. In doing so, we reduce the researchable elements to those which will more likely yield discrete results. The reduced researchable elements will then yield recognizable results, but those results will often not independently solve the problems that prompted the research.

Conclusion

The biodiversity issue has matured to the point where policymakers are beginning to understand the existence of, and perhaps even the essence of, the problem, and they recognize and are frustrated by the appalling complexity of the social, economic, and political forces that drive it. Exercises such as those listed above indicate that the scientific community recognizes the problem, and has begun to treat it as a research opportunity. The challenge will be to make a clear *prima facie* case for the utilitarian value of the work proposed.

If we view biodiversity as intrinsically important, then what are the questions unique to biodiversity that we should be able to call upon science for answers to, which in turn will be of arguable use in addressing the problem? Will these answers come from traditional scientific disciplines such as ecology, systematics, population biology, or demographics? Will they come from developments in methodology or from the examination or development of management insights or techniques? This is the challenging question. One priority-setting approach, the "Biodiversity Uncertainties and Research Needs" project, sponsored by the U.S. Environmental Protection Agency, is attempting to deal with this problem. Biodiversity decision makers have been identified and interviewed, and their information needs are being defined. Scientists will then examine the needs and attempt to structure a responsive research agenda (EPA 1992).

We recognize the abundance and variety of life on Earth as a valued resource for reasons that range from the pragmatic; such as the provision of life-giving medicines, to the sublime: the satisfaction we find in contemplating the beauty, mystery, and variety of life about us. We simply must realize that in order to deal with this assemblage of concerns generalized under the name of biodiversity, science will likely not provide short term solutions. If we wish to effect change, whatever else we do, we must effect social changes—changes in the way we live, changes in our expectations of our institutions, of each other, and of ourselves—changes in our values (see Egerton, Chapter 2; Hall, Chapter 5, this volume).

To the extent that biodiversity has become a word fraught with political baggage, it presents us with a challenge: we must identify what we can do as scientists in improving understanding of the issue and in helping cope with it. We we must find honest and effective ways of conveying the need for that research, and we must encourage the intelligent application of its results.

Recommended Readings

Committee on Global Change. (1988). *Toward an Understanding of Global Change: Initial Priorities for U.S. Contributions to the International Geosphere-Biosphere Program.* National Academy Press, Washington, D.C.

Committee on the Applications of Ecological Theory to Environmental Problems. (1986). *Ecological Knowledge and Environmental Problem-Solving.* National Academy Press, Washington, D.C..

Murphy, D.D. and S.B. Weiss. (1992). Predicting effects of climate change on biological diversity in western North America: species losses and mechanisms. In: R.L. Peters and T.E. Lovejoy, eds. *Consequences of Greenhouse Warming to Biodiversity.* Yale University Press, New Haven, Connecticut.

Soulé, M.E. and K.A. Kohm. (1989). *Research Priorities for Conservation Biology.* Island Press, Washington, D.C..

Wilson, E.O. (1985). The biological diversity crisis. *BioScience* 35:700-706.

20
"Natural" or "Healthy" Ecosystems: Are U.S. National Parks Providing Them?

Frederic H. Wagner and Charles E. Kay

Introduction

Ecosystem ecologists are now generally agreed on the desirability of having areas of natural biota minimally disturbed by technological societies to serve as reference points for understanding the structure and function of ecological systems. Systems, of course, exist along a continuum from little or no human disturbance to complete anthropogenic alteration (cf. McDonnell et al., Chapter 15, this volume). An in-depth understanding of how they respond to different intensities of human perturbation, what constitutes ecosystem sustainability, and how profoundly altered systems can be restored, is facilitated by a knowledge of structure and function along the entire continuum.

A number of different kinds of areas, acquired and administered under a variety of arrangements, are expected to provide relatively undisturbed reference systems. These include state natural areas, as in the Wisconsin system, privately owned areas like those of The Nature Conservancy, and wilderness areas under the aegis of several federal agencies. But most important in terms of their age, size, prohibition against consumptive use of natural resources, and public recognition, are those national parks and monuments that were established to protect exceptional biotas minimally affected by post-industrial humans. This chapter examines how effectively the U.S. national parks are serving this reference-point role, but raises ecological and policy considerations that also apply to the other types of areas.

Some Conceptual Issues

National Park Service policies generally adopt this reference-point role for those areas in the National Park System that contain significant biological resources. Indeed, the legislation establishing Yellowstone National Park, the world's first national park, ordered retention of the "...natural curiosities, or wonders...in their natural condition..." (16 USC 21-22). Since that time numerous documents and policy statements have reiterated this general theme:

- "As a primary goal, we would recommend that the biotic associations within each park be maintained...as nearly as possible in the conditions that prevailed when the area was first visited by the white man" (Leopold et al. 1963).

- "The use of national parks for the advancement of scientific knowledge is also explicit in basic legislation. National parks, preserved as natural, comparatively self-contained ecosystems, have immense and increasing value to civilization as laboratories for serious basic research. Few areas remain in the world today where the process of nature may be studied in a comparatively pure natural situation" (Anon. 1968:43).

- "The primary purpose of Yellowstone National Park is to provide present and future visitors with the opportunity to see and appreciate the natural scenery and native plant and animal life as it occurred in primitive America" (Anon. 1977).

Although the words "nature" and "natural" feature prominently in these and other similar statements, they provide no criteria of ecosystem structure and function by which "natural" ecosystems can be characterized, or which can be used as goals to be achieved by management programs. But they do generally state or imply, in many cases ambiguously, two characteristics:

- "A portrayal of primitive America is defined as having natural conditions where scenery and 'balance of nature' in ecosystems are not altered by man" (Anon. 1967).

- "The natural resource policies of the National Park Service are aimed at providing the American people with the opportunity to enjoy and benefit from natural environments evolving through natural processes minimally influenced by human actions" (Anon. 1988).

- "National parks were founded on the principal of...natural processes, and the importance of minimizing human interference...management's primary purpose is to maintain the area's pristine condition to the fullest extent possible...This includes the perpetuation of natural processes in the absence of human interference—processes essential to the existence of a healthy ecosystem" (Varley 1988).

The ambiguities in these statements lie in the variations between "minimal" and no human actions, and in whether they refer implicitly to the effects of post-Columbian Europeans or Native Americans or both. We are inferring that the references are to Europeans, and that any pre-Columbian human effects are implicitly assumed to be inconsequential. We base this inference on the 1967 statement above in which "...primitive America is defined as having natural conditions...not altered by man..." and the last which defines "pristine conditions" as having "...natural processes in the absence of human interference...

The second ambiguity lies in the evolutionary stipulation of the second of the three statements. The same document expands on the point further:

- "Managers...will try to maintain all the components and processes of naturally evolving park ecosystems" (Anon. 1988:4.1).

- "The National Park Service will strive to protect the full range of genetic types (genotypes) native to plant and animal populations in the parks by perpetuating natural evolutionary processes and minimizing human interference with evolving genetic diversity" (Anon. 1988:4.10).

Here again we infer that the explicit reference is to post-Columbian Europeans, and that Native American effects are implicitly assumed, consciously or subconsciously, to be unimportant (see also Egerton, Chapter 2; Williams, Chapter 3; Russell, Chapter 8; Foster, Chapter 9, this volume).

In the remainder of this chapter, we will develop the thesis that:

1) There is growing archaeological and anthropological evidence worldwide that pre-industrial humans significantly modified the structure and function of ecosystems, including those now in national parks;

2) Ecosystems of national parks in the United States are being materially altered by burgeoning, ungulate herds which may be increasing, in part, because they are no longer being constrained by aboriginal hunting; and

3) Park Service policies do not face up realistically to either the pre-industrial role of humans or the role of unconstrained animal populations. These two circumstances need to be given serious consideration in more explicitly defining the ecological purposes of national parks, and tailoring management programs to those purposes.

Ecological Effects of Pre-Industrial Humans

Until quite recently, North American ecologists have tended subconsciously to dismiss the influence of pre-Columbian humans on their continent's ecosystems, and commonly have sought to study systems supposedly free of any anthropogenic effects. But archaeologists, anthropologists, and other social scientists worldwide are forcing on ecology the realization that such effects have been dismissed too lightly (e.g., Turner and Meyer, Chapter 4; Vayda, Chapter 6; Boyden, Chapter 7; Richerson, Chapter 11, this volume). Effects on animal populations, influence of fire, and alteration of vegetation by pre-industrial cultures are being reported from every continent with increasing frequency.

Evidence from Outside the Western Hemisphere

There is now wide familiarity with the hypothesis and evidence that many Holocene faunal extinctions worldwide followed the first human contact with unadapted faunas. Best known are the North American and Australian megafaunal extinctions. Less well known are numerous insular extinctions (Martin and Klein 1984; Steadman and Olson 1985; Diamond 1988). Diamond (1988) commented that "on every oceanic island for which we have adequate knowledge, the first arrival of humans was quickly followed by the extermination of all or most large animals, the best known victims being the moas of New Zealand, the giant lemurs and elephant birds of Madagascar, and the flightless geese of Hawaii..."

Still less well known, and of greater interest here, are the effects of hunter-gatherer cultures on population densities and community structure of persisting animal species. Thorbahn (1984) investigated the archaeology and ethnohistory of the Tsavo National Park region of central Kenya in an effort to relate precolonial human activities to the character of the biota. He concluded that heavy ivory trade to India and the Middle East by indigenous people between A.D. 155 or earlier and the end of the nineteenth century significantly reduced elephant populations. Prior to this period, the vegetation was a more open grassland, and human populations were more dense. Today, following reduction of the browsing elephants, the vegetation has changed to a dense thorn-scrub called *nyika*, which means in Swahili "wilderness."

Similar vegetation changes have occurred in the southern African countries. By 1900 there were only 4,000 elephants in Zimbabwe, a legacy of the ivory trade which had been underway for several centuries (David Cummings, pers. comm.). Vegetation here, too, has grown up to a dense thorn scrub. Today, following protection, elephant numbers have increased to 50,000.

Klein (1979) traced the species composition of ungulate remains in South African archaeological sites through the Middle Stone Age (roughly 130,000 to 30,000-40,000 years B.P.) and Late Stone Age (ca. 40,000 B.P. to the colonial era). At about 10,000-12,000 B.P., there was a pronounced improvement in the weaponry and hunting techniques. Coincidentally, changes in ungulate species composition and age ratios reflected changing exploitation rates. Some species present in the earlier strata disappeared.

Legge and Rowley-Conwy (1987) traced the species composition of ungulate remains in archaeological sites of hunter-gatherer cultures occupying an area of northern Syria between 11,000 and 8,000 B.P. From 11,000 to shortly after 8,500, goitered gazelle (*Gazella subgutturosa*) made up 80% of the large animal remains. Sheep, goats, wild asses, wild cattle, wild pigs, and deer made up the remainder. At roughly the seventh century B.P., there was an extensive proliferation of drive traps, termed "desert kites," in northern Jordan on the migration routes of the gazelles summering in northern Syria. Within a few hundred years thereafter, gazelles declined to 20% of animal remains in the sites, and sheep and goat remains rose to the gazelles' previous 80%. These authors concluded that the extensive increase in gazelle trapping reduced their populations and forced a dietary shift to domestic animals.

Koike and Ohtaishi (1985) analyzed the age composition of sika deer (*Cervus nippon*) in archaeological sites from the Jomon Period (5,500-3,000 B.P.) on the island of Hokkaido. Age composition of animals from the Early Jomon Period contained a large proportion of older animals, as would occur in a lightly exploited population. But samples from the Late and Latest Jomon Periods were mainly comprised of animals younger than 5 years of age, an age distribution similar to those of contemporary, heavily hunted populations on Hokkaido. The authors concluded that hunting pressures by Jomon peoples increased over time and reduced deer densities.

Although declines in game animals are most often cited, reduction of plant species used for food (Jochim 1981) and firewood (Baksh 1985) have been reported. Feely (1980) and Hall (1984) cite evidence indicating that pre-colonial agriculture dating back at least 1,500 years significantly shaped the character of vegetation in southern Africa through cultivation, grazing, and burning. Vestiges of these effects are still evident today. Flenly and King (1984) documented the deforestation of the Easter Islands by early Polynesian immigrants.

Assigning ecological meaning to archaeological data always depends on inferences about past events and processes that cannot be observed. Hence, they always leave an element of uncertainty. However, the hypotheses proposed by archaeologists concerning the influences of subsistence cultures on animal populations and communities are gaining support from anthropological investigations on contemporary hunter-gatherer cultures. These, too, are finding evidence of significant impacts on the fauna. Resource depression around camps and villages has frequently been reported from the historic literature and from studies of hunter-gatherer cultures. Hunting yields decline as a function of village age, and increase as a function of distance from settlement (Hames and Vichas 1982).

The ubiquity of these influences prompted Simms (1992) to generalize that "...all human societies have played a role in shaping ecosystems, [and] there are many cases of hunter-gatherer behavior that...would be classified as having negative impacts upon wilderness environments. The evidence suggests that simple societies may be as susceptible to causing significant environmental damage as more complex societies. To be sure, there is a difference of scale..." And Redford (1990), referring to what he called "the myth of the noble savage," commented that "The recently accumulated evidence...refutes this concept of ecological nobility...Paleobiologists, archaeologists, and botanists are coming to believe that most tropical forests have been severely altered by human activities before European contact...These people behaved as humans do now: they did whatever they had to to [sic] feed themselves and their families."

New World Evidence

The interactions between Western Hemisphere pre-industrial humans and other components of their ecosystems differ from those of Eurasia, Africa, and Australia in three historical ways: 1) the New World interactions have taken place over a much shorter time period; 2) humans arrived in the Western Hemisphere at a relatively late stage of neolithic evolution in the sophistication of weaponry and foraging behavior; and 3) the biota did not coevolve with humans before the latter's recent arrival. Hence, there were circumstances that predisposed the Western Hemisphere biota to especially significant human impacts, and, indeed, nowhere in the world have the megafaunal extinctions been as sweeping as those of this half of the world.

Skeptics of pre-European impacts have suggested that North American human populations were too sparse to effect any major ecological alterations. But these judgments are based on the assumption that Native American populations at the time of early European contact were similar to those of preceding centuries. Recent investigations (cf. Dobyns 1983; Ramenofsky 1987) present evidence that European diseases were introduced by the first Spaniards arriving in the West Indies. These were carried to the mainland by the natives, whereupon the epidemics swept across the continent and decimated populations up to two centuries in advance of European contact on the mainland. Pre-Columbian human populations are thus thought to have been anywhere from 4 to 10 times as dense as those encountered by the eighteenth and nineteenth century European explorers, trappers, and settlers.

In an analysis of the ecological impacts of North American pre-industrial cultures on wild ungulate populations, Kay (1990a) quantified historical and archaeological evidence on the abundance and species composition of large mammals in the pre-Columbian western United States. Kay (1990a) tabulated the number of occasions on which large mammals were recorded in the journals of 19 known expeditions conducted in the region of what is now Yellowstone National Park during the early and mid 1800s. The number of individuals in each party varied from 3 to 60, with over half of the groups containing 17 or more. The total time spent was 765 party days. Ungulates were seen on only 121 occasions, or approximately once every 6 party days. Elk (*Cervus elephas*), by far the most abundant and frequently seen ungulate in the region today, were seen only once every 18 party days. Moreover, not one single wolf or mountain lion was reported to have been seen by the 19 expeditions. This scarcity of large carnivores would be expected if prey animals were scarce.

Kay (1990a) analyzed the ungulate species composition in the remains of more than 200 archaeological sites in Washington, Oregon, Idaho, Montana, Wyoming, Utah, and Nevada. These contained over 52,000 identified ungulate bones. Yet, elk remains made up only 3% of that total. Elk were similarly rare in sites near Yellowstone National Park where today they comprise roughly 80% of all ungulates.

From these and other sources of evidence, Kay (1988, 1990a,b) concluded that the low ungulate densities and very different species composition in the pre-Columbian western United States were maintained by aboriginal hunting, possibly in synergy with carnivorous predators.

An extensive literature reports the role of fire in North American ecosystem structure and function, and implicates aboriginal burning (cf. Lewis 1982). Two studies in the western United States exemplify the significant control of vegetation structure by fire. Barrett and Arno (1982) conclude that Native American fires maintained the open, park-like structure of ponderosa pine (*Pinus ponderosa*) forests in the northern Rocky Mountains, which the first European settlers encountered. Similarly, Bonnicksen and Stone (1981, 1982a) describe the patchwork, mixed-age character of the giant sequoia-mixed conifer forests of Kings Canyon National Park, which prevailed near the end of the last century. They ascribed this character, and the shrub and hardwood understory, to periodic burning by Native Americans.

In all of these cases, the authors recognize that Indian burning and lightning fires interacted. Consequently, the effects of aboriginal burning were difficult to separate out and quantify. Greenlee and Langenheim (1990) attempted this separation by using both the empirical evidence of fire scars on coastal redwoods (*Sequoia sempervirens*), and modeling approaches. These investigators inferred five successive, historical fire frequencies in the coastal vegetation of Santa Cruz and Monterey Counties of California: 1) lightning (up to 11,000 B.P.); 2) aboriginal (11,000 B.P. to A.D. 1792); 3) Spanish (1792 to 1848); 4) Anglo (1848 to 1929); and 5) recent (1929 to present). Thus, by subtracting the lightning frequency from the aboriginal, the difference, in these authors' opinion, constituted the effects of early human-set fires in the locale. The separation was made possible by the datable arrivals of the first humans in the region, and of the Spaniards.

There is also evidence of significant modification of North American landscapes through deforestation. Betancourt and coworkers (Betancourt and Van Devender 1981; Betancourt et al. 1986) traced the vegetation changes in the Chaco Canyon region of northern New Mexico between the early tenth and twelfth centuries. The Anasazi peoples who occupied the area during this period abandoned it, apparently in part because they had exhausted woody vegetation used for building construction and firewood over a radius of 75 km.

In total, the evidence for alteration of North American ecosystems by pre-industrial Native Americans appears as extensive as that for other parts of the world. The North American landscape was already significantly shaped by human action when the first Europeans arrived.

Recent Changes in National Park Ecosystems

Given the prevalence of pre-industrial human ecological influence on the landscape, including areas now in national parks (e.g., Yellowstone and Kings Canyon), we may now return to the meaning of such terms as "natural," "healthy," and "pristine," and consider what purpose parks should serve. Although the early human influences have rarely been considered or incorporated in park objectives for managing natural resources, the Park Service has consistently held to the policy of preserving examples of "natural" or "pristine" (i.e., commonly, pre-European) biotas. This is implied in the legislation establishing the early parks, and in the 1916 Organic Act. In 1963, a National Parks Advisory Committee, appointed by the Secretary of the Interior, and generally termed "The Leopold Committee" after its chair, A. Starker Leopold, recommended that the purposes of parks should be "to preserve vignettes of primitive America" (Leopold et al. 1963).

The Service has vacillated over the years on the question of whether it should engage in active resource management to maintain ecosystems in the conditions thought to prevail in pre-Columbian times. The Leopold Committee firmly recommended such management procedures as controlling excessive ungulate numbers and use of fire to maintain seral stages. Indeed, Yellowstone Park controlled predators and fed ungulates in the first half of this century (Houston 1982). It then entered a stage of controlling elk through shooting and trapping, and controlled fires up to the early 1970s. Wind Cave National Park currently controls elk, prairie dogs, and bison, and engages in controlled burning. Current Park Service policy (cf. Anon. 1988) advocates these kinds of management where deemed necessary.

However, in 1967, Senator Gale McGee of Wyoming, then on the Appropriations Subcommittee which funded the Park Service, threatened to cut off Yellowstone funding if the Park did not discontinue shooting elk

(Chase 1986). Promptly thereafter, the Park adopted the natural-regulation paradigm as a conceptual basis for a new management policy.

Without the support of empirical evidence, and based only on an evolutionary rationale, the paradigm concluded that elk and other ecosystem components had coexisted since the Pleistocene without predatory (and implicitly, human) constraint (Cole 1971). Elk numbers were considered to be "regulated" by food resources, and given more than 10 millennia, the system must have arrived at some equilibrium over this period. The Park had been established early in the European settlement of the West (1872) when there presumably had been little human ecological alteration, and by the late 1960s had been protected from significant human intrusion for the better part of a century. For these reasons, the contemporary, large elk herd and character of the remainder of the system must be presumed to typify the pre-Columbian conditions. Hence, there should be no need to intervene with advertent management.

The natural-regulation concept became a strong influence in the Park Service despite the recommendations of the Leopold Committee, the language in the policy documents approving management where desirable, the practice of advertent management in some parks, and strong criticism from a number of biologists (cf. Erickson 1981; Gale 1987; Gilpin 1987). Most parks have refrained from controlling ungulate herds. And as Bonnicksen (1989a) has pointed out, the philosophy's manifestation in park fire management is the "let burn" policy: the practice of not controlling naturally set fires.

It is now of interest to examine whether any changes have occurred in park ecosystems during the approximately 20 years of natural-regulation management. The natural-regulation concept implicitly dismisses any pre-Columbian human influences on park ecosystems. If there have been any ecological changes during the history of the parks, they have occurred in the absence of pre-European anthropogenic effects. Such changes provide a partial test of the natural-regulation paradigm which holds that the systems have arrived at some equilibrium, implicitly without human intervention. It is true that large, nonhuman predators have been removed in many cases. Hence, aboriginal influences are not the only variable removed. But as shown below, simulation modeling of wolf predation on Yellowstone elk herds does not project much effect.

Kay's (1990a) study of Yellowstone's northern range examined nearly 50,000 early photographs in ten photo archival collections to reconstruct the character of the vegetation at, or in the early-decades after, Park formation in 1872. Aspen (Populus tremuloides) woodlands today: 1) consist only of older trees without young saplings; 2) have black, scarred bark up to about 2 m induced by elk chewing; and 3) have a lawn-like understory dominated by exotic grasses and low forbs. Early photographs show aspen stands of mixed age classes, no bark scarring, and dense understories of shrubs and tall forbs. The early stands resembled the character of contemporary aspen stands in exclosures in the Park, and outside but near Park boundaries where elk numbers are held at lower levels by public hunting. The area of aspen woodlands, as a type, has declined between 50 and 95% in the Park's north range since Park formation.

Kay (1990a) also assembled repeat photosets of riparian zones. Early photographs show dense shrub strips along streams on the northern range. Today, these are totally eliminated, the banks bare, and in many cases, sloughing into the streams. Once again, willows (*Salix* spp.) and other shrubs flourish inside Park exclosures and along streams outside the Park boundaries.

Kay (1990a) unearthed a number of historical accounts describing an abundance of berry-bearing shrubs (*Prunus virginiana, Amelanchier alnifolia, Shepherdia canadensis*) and extensive use of the fruit by the local Native Americans. Today, these species seldom exceed a height of 0.3 m and rarely produce fruit in Yellowstone except in exclosures. They reach heights up to 3 m and produce abundant fruit outside Park boundaries where ungulates are not concentrated.

The vegetation changes are accompanied by faunal changes. Kay (1990a) reviewed early accounts of abundant beaver (*Castor canadensis*) in the Park, large numbers being trapped, and one estimate placing the number at 10,000 in the early 1900s. Today, beaver are infrequently encountered and are absent from areas in which they were historically numerous. Sally Jackson (pers. comm.) has found avian densities and diversity much lower in heavily browsed riparian zones in and near the Park than in relatively unused stands outside the boundaries. White-tailed deer, once present in the Park, no longer occur there (Kay 1990a).

In sum, deciduous woody vegetation and its associated fauna in the ecosystem of the Park's north range have declined sharply during the Park's history. Understory vegetation in aspen stands inside the Park is much less diverse than that in aspen exclosures and in aspen stands outside the Park. In total, the system has experienced a marked decline in plant and animal diversity. The overwhelming evidence indicates that this decline has occurred as a result of the pressure of elk on the north range, where today they number around 20,000. The unbrowsed condition of the vegetation, shown in the early photographs, also supports the inferences drawn above from the historical and archaeological data, that elk numbers in the region (and other ungulates as well) were low at the time of European contact. And the collective evidence challenges the natural-regulation paradigm which holds that modern abundance of elk and condition of the rest of the biota are what prevailed in pre-Columbian times.

It is true that aboriginal hunting is not the only pre-Columbian population constraint on elk that is no longer present. There are, today, mountain lions but no wolves. However, recent simulation modeling of the effects of wolf reintroduction into Yellowstone National Park (cf. Boyce 1992) predicts only a minor reduction in elk numbers.

Braun et al. (1991) observed an increasing elk population in Rocky Mountain National Park, which winters in the subalpine and alpine zones, an unprecedented behavior pattern. The elk browse on, and are markedly reducing, the willows in the wintering areas. The willows are also staple food for white-tailed ptarmigan (*Lagopus leucurus*) in the Park. Ptarmigan censuses over 25 years have shown a population decline, and the authors hypothesize that this may be associated with elk-induced reduction in willow abundance.

Warren (1991) reviews the extensive literature describing the effects of white-tailed deer on eastern North American forest vegetation. Deer populations have increased to high densities throughout much of the eastern United States in the twentieth century, and negative effects on vegetation are widely observed. Warren also reviews studies in which vegetation alteration by deer has been accompanied by shifts in species composition of rodents, lower abundance of snowshoe hare (*Lepus americanus*), and lower avian species richness.

Warren cites evidence from several studies, including his own, which have measured some of these effects in Great Smoky Mountain National Park (GSMNP), several other national park areas in the eastern United States, and Catoctin Mountain Park in Maryland. Effects in GSMNP include reduction in number of plant species, loss of hardwood species, and heavy grazing on and reduction of potentially threatened plant species.

Pastor et al. (1987, 1993) have measured the effects of moose (*Alces alces*) browsing on entire nutrient cycles in Isle Royale National Park. Moose arrived in Isle Royale somewhere around the turn of the century, and, hence, are a relatively new influence in the island ecosystem. Wolves arrived on the island somewhat later, and since then have entered into what may in part be a predator-prey oscillation (Peterson 1988), although there is some evidence of a vegetation oscillation as well.

As the moose reach the peaks of their oscillation, they browse heavily on the deciduous species, particularly aspen, and convert the vegetation to a coniferous type. Nitrogen mineralization under deciduous litter proceeds at a faster rate than under coniferous litter, and primary production is higher in soils with rapid nitrogen-mineralization rates. Hence, the effects of the moose browsing in the Park are: 1) to convert the vegetation to coniferous growth, thereby reducing diversity; and 2) to slow nitrogen mineralization and thereby reduce primary production.

Clearly, there are profound and unprecedented changes taking place in national park ecosystems. There is a substantial basis for inferring that these are taking place because of the removal of pre-Columbian ecological constraints. Aboriginal hunting may well have been one of the most important of those constraints. If the purpose of national parks is to preserve "pristine" or "healthy" ecosystems, with pristine and healthy defined as the conditions prevailing at the time of European settlement, many of the parks are clearly not achieving this purpose.

What Needs to Be Done?

In one sense, this question can be construed as a query about what management protocols need to be instituted. Should there be: no management, as at present? Active, advertent management? Management to simulate early human effects?

But management programs are designed to achieve goals, and effective management protocols cannot be designed until goals are clearly defined. Numerous authors (cf. Bonnicksen and Stone 1982b,c; Agee and Johnson 1988; Johnson and Agee 1988; Bonnicksen 1989a; Gordon et al. 1989) have commented on the ambiguity of park goals, and the inconsistency with which the existing ones are carried out: ungulate control in Wind Cave, not in Rocky Mountain; prescribed burning in Kings Canyon, not in Yellowstone; and elk control in Yellowstone up to 1967, no control from 1967 to the present. National policy documents advocate ungulate control where needed, yet most parks rigidly avoid control. Hence, as the above authors concur, the most immediate need is a precise definition of the ecological purposes or goals of the national parks. Once this is achieved, the appropriate management goals can be decided upon.

Ecologists, natural-resources managers, and policy analysts are increasingly converging on the view that management policies for public resources are set to satisfy social values (cf. Hendee 1974; Giles 1978; Kennedy 1985; Bonnicksen 1989a; Kania and Conover 1991; Wagner 1993). As Bonnicksen and Stone (1985) state, "Goals are value judgments that describe the ideal or preferred condition. Therefore, goal setting is a social or political decision, not a professional decision."

Park goals were/are appropriately set by social and political action when Congress passed the Park Service Organic Act in 1916, and passed the enabling legislation establishing each park. And as discussed at the beginning of this chapter, the biological goals have generally been the provision of ecological reference areas. But legislation is characteristically abstract and vague, and that ambiguity provides the interpretative leeway within which Park Service officials can employ the diversity of management approaches described above, or simply the inaction of the natural-regulation paradigm.

What is urgently needed from the ecological community, as pointed out by other authors (cf. Bonnicksen and Stone 1982b,c, 1985; Agee and Johnson 1988; Bonnicksen 1989b), is an ecologically explicit and substantive definition of the reference-area goals. Bonnicksen (1989b) advocates setting quantitative and measurable standards of ecosystem structure and function as part of such goals. This would require the most insightful and sophisticated ecological understanding of each ecosystem, and should be carried out collaboratively by ad hoc groups of a wide variety of ecologists familiar with each area. The task is too large and complex to be carried out internally by the small cadre of Park Service biologists, as such efforts to date indicate. And care should be taken to include scientists with differing points of view. Too often, Service officials seek advice from outside scientists who are known to concur with their views and policies, and ignore scientists who draw contrary inferences from the available evidence.

In this definition process, the options of simulating and not simulating pre-industrial human influences on park ecosystems would be kept open. If such influences were not included in the goals, it would be incumbent upon the deliberation process to state what scientific purpose would be served by allowing these systems to seek some end determined by the absence of ecological factors with which they have co-evolved for ten millennia. If they were included, that too should be based on a scientific rationale and accompanied by precise definition of management protocols.

More generally, goal definition would be accompanied by prescriptions for management policies, with the overriding alternatives of managing or not managing to compensate for the changes these systems have experienced since settlement (cf. Wagner 1991). Current absence of the ecological conditions established by early human influence is only one such change. At present, the systems are all changed and are changing to varying degrees from their pre-Columbian structure, and no one knows how they will develop, or whether they will irrevocably alter themselves without judicious and ecologically enlightened management assistance. Advisory committees have repeatedly recommended such management (cf. Leopold et al. 1963; Gordon et al. 1989). To continue the no-management policy, with its risk of ecosystem self destruction, is a goal option. But that would need to be supported with a well considered and explicated ecological rationale. Self destruction would eliminate the reference-area value.

Epilogue

Historical, archaeological, and anthropological research is providing us with new perspectives on ecosystem structure and function. Ecosystems analysis and natural-resources management will be increasingly enhanced by including these disciplines in the research efforts. Humans have been components of ecosystems for millennia, and the systems that we encounter today are significantly shaped by their presence. It is the "natural" systems—i.e., those completely free of any human influence—that are the exceptions. We can see a rich future for the social-scientific disciplines working collaboratively with ecology.

Recommended Readings

Agee, J.K. and D.R. Johnson. (1988). A direction for ecosystem management. In: J.K. Agee and D.R. Johnson, eds. *Ecosystem Management for*

Parks and Wilderness, pp. 226-232. University of Washington Press, Seattle.

Bonnicksen, T.M. and E.C. Stone. (1985). Restoring naturalness to national parks. *Environ. Manage.* 6:109-122.

Boyce, M.S. (1992). Wolf recovery for Yellowstone National Park. In: D.R. McCullough and R.H. Barrett, eds. *Wildlife 2001: Populations*, pp. 123-138. Elsevier Science Publishers Ltd., Essex.

Chase, A. (1986). *Playing God in Yellowstone: The Destruction of America's First National Park*. The Atlantic Monthly Press, Boston.

Simms, S.R. (1992). Wilderness as a human landscape. In: S.I. Zeveloff, M.L. Vause, and W.H. McVaugh, eds. *A Wilderness Tapestry: An Eclectic Approach to Preservation*. University of Nevada Press, Reno.

21
Restoration as a Technique for Identifying and Characterizing Human Influences on Ecosystems

William R. Jordan III

While the obvious way to detect human influences on a natural landscape is simply to compare descriptions of the landscape in the presence and absence of this influence, this approach has certain obvious limitations.

For one thing, it is frequently impossible to make a clear comparison, either between two similar systems or of the same system before and after it is subject to human influence, and even when such comparison is possible it is usually both difficult and expensive. Even more fundamentally, however, description *by itself* produces only a mass of detail. It provides no way of discriminating what is significant from what is not. Equally to the point, it provides no way of establishing causal relationships. It may reveal correlations, but can only suggest causes. This may be sufficient in the case of more or less obvious relationships, but it will generally fail to reveal those subtler phenomena and relationships that are of special interest to us here.

Elsewhere, my colleagues and I have explored in some detail the idea that one way to overcome the limitations of comparison is to adapt what we have termed a synthetic approach to ecological research involving attempts to repair or reassemble ecological systems, or to reverse human influences on them (Jordan et al. 1987). We have pointed out that the value of this approach is clearly illustrated by the experience of those who have attempted the wholesale restoration or repair of entire ecosystems, but that this is only the tip of a heuristic iceberg, and that, in fact, synthesis—the actual construction of ecosystems or assembly of ecological communities—in various forms and in varying degrees is a normal part of the process of ecological research, and has been since the time of Charles Darwin, or even earlier.

In this chapter I will explore the implications of this approach for the discernment and characterization of subtle human influences on natural ecosystems. What I want to suggest, briefly, is the idea that description, though crucial, is only the first step in the discovery of any kind of influences on ecological systems. Real insight will invariably depend not merely on descriptions, however meticulous, nor on comparisons, however imaginative, but on the attempt to refine descriptions by reducing them to practice by using them as blueprints, recipes, or sets of specifications for the management and the restoration or re-creation of the system.

More specifically, I suggest that, ultimately, the best way to discern the precise nature of human influence on a natural system, is actually to attempt to reverse that influence by taking steps calculated to compensate for the effects of specific human activities. Since any such compensatory or restorative effort is based on an idea—or hypothesis—of which activities are involved, and also on the precise way they influence the system, any such effort serves as a test for conceptual or quantitative models of the human role in ecosystems. In this way, it is possible not merely to compare descriptions of pristine and "influenced" systems, guessing what might be related or significant and what not, but actually to discern among all the differences that emerge which ones are related to particular human activities and, more than this, to deduce something about the precise nature of this influence in ecological terms. Clearly, a single restoration experiment will often be inadequate to answer these questions. But, as in other forms of experimental science, it should often be possible to design restoration experiments properly to provide the controls needed to isolate and characterize the factors influencing the system.

I will illustrate the value of restoration for this purpose with accounts of two experiences, both from the grasslands of the upper Midwest, USA. One of these—the "discovery" of the role of fire in the tallgrass prairies of the prairie-forest border—provides an early, classic case. The other, which has resulted in some new ideas regarding the composition and dynamics of the oak openings of the same area, is a more recent experience. Both point directly at the value of the attempt to reverse influence—restoration—as a way of analyzing and coming to understand humans as components of ecosystems.

The Discovery of Fire

When I first began thinking about the relationship between ecological research and the practice of ecological restoration, and about the role of restoration in the testing and refining of ecological ideas, the first example that occurred to me was what people now refer to somewhat jokingly as the "discovery" of fire on the tallgrass prairies of the upper Midwest. What is meant by this, of course, is the formal recognition of the role of fire in the ecology of these prairies, in a sense the re-discovery of a technology for environmental management that had been practiced by indigenous inhabitants of central North America for thousands of years before the arrival of Europeans. This is obviously an important insight into prairie ecology, but it is also an important insight into the ways humans have influenced the tallgrass prairies historically through activities that influenced the timing and frequency of fires—a crucial insight, in fact, into several more or less subtle human influences on the natural landscape.

From a purely ecological point of view, this is an old story (see, for example, Curtis 1959; Curtis and Partch 1948). The early insights achieved at Wisconsin and elsewhere beginning in the 1940s have long since been confirmed, and continue to be refined through basic research. It is now firmly established that the tallgrass prairies are fire-dependent ecosystems, and their story makes up just one chapter in the story of the working out of the role of fire in a wide variety of community types during the past half-century. My purpose here is not to summarize the results of this effort, but to draw attention to certain aspects of the way it has been carried out, concentrating my attention especially on the classic work at the University of Wisconsin-Madison Arboretum, with which I am most familiar.

Briefly, what happened was this. Beginning in 1934, a group of scientists at the UW-Madison under the leadership of Norman Fassett and Aldo Leopold undertook restoration of a tallgrass prairie on sixty acres of derelict farmland recently acquired by the University for an arboretum. The prairie was to be the centerpiece of an extensive collection of ecological communities, many of which had never existed or no longer existed on the site, and so would have to be created slightly outside their historic range, or, as in the case of the prairie, restored more or less from scratch.

To carry out the restoration of the prairie, the Arboretum team brought in Dr. Theodore Sperry, who had recently completed a doctorate at the University of Illinois at Urbana, specializing in prairie ecology. Though intimately familiar with the prairies of the area, Sperry had no better idea than anyone else how to restore or create a prairie. He began by doing the obvious—bringing in transplants, and later seeds, collected from relict prairies in the area, and planting them in a patchwork-quilt pattern on the restoration site.

At least in a rough sense, what Sperry was doing, of course, was attempting to reduce to practice his own conception of prairie, based on his earlier experiences and descriptive studies. The results, after the first few years, were mixed. On the one hand, the prairie plants that Sperry managed to establish, despite the severe droughts of the late 1930s, had for the most part flourished. On the other hand, so had the exotic grasses and weeds on the site, part of which had formerly been in pasture, part in corn. As a result, what Sperry had after several years of work was a good sprinkling of some three dozen species of prairie plants growing and flowering in a matrix of weedy grasses and forbs.

Figure 21.1. The discovery of fire as a key element in prairie ecology was one result of early prairie restoration attempts, illustrating the value of restoration as a way of elucidating subtle human influences on ecosystems. Here, managers conduct one of the earliest experimental prairie burns at the UW-Madison Arboretum sometime in the late 1940s. (Photo by Max Partch. Courtesy UW-Arboretum.)

At this point there must have been a serious question in Sperry's mind as to the feasibility of restoring prairie on an ecologically significant scale. What was needed, of course, was a way to tip the balance between natives and exotics growing together on the heavy, relatively rich soil of the restoration site. Groping for a solution, those working on the project introduced fire into the trial plantings sometime during the early 1940s. Exactly who initiated these experiments is unclear, but they were eventually carried out systematically and completed by John Curtis, who reported in 1948 that burning did indeed favor some prairie natives in their competition with Kentucky bluegrass, one of the most persistent and troublesome of the non-prairie species infesting the sites (Curtis and Partch 1948). Fire, in fact, turned out to be a key to the restoration and maintenance of prairies in the prairie-forest border area. Within a few years, frequent burning of the prairie was standard practice at the Arboretum, and it has since become part of the routine and the ritual of maintaining prairies throughout the region (Fig. 21.1).

The ecological implications of this work are widely appreciated. What I want to emphasize here are the implications for our subject of the way this work was carried out —the discovery of subtle human influences on ecosystems. The story I have outlined above illustrates perfectly the pattern of discovery and realization I described at the outset. The point here is not that this work led to an important insight into the ecosystem, though it certainly did that. The point is that it had what might be called an anthropological aspect as well—that is, it revealed something vitally important about the relationship between human beings and the prairies, pointing in fact to a number of subtle, previously unappreciated, influences of human cultures on the prairies (see also Russell, Chapter 8, this volume). The salient point is that knowledge of the burning of the prairies was not new. That fire was a common event on the midwestern prairies was well known even to the earliest settlers. Some people, including Thomas Jefferson, had even speculated that it was fire that kept the prairies free of trees. Nevertheless, the exact role of fire in the ecology of the prairies remained unclear until Sperry, Curtis, and their successors reduced the descriptive information available to them to practice, and actually tried the experiment of making a prairie first without and then with fire.

The result was a new appreciation of the precise role of fire in the ecology of the prairie. This is obviously of great ecological importance. But it is also related to our topic here, since this experience also revealed the subtle role humans had played both in the shaping and in the destruction of the prairie. It illuminated, first of all, the influence of Indians, for whom fire had been a powerful tool for managing both vegetation and game (see also Williams, Chapter 3, this volume). But it also shed light on an even subtler influence—subtler because it was the consequence not of action, but of inaction. That was the influence of European settlers in the nineteenth century who, more or less inadvertently, virtually eliminated fire from the landscape, not only because they did not burn prairie themselves, but also, indirectly, by opening up land with the plow, interrupting the vast, seamless expanses of the prairie. This limited the spread of fire, and so dramatically reduced the likelihood of any particular bit of landscape being visited by fire in a particular year. As a result, millions of acres of prairie and oak opening began growing up into oak forests. Each of these, crowning hilltops throughout southern Wisconsin and Minnesota, northern Illinois and Indiana, represented a prairie lost not to the obvious agency of the plow or the cow, but to the subtle influence of merely not burning. This is a subtle, but powerful influence. By not recognizing it for a century, we nearly lost the prairies entirely. And the way we finally realized this, opening up the possibility of a future for the prairies, was by trying to reverse the processes that had led to their decline, and very nearly to their extinction.

A Question of the Oak Openings

The story of the role of fire in prairie clearly illustrates how the process of restoration, the attempt to reverse or compensate for human activities and choices, can contribute to understanding the ecological effects of those activities. The story shows how the experience of restoration can lead to the revision of an existing model of an ecosystem or, more accurately, to a clearer understanding of the role and importance of the various elements in the description—in this case fire.

A more recent experience, in this case in the derelict and degraded oak openings of southern Illinois, reveals a similar pattern and is, if anything, an even more striking example of the process of discovery through restoration—of discerning human influence by attempting to reverse it (Packard 1988). In this case, the objective was re-creation of examples of the oak opening or tallgrass savanna communities that had once been widespread in the prairie-forest border region. Interestingly, however, while a few remnants of original tallgrass prairie have survived even down to the present, those oak openings that still exist are so degraded, principally by invasion by exotic trees and shrubs such as buckthorn (*Rhamnus cathartica*) and honeysuckle (*Lonicera tatarica*), that the system has generally been regarded as ecologically extinct.

With this in mind, Steve Packard, Director of Science and Stewardship for The Nature Conservancy in Illinois, began attempting to restore the understory of some of the surviving oak groves during the early 1980's. Like Sperry a generation earlier, Packard began by following existing descriptions of the oak opening vegetation, notably that provided by John Curtis (1959). In general, Curtis described the oak openings as prairies with trees, downplaying the importance of any species or other features that might distinguish them from prairies on the one hand or oak woodlands on the other. Following this description, Packard began his oak opening restoration efforts by simply clearing the exotic understory from beneath existing oaks and trying to introduce prairie species by seeding in the cleared areas.

Somewhat surprisingly, however, this failed. Weeds, rather than prairie plants, began taking over the groves, and so Packard began searching for other species that for one reason or another he suspected might flourish in the partly-shaded conditions peculiar to the oak groves. He eventually developed a list of more than a hundred putative "oak opening" species. When he introduced these into the cleared areas under the old savanna oaks, he found they quickly took over, filling the space and crowding out competing weeds. In the meantime, a colleague drew his attention to a list of species of the "barrens" (the name given to the oak openings by pioneer settlers) compiled by a pioneer doctor, S.B. Mead (1846). Remarkably, Mead's list included many of the species with which Packard was experimenting. Moreover, many of these species shared characteristics, such as fleshy fruits and an evergreen habit, which distinguished them from prairie species, and could be construed as adaptations to the variable, edge-like conditions of the oak groves. This lent further support to the notion that Packard's list, though obviously far from definitive, might be closer to the real composition of the historic oak groves than were Curtis's descriptions, based of course on severely degraded remnant examples. And this line of reasoning soon gained further support when a colleague pointed out that he knew of a community in the Chicago area with an understory closely matching Packard's description. Managers had known of the site before, but since it hadn't matched any of the published descriptions of plant communities of the area (Curtis 1959), they hadn't known what to make of it and had more or less overlooked it. Not matching the existing description for any recognized ecosystem, it had very nearly fallen through the cracks in ecologist's knowledge—into ecological oblivion. Today this single site in the suburbs of Chicago is the only natural, high quality example of an entire ecosystem regarded by The Nature Conservancy as one of the most endangered ecosystems on earth.

Packard's work provides another example of the value of the process of restoration as a way of testing and refining ideas about ecological communities and the factors that influence them. It is an especially dramatic example both because the community in question is vanishingly rare, and also because the restoration effort led not merely to the refinement of the existing description, but to its radical revision (or at least to the notion that a radical revision is in order, since Packard's work is not conclusive, and is still being debated even as restorationists and managers are applying and experimenting with it throughout the oak opening region) (Fig. 21.2).

In the oak openings we encounter yet another natural or classic ecosystem subject to human influences that has changed dramatically during the past century and a half. In the oak groves, as on the prairies, the consequences of these changes have not been clearly understood. Here, as on the prairies, the problem is complicated by the fact that ecological change, as is usually the case, is due not to one but to a combination of factors. Fire is clearly a factor, and one of the most powerful influences on the oak openings, as on the prairies, has undoubtedly been protection from fire—mainly inadvertent protection, another subtle influence. This, however, is complicated by still other factors such as the introduction of exotic plants like the ubiquitous buckthorn, possibly by subtle changes in climate, certainly by isolation of areas by cornfields and developed landscapes, and probably by the replacement of native herbivores such as bison, elk, and turkeys by cows.

Such influences may be subtle in themselves, but they typically interact in ways that are even more subtle. In a surprising twist, one restorationist has recently presented evidence that the worst thing to do to an oak opening is to *remove* cattle from it following a period of grazing in the absence of fire (Bronny 1989). Taking away fire is bad, Bronny believes, but light grazing may retard deterioration in the absence of fire. It is the cessation of fire followed by the cessation of grazing—two subtle influences, including one likely to be carried out in an attempt to rescue and restore the system—that together constitute what Bronny calls the "one-two punch" that will finally obliterate the oak opening.

Figure 21.2. Efforts to restore native ecosystems continue to lead to new insights into their ecology and the nature of human influence on them. These volunteers clearing understory brush from a grove of oaks in a forest preserve in Chicago are participating in restoration efforts that have recently raised fundamental questions, challenging existing ideas about this ecosystem and its historic interaction with human culture. (Photo courtesy of The Nature Conservancy of Illinois.)

Again, as Bronny's work shows, the way to get at this, to discern the subtle influences at work on and within the system, is to attempt to counteract and reverse them. Only in this way, perhaps, will we learn enough about ourselves and our relationship with the natural landscape to be able to ensure a place for communities such as the prairies and oak openings in the landscape of the future.

Recommended Readings

Bronny, C. (1989). One-two punch: grazing history and the recovery potential of oak savannas. *Rest. Manage. Notes* 7:73-76.

Curtis, J. T. (1959). *The Vegetation of Wisconsin*. University of Wisconsin Press, Madison, Wisconsin.

Curtis, J. T. and M. L. Partch. (1948). Effect of fire on the competition between bluegrass and certain prairie plants. *Am. Midl. Nat.* 39:437-443.

Jordan, W.R., III, M.E. Gilpin, and J.D. Aber. (1987). *Restoration Ecology: A Synthetic Approach to Ecological Research*. Cambridge University Press, Cambridge, U.K.

Packard, S. (1988). Restoration and the rediscovery of the tallgrass savanna. *Rest. Manage. Notes* 6:13-22.

22
Biosphere 2 and the Study of Human/Ecosystem Dynamics

Mark Nelson, Edward P. Bass, and Linda Leigh

Introduction

Biosphere 2 is a materially closed, 1.28 hectare facility in Arizona designed for research ranging from organismic to systems ecology, including opportunities for management by humans of a total ecological system. It was also designed to serve as an educational tool toward the understanding of Earth processes and the consequences of human actions, and for the development of technologies, products, and know-how for commercial application and exploitation. Biosphere 2 is financed through venture capital investment, and it is expected to spawn many discoveries with application not only to environmental and ecological problems facing our planet today, but also to future human activities in space.

Biosphere 2 contains seven biomes within its 180,000 m^3 volume—rainforest, savanna, desert, ocean, and marsh ("wilderness biomes"); intensive agriculture and human habitat ("anthropogenic biomes"); and an extensive and complex mechanical system to assist in such functions as atmospheric and water circulation and heat exchange. The more than 3,000 species (excluding soil microbiota) introduced to Biosphere 2 were selected over a period of five years of research and development. After closure in September 1991 for an initial two year experiment trial, there has been very little material exchange between Biosphere 2 and the Earth (Biosphere 1). Currently the air exchange (atmospheric leak rate) is estimated to be less than 10% per year—far tighter than any experimental life support facility ever constructed (Dempster 1992). The facility is open to energy exchange, principally the input of solar and electrical energy and the removal of excess heat, and to a continual and substantial information exchange. A crew of eight people resides in Biosphere 2, fully supported by and essential to the operation and research of the system. All of the crew's food is

Table 22.1.   Areas and volumes of Biosphere 2.

Biosphere 2 areas	Square feet	Square meters	Acres	Hectares
Glass surface footprint	170,000	15,794	3.90	1.58
Intensive agriculture	24,020	2,232	0.55	0.22
Habitat	11,592	1,077	0.27	0.11
Rainforest	20,449	1,900	0.47	0.19
Savannah/ocean	27,500	2,555	0.63	0.26
Desert	14,641	1,360	0.34	0.14
West lung (airtight portion)	19,607	1,822	0.45	0.18
South lung (airtight portion)	19,607	1,822	0.45	0.18
Total airtight footprint	137,416	12,766	3.15	1.28
Energy center	30,000	2,787	0.69	0.28
West lung (weathercover dome)	25,447	2,364	0.58	0.24
South lung (weathercover dome)	25,447	2,364	0.58	0.24
Ocean water surface area	7,345	682	0.17	0.07
Marsh surface area	4,303	400	0.10	0.04

Biosphere 2 volumes	Cubic feet	Cubic meters
Intensive agriculture	1,336,012	37,832
Habitat	377,055	10,677
Rainforest	1,225,053	34,690
Savannah/marsh/marine	1,718,672	48,668
Desert	778,399	22,042
Lungs (at maximum)	1,770,546	50,137
Total	7,205,737	204,045

grown, and air, water, and waste are recycled inside the facility.

The material closure of Biosphere 2 builds on such ecological tools as micro- and mesocosms and the study of natural watersheds, and affords the important opportunity of tracking such aspects of an ecosystem as nutrient and atmospheric balances and interactions more comprehensively than can be achieved in natural systems which are open atmospherically and linked with the vast buffers and reservoirs of Earth's environment. Material closure has necessitated the development of technologies to recycle organic waste products, purify air of potential toxins, and to provide a non-polluting intensive agricultural system. Biosphere 2, with a design goal of one hundred years of operation, is intended to afford the opportunity for significant long-term studies of the maturation and dynamics of its ecosystems, and to develop intelligent human management and technological practices.

System Description

Biosphere 2 is an ecosystem laboratory of a new kind. It is essentially materially closed to exchange with the outside atmosphere and edaphosphere by a stainless steel liner and a glass and spaceframe superstructure. Within its 1.28 ha air-tight footprint are two wings of the main structure (Table 22.1). The east wing contains rainforest, savanna, and desert terrestrial ecosystems, and estuarine marsh and ocean aquatic ecosystems. The west wing contains the intensive agricultural area for human food production and wastewater recycling, and the human habitat which contains offices, residences, workshops, laboratories, and recreation/relaxation areas. Two variable volume chambers ("lungs") allow for expansion and contraction of the internal air volume (Fig. 22.1). An energy center external to Biosphere 2 generates electricity and provides thermal control of the facility's tropical biomes through energy exchange using an isolated closed-loop piping system (Allen 1991; Allen and Nelson 1988; Nelson and Soffen 1990; Nelson et al. 1993).

Biosphere 2 is the first attempt to create an indefinitely operating life support facility using bioregeneration. All of the eight person crew's food, air, and water is produced and recycled inside the facility. A major aspect of the experiment of Biosphere 2 is to test whether ecosystems capable of stable operation and evolution can be created using collections of biotic materials from the natural analog ecosystems. Temperature ranges in the internal ecosystems are maintained through a series of air handler units in the basements of Biosphere 2, which exchange heat with cooled or heated water passed through the facility. The air the air handlers emit over the ecosystems can be adjusted depending on day/night and seasonal requirements, and the air flow of up to 4-5 km·hr permits wind pollination and

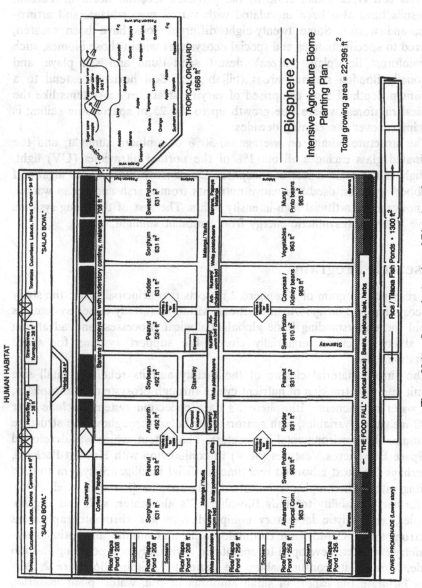

Figure 22.1 Isometric drawing of Biosphere 2.

facilitates general air circulation within the enclosed structure (Dempster 1989; Nelson 1990).

Soils have been made for the various ecosystems predominantly out of Arizona soil types found close to the project's location north of Tucson. The soils have also been inoculated with varied mycorrhizae, soil arthropods, and worms. Some twenty-eight different soils have been created, tailored to specific biomes and special ecosystems within those biomes, such as rainforest floodplain (varzea), desert sand dune and salt playa, and seasonally flooded savanna areas (billabongs). Soil horizons extend to a maximum depth of 5 m, comprised of varying strata, in ecosystems like the tropical rainforest, where tree growth up to the 25 m spaceframe ceiling is anticipated over the coming decades.

The structure admits on average 40-50% of ambient sunlight, and the laminated glass excludes all but 1% of the normal ultraviolet (UV) light. Sunlight is the main driver of Biosphere 2's photosynthesis. The "algal turf scrubber" systems, used to remove nutrients from marsh and ocean waters, are powered by artificial high-intensity lights. The rest of the living systems derive their photosynthetic energy from ambient sunlight.

Research Program

The research program of Biosphere 2 reflects two principal goals: the study of ecosystem and biospheric interactions within materially closed systems as a tool for understanding basic global ecological processes; and advancing our ability to create materially closed life support systems for space habitation.

The virtual material closure of the facility and its relatively small size permit intensive tracking of nutrient cycles through ecosystems, atmosphere, and water components. Biosphere 2's data collection systems include over 2,000 measured variables, with sensors distributed throughout the structure to monitor key environmental parameters. The system, which was developed by Space Biospheres Ventures (SBV) in conjunction with Hewlett-Packard, is perhaps the most advanced real-time artificial intelligence system to date dealing with both complex technical and life systems (Petersen et al. 1991). To further the ability to study Biosphere 2's air, water, soil, and organic samples, an analytic laboratory equipped with gas chromatograph/mass spectrometer (GC/MS), ion chromatograph, and atomic absorption spectrometry has been developed to operate in a virtually non-polluting fashion inside. Most of its consummables are generated within Biosphere 2. Airlocks facilitate the export of small amounts of soil, water, plant material, and air samples as required for outside analysis and research with a minimum exchange of air. At the end of the initial two year closure, and at subsequent crew exchange intervals, researchers can use the airlocks to enter Biosphere 2 to conduct more detailed research projects.

Biogeochemical Cycling Studies

Carbon cycling is a major factor in ecosystem dynamics, and because of the increase of carbon in our planet's atmosphere, an issue of overriding environmental concern. Its management inside Biosphere 2 is particularly important because of the rapidity of cycling occasioned by the small reservoirs and high ratio of biomass to atmosphere of the system. The concentration of living biomass inside Biosphere 2 is some 100 times greater in relation to the atmosphere than in the global environment. Soil organic carbon in proportion to the atmosphere is thousands of times greater This has reduced mean residence time of CO_2 in the atmosphere from an average of about three years for Earth's biosphere to about four days in Biosphere 2. The diurnal and seasonal fluxes of CO_2 are also much greater. During the first year of closure, daily fluxes were as great as 600-800 ppm; and while December 1991 saw a monthly average CO_2 concentration of about 2,400 ppm, in June 1992, during the maximum sunlight period, the monthly average dropped to about 1,050 ppm.

SBV and researchers at the Yale School of Forestry and Environmental Studies have begun collaborative work aimed at documenting the carbon budget of Biosphere 2—the distribution of carbon living and non-living elements at initial closure throughout the system (Petersen et al. 1992). Fluctuations are being tracked by periodic biomass measurements made by the crew during the first two-year closure experiment and a more detailed examination on occasions when other scientists enter the system, for example on occasions of crew change, during the facility's designed hundred-year life. An unanticipated decline in atmospheric oxygen, from an initial level at closure of about 20.9% to 16.25% (August 15, 1992) has sparked a collaborative investigation with geochemists from the Lamont-Doherty Laboratory to determine the soil reactions responsible through analysis of oxygen isotopes and other methods. Similar studies are planned for other major nutrient cycles, such as nitrogen, phosphorus, etc., along with studies examining specific mechanisms and dynamics.

Restoration Ecology

The study of Biosphere 2's micro-ecosystems include questions relevant to human impact and possible restoration of natural ecosystems (see Jordan, Chapter 19, this volume). For example, the half-acre tropical rainforest is testing a strategic plan to minimize the problems associated with harsh sunlight, a problem faced in natural areas after road-building or timber clearcutting. With consultation from Dr. Ghillean Prance, director of the Royal Botanic Gardens at Kew, a strategy of planting fast-growing trees and a border of sun-loving plants from the Zingerberacae (Ginger) family is providing the shade required by the lower canopy rainforest species. Thus, there is the opportunity for study of a planned succession from species which will form the first canopy, to those which will dominate when the biome reaches maturity. This long-term study of Biosphere 2 will also facilitate examination of the viability of small plant and animal communities, to see what introductions of genetic material will be required for plant and animal communities to retain sufficient genetic diversity. Baseline studies of a number of key plant and animal species are currently being planned after a genetics workshop at Biosphere 2 in order to follow founder effects in these small gene pools.

Atmospheric Regulation

Given the recent perception that human activities are severely affecting atmospheric parameters such as CO_2 and other greenhouse gas buildups (e.g., smog, acid rain) there is considerable theoretical interest in how atmospheric processes are regulated. Biosphere 2 is a laboratory where the biological impact on air parameters can be closely examined. This is especially true since many other factors that influence the global environment are absent. For example, Biosphere 2 will lack the high stratosphere where incoming UV light energizes a number of chemical reactions. In addition, the high ratio of biomass to atmosphere, and small reservoirs insure a most rapid biological impact on air composition. During a calendar year in Biosphere 2, there will be almost a hundred cycles of CO_2 from the atmosphere into living or water systems, and back into the air again. This ability to track cycles which have an atmospheric component is a fundamental difference between studies in natural ecosystems, which are atmospherically open, and those in materially-closed ecological systems.

Carbon dioxide and methane are leading choices for early study because of their role in creating the greenhouse effect. Response of varying ecosystems and individual species to heightened levels of such gases will be important in predicting possible feedback mechanisms. Biosphere 2 includes several systems that are known sources for methane generation—marsh ecosystems, rice paddies, and aquatic plant waste-treatment areas. But, as opposed to the current global situation, the crew of Biosphere 2 will be responding and actively managing any health-threatening buildups of atmospheric components. This is done in several ways. One is through employment of the "soil bed reactor" technology developed for air purification in Biosphere 2.

Soil bed reactors function by forcing air through the microbial communities of actively functioning soil (Bohn and Bohn 1986; Carlson and Leiser 1966; Bohn 1972). These beds are constructed, prepared, inoculated, and maintained so as to optimally expose trace gases to enhanced and diverse microbial populations. In preliminary studies conducted for SBV at the Environmental Research Laboratory of the University of Arizona and in the

Biosphere 2 Test Module, soil bed reactors were shown to be effective at controlling levels of gases from both biological and material sources (Hodges and Frye 1990; Alling et al. 1993; Nelson et al. 1991). Literature searches (Hodges and Frye 1990) and previous research indicate that a wide variety of gases can be metabolized by soil microbes. The entire air volume of Biosphere 2 can be pumped through the soil bed reactor in less than a day's operation (during the first 11 months of operation, trace biogenic and technogenic gases have remained at low enough concentrations that use of the soil bed reactors has not been required). Soil bed reactors are one of a number of innovative technologies in which SBV holds patents as a result of its research and development for the Biosphere 2 project.

In addition to deciding on the operation of the soil bed reactors, the crew manages the growth rates of the biomes in response to atmospheric needs. For example, when carbon dioxide levels are high, as occurred during the first fall/winter, the savanna and desert biomes are kept active longer than their normal seasons. Cutting of grasses or annuals in these biomes also encourages continued plant growth, while cut materials are stored dry to slow their decomposition and release of CO_2. Should CO_2 levels be low, the opposite strategy may be employed by limiting rainfall in those systems or disturbing the soil to encourage soil respiration. Manipulation of day/night temperatures is another technique for managing rates of photosynthesis and soil and phytorespiration. Thus, the human crew can help manage the systems of Biosphere 2 in response to air quality conditions.

Fluxes of CO_2 and predominance of system respiration over photosynthesis are anticipated to be greatest during the first years of Biosphere 2's operation because the ecosystems are still growing to maturity, while there is increased respiration from the recently installed soils. To help buffer these fluxes and seasonal swings in CO_2, a recycling physio-chemical system was also developed for Biosphere 2. It operates by reacting CO_2 from the air in a sequence of reactions with sodium hydroxide and calcium oxide to precipitate limestone ($CaCO_3$). It is possible to regenerate the starting chemicals, and heating the limestone re-releases the CO_2 back into the atmosphere when required. Failure of the heating element in the furnace has prevented the release of the CO_2 from the $CaCO_3$ made during this first two-year experiment. This material accounts for almost 1% of the decline in atmospheric oxygen of the system.

Water Systems

Biosphere 2 offers a chance to expand on the kind of watershed study pioneered in the Hubbard Brook Experimental Forest in relation to water systems (Likens et al. 1977; Bormann and Likens 1979). There are a number of water subsystems in Biosphere 2, including streams in savanna and rainforest, marsh and ocean tides, potable water, wastewater treatment, and irrigation water. The latter is principally collected from condensation, which is then combined with leachate containing subsoil drainage collected separately from and returned to each of the biomes.

It is calculated that an additional 757,060 liters of water will be incorporated in living biomass as Biosphere 2 progresses from a relatively juvenile to a mature system. It is estimated, for example, that the rainforest at initial closure contained only some 5-10% of the biomass it will contain when it reaches maturity. So, a reservoir of water for that purpose is being held in a tank incorporated into the south lung.

To mitigate the kind of perturbation of aquatic systems that human and domestic animal waste generate, SBV has developed a bioregenerative method for recycling and reclaiming nutrients from domestic habitat water. This method builds on previous work in the field which has shown the ability of such systems to concentrate toxins and heavy metals, and to purify human domestic graywater and sewage (Hammer 1989; Wolverton 1986; Wolverton and McDonald 1979). In Biosphere 2's system, regeneration is accomplished by first holding such waste water in tanks for anaerobic digestion, followed by circulation through aquatic plant/microbial lagoons (Fig. 22.2). The biomass increase of the lagoons is harvested for forage or inclusion in the composting process. Thus, nutrients derived from the agricultural soil are returned there. A similar system will handle any chemical/heavy metal emissions should accidents occur in workshops or biomedical and analytic laboratories inside the facility (Dempster 1988).

To help balance the water systems of the wilderness biomes, human technology must complement some natural functions. For example, because of its small surface area, evaporation from the ocean surface returns only 10% of the fresh water input from the estuarine stream. Therefore, a flash evaporation desalinator unit is operated to restore the required fresh water to the system. The Smithsonian Marine Systems Laboratory (MSL) consulted on the design and engineering of Biosphere 2's marsh and ocean systems. To maintain the low nutrient levels in the ocean required for the health of the coral reef community, SBV and MSL have designed algal turf "scrubbers" powered by 16 h·day^{-1} of high intensity lighting and continuous water pumping to complement the natural algae found in the ocean (Fig. 22.3).

Genetic Diversity

Biosphere 2 may yield interesting data on the importance of diversity in ecosystem operations. Its design has anticipated a probable loss of species over time and, therefore, followed a strategy of "species-packing" and deliberate redundancy of ecological function. All introductions have been documented in a computer bio-accessions system, and baseline work has

Figure 22.2.   Aquatic/plant microbial lagoons for wastewater recyclying in basement of Biosphere 2. Photo by D.P. Snyder

Figure 22.3.   Algal scrubber room, with high intensity artifical lighting, in Biosphere 2. Photo by Gill Kenny.

included the mapping and measurement of all plantings in the biomes. Studies of species loss, expansion, and phenology will be relevant to determinations of the role of ecological self-organization in shaping what the Biosphere 2 systems will become over time. Within biomes, how well the planned landscapes persist, and the functioning and persistence of food webs will also be foci of research. Biosphere 2 offers a unique opportunity to observe to what extent "island" effects lead to ecosystem and species simplification and loss in terrestrial and aquatic ecosystems.

Interaction of the Biomes: Human Role

Biomes (also called biogeocoenoses) play a key role in the structural organization of the biosphere. The Russian biologist M.M. Kamshilov recognized the

"ability [of biomes] to withstand various external effects...[due to their] homeostasis or buffering power. There seems to be a direct relationship between the complexity of biocoenosis and its ability to withstand diverse external effects...greater resistance not only to intrusion of individual species from different ecosystems but also to abiotic factors...The stability of the biosphere as a whole, and its ability to evolve, depend, to a great extent, on the fact that it is a system of relatively independent biogeocoenoses...[which] compete for habitat, substance, and energy provides optimal conditions for the evolution of the biosphere as a whole" (Kamshilov 1976).

Biosphere 2 will afford a unique laboratory for studying the interrelationship of biomes, and how their interplay affects overall system balances. In addition, the influence of eight human beings, including their technology and agriculture, can be studied for their impacts on individual ecosystems and overall system dynamics.

The link between the health of the crew and the health of their enclosed environment will be much more immediate and quantifiable in Biosphere 2 than in the global environment. For example, studies are underway of the crew's blood chemistry and toxicology using pre-closure and post-closure samples. Since there is careful monitoring of the water and air quality the crew is exposed to inside Biosphere 2, it will be possible to see the effects on their blood systems of chemicals they have been previously exposed to as well as trace compounds presently found in the facility.

The crew members of Biosphere 2 are interacting with their biosphere in some markedly different ways compared with the global situation. First, care has been exercised to eliminate potential problems, such as pollution from agrichemicals, and to minimize others, such as trace gas emissions, by selection of natural or less toxic materials whenever possible. The quantity of paper and other potential solid wastes is restricted; fire and open flames are also forbidden by Biosphere 2 Mission Rules to minimize air pollution. The humans are restricted from expanding into the wilderness biomes. Apart from an occasional bowl of fruit or some specialty items like coffee beans from rainforest trees, human food will come from the agricultural crops. Even such subtle impacts as the creation of paths must be carefully regulated by the crew, given the small size of the ecosystems being traversed. The goal will be for the crew to function as discretionary top level "predators" (keystone predators; Paine 1966) when necessary in the biomes. This will include culling when predators are overtaxing their prey (as has already been done with parrotfish and lobsters in the ocean) and intervening when needed to weed invasive vegetative growth or to prevent vegetation from interfering with mechanical or structural systems. However, should a fish cull from the ocean go to the kitchen, or a grass cutting from the savanna go to domestic animals, a restoration of the biomass will be required by returning nutrients to the wilderness in an appropriate fashion. Similarly, the crew will regularly use soil and water tests in the biomes to determine where to distribute nutrients removed from marsh and ocean waters by the algal scrubber systems. The fundamental currency in the economy of Biosphere 2 is its material cycles. The humans must intelligently intervene to restore and maintain biogeochemical cycles if Biosphere 2 is to continue to function well over time.

Biosphere 2 as Cultural Landscape: Determining Ecosystem Response to Unusual Given Conditions

Biosphere 2 is in an important sense a human-apexed system but not a human dominated ecosystem. The areas patterned on natural ecosystems have been included to study their role in maintenance of overall biogeochemical cycles, and to investigate to what extent they will function comparably to their natural analogues. These ecosystems play crucial roles in the persistence of Biosphere 2, although they are dependent on technology to provide appropriate environmental parameters. In this sense, Biosphere 2 represents a unique symbiosis of ecology and technology. While it provides a laboratory for studying how ecological systems self-organize, testing the relationship between biodiversity and stability, it also offers a chance to further our capacity to develop technologies that support rather than degrade the environment. Perhaps Biosphere 2 indicates pathways, new types of cultural landscapes that may resolve current dilemmas posed by an apparent conflict between humans and their technologies, and our global biosphere.

Part of the ecological interest in Biosphere 2 stems from its attempt to create ecosystems that parallel natural ones, with some environmental vectors that are significantly different. What effect will the exclusion of certain ultra-violet radiation inside Biosphere 2 have on biological systems? What will phenology studies show about the adaptation of species to conditions which vary from their natural limits? Biosphere 2 has already transplanted Everglades marsh and Caribbean/Yucatan coral reef with very small losses despite the sharply reduced overall light levels. The Amazonian species now housed in Biosphere 2's tropical rainforest customarily reside under an equatorial diurnal light cycle rather than that found at 33° N. Many of the insects and other animals of the terrestrial systems will have access to biomes and plant species far distant from their native habitats; how will their behavior change in response to a new range of ecological opportunities? (For instance, the normally arboreal prosimian galagos, *Galago garnetii*, have ventured past the savanna's gallery forest to explore the treeless lower savanna and into the desert area in response to food sources or simple curiosity, and have even been seen exploring technical basement areas of Biosphere 2.)

While precise information on plant pollination has been sought through literature searches and consultants as a part of the process of plant selection, there are quite a number of plant/pollinator relationships in Biosphere 2 about which little is yet known. Close observation will clarify which of the predominantly generalist insect and bird pollinators selected for Biosphere 2 perform these functions for selected species.

The capacity for people to intelligently create ecosystems, and to manage and mitigate impacts upon natural ecosystems, may be enhanced by experience from Biosphere 2.

Sustainable, Non-polluting Agriculture and Micro-City

While the designers of Biosphere 2 have sought to model the wilderness biomes on analog natural areas, it has been necessary to avoid imitation of current predominant agricultural and urban waste-disposal practices in the design of the agricultural systems and micro-city (human habitat) of Biosphere 2. Typical high-input chemical agriculture would be untenable because of its pollution of air, water, and soil, and monocultural bias. Peasant agriculture would be untenable because of the long hours required. Similarly, the pollution and sewage dumping of modern urban areas would soon threaten the health and nutrient cycles of Biosphere 2.

The requirements of Biosphere 2's intensive agricultural biome are to remain indefinitely sustainable and highly productive (complete nutrition for eight people year round) without the use of chemical products. This requires an agricultural system that maintains soil fertility by returning nutrients to the soil, and one which can cope with a wide variety of potential pests and diseases through methods such as beneficial insects, crop rotation, intercropping, etc. Since the internal climate of Biosphere 2 is

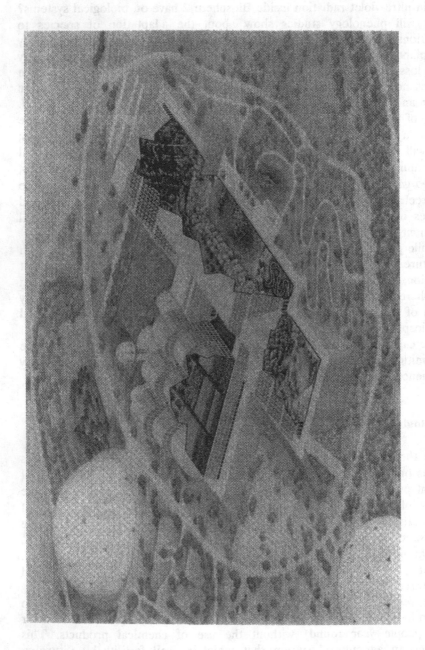

Figure 22.4   Typical planting plan of Biosphere 2 agriculture. Drawing by Elizabeth Dawson.

semi-tropical (13-35°C except in the desert, where winter temperatures mayget as low as 4.4°C), such a system may be of greatest applicability to developing countries in the tropics, which at present are food-impoverished, least able to afford high-input agriculture, and which have great problems dealing with the recycling of human wastes.

Biosphere 2's agriculture employs a total of over a hundred crop cultivars (Figure 22.4). Some of its major elements are: the Tilapia fish-rice-*Azolla* (water fern) paddies; fodder intercropping of *Leucaena*-elephant grass and siratro; a variety of tropical fruit crops; root crops like white and sweet potatoes, taro, and peanuts; a variety of grains including wheat, barley, oats, amaranth, and sorghum; and temperate and tropical vegetables. For milk, meat, and eggs, there are pygmy African goats, Ossabaw feral swine, and jungle fowl. A variety of mechanical equipment assists in harvest and food processing to minimize labor requirements (Glenn et al. 1990; Leigh et al. 1987).

The detailed study of these cropping techniques, waste recycling systems, air purification, and biological control of pests will also be of interest to the scientific community, especially in the light of the reexamination of "alternative agriculture" systems called for in the recent U.S. National Research Council report (NRC 1989). The development of a non-polluting, sustainable agricultural system is one of the major differences between Biosphere 2 and the global arena, where agriculture expands to cover wilderness areas to feed urban populations, sewage pollutes aquatic and marine environments, and agrichemicals cause environmental problems in farm areas.

Summary

In the operation of a closed biospheric system such as Biosphere 2, the limits of the system in terms of such aspects as energy exchange, productivity and the capacity to sustain a variety and quantity of life are dramatically evident and pose constraints which must be dealt with in a practical and immediate fashion. Limits and constraints inherent in our planet's biospheric system are no less real, but are less immediately and readily apparent. The role of the human species and impact of our activities within Biosphere 1 (Earth) is a matter of substantial consequence, and warrants the highest level of attention in terms of both science and policy.

Biosphere 2 affords a valuable laboratory in which to develop our understanding of these issues. In addition, it can function as a test bed for mitigation of human impacts through the development of appropriate bioregenerative technologies, and provide insights into the dynamics of ecosystem and biospheric processes. It is crucial to understand the impacts of humans upon ecosystems and our overall biospheric life support, whether those impacts be deleterious or beneficial, inadvertent or intentional, irresponsible or through intelligent management practices. Biosphere 2 may also be a test bed for viewing the impacts of ecosystems on humans—since in this miniature world the close coupling of ecosystem/biosphere health and human health and well-being is dramatic and immediate.

Recommended Readings

Allen, J. (1991). *Biosphere 2: The Human Experiment*. Penguin Books, New York.

Glenn, E., C. Clement, P. Brannon, and L. Leigh. (1990). Sustainable food production for a complete diet. *HortScience* 25:1507-1512.

Hammer, D., ed. (1989). *Constructed Wetlands for Wastewater Treatment: Municipal, Industrial, and Agricultural*. Lewis Publishers, Boca Raton, Florida.

Kamshilov, M.M. 1976. *The Evolution of the Biosphere*. Mir Publishers, Moscow.

Nelson, M. and G.A. Soffen, eds. (1990). *Biological Life Support Systems*. Synergetic Press, Oracle, AZ. Reprinted from NASA, Office of Management, Scientific, and Technical Information Divison, Proceedings of the Workshop on Biological Life Support Technologies: Commercial Opportunities, M. Nelson and G.A. Soffen, eds. (1989).

Section IV   Overview

The book has addressed a complex, problematic, and assumption-laden topic, the great question of George Perkins Marsh (1864), "Whether man [sic] is of nature or above her?" Because the question is so fundamental and has stimulated the founding of new disciplines, it is not surprising that it ripples down to our own time. To put the question in the terms we introduced in the title of this book, what is the human role in ecosystems? What are the subtle effects of humans on ecological systems (individuals, populations, communities, ecosystems, landscapes, etc.)? What is the nature of ecosystems in human populated areas? Because mainstream ecology has so often ignored all but the most blatant human effects, or avoided studying populated areas altogether, the question is still an open one for the discipline of ecology. The problem is large enough that we have provided two kinds of overview. The first (Chapter 23) is a compilation of conclusions from the 1991 Cary Conference, where this book was conceived. The second (Chapter 24) is the editors' collection of highlights and presentation of themes that emerge from the first 22 chapters of the book.

The diverse contributors to the 1991 Cary Conference brought a diversity of perspectives and data to bear on Marsh's great question and its answer. We asked four researchers to share their insights from the conference. Two are trained in the social sciences, and two in "biological" ecology. Padoch and Borden each represent the perspective of human ecology or ecological anthropology. Both commentators are social scientists by early training, but have been active in the relatively new "bridge" discipline of human ecology. The remaining two views are ecological. Castilla, a marine ecologist, clearly understands the traditional ecological viewpoint, but has, in addition, interacted with researchers in other disciplines to deepen and broaden the understanding of ecosystems by taking humans into account. The final commentary is provided by Pacala, a theoretical ecologist, who has also pursued empirical work in the real world, where humans exist. The four commentaries comprising Chapter 23 are purposefully personal and impressionistic accounts.

The other essay (Chapter 24) is our own attempt to abstract some of the larger ideas that emerge from or run through the book. We hope that some of the excitement and sense of cross disciplinary co-discovery and promise that prevailed at the 1991 Cary Conference comes through in the overviews. There is great opportunity for ecology to grow by understanding of subtle human effects and the effects of human populations on the landscapes and ecosystems they inhabit or visit.

23
Part I: A Social Scientist's Perspective

Richard J. Borden

This important and timely fourth Cary Conference on "Humans as Components of Ecosystems" addresses some of the most challenging issues of our time. Its main purpose has been to enlarge our understanding of human affairs within their relevant ecological contexts. By combining both human and ecological themes, of course, it assured that our exchanges would inspire a special openness to complexity, updated with contributions by leading researchers and practitioners from diverse backgrounds. This unique gathering of ecological science, human studies and policy, and planning and decision-making perspectives is a necessary condition for a *bona fide* understanding of living systems. The combination is a rare opportunity to coalesce the contributions of the various disciplines represented in this volume.

I am honored and delighted to be a participant and summarizer. Before I make some short comments, I would like to briefly clarify my prior involvement with the general topic of this meeting.

By training, I am a psychologist with specialties in the structures of thought, identity formation, and personality change. My post-doctoral studies were taken in an entirely different field—animal behavior and ecology. I have combined these interests through research on the development of an ecological perspective, and especially the relationships between ecological understanding and action.

Bringing these two approaches together within a conventional academic setting was always a challenge. About ten years ago, these conflicts were relieved when I took a position at College of the Atlantic (COA), a small private college located on the Maine coast. The college was founded in the late 1960s to provide interdisciplinary studies of complex human and environmental problems within a human ecological perspective. Its academic program relies on a non-departmentalized faculty as a resource to students who design individualized programs leading to the B.A. and M. Phil. in Human Ecology. In addition to my teaching and research, I am also responsible for academic coordination as Academic Dean.

In recent years, human ecology education and research programs have developed in many other places. On behalf of COA and also as President of the Society for Human Ecology, I have organized meetings and workshops and visited many institutions around the world. In these various roles, I have witnessed numerous attempts to put something around the ideas which are central to this Cary Conference. It is against this backdrop I will extend a few observations. And in keeping with the request of the editors, my reflections are impressionistic and extemporaneous.

Overall, the meeting and resultant volume have brought together a remarkable range of viewpoints, combining detailed scientific studies with social science and humanistic understanding. For myself, there were several presentations of extended scope that got squarely at the heart of the issue. Jim Kitchell's and Steve Carpenter's review (Chapter 10) of inland fisheries was a captivating illustration of the interaction of ecological processes and human intentions. It disclosed the inherent unpredictability of these two realms and at the same time set the attitude of understanding needed for intelligent management and planning. Wolf Grossmann's use of multiple methodologies (Chapter 18) was likewise impressive. He showed the value of working directly with people on complex systems, not only in the role of scientist-as-observer, but also as an active catalyst for regional self-understanding.

Some additional studies that paid attention to special human qualities in other settings included: 1) Bill Turner's and William Meyer's emphasis (Chapter 4) on the role of meaning, attitudes, and values; 2) Pete Vayda's focus (Chapter 6) on individual intentions and actions—and the actual consequences of those behaviors—as a rigorous basis for contextual analysis; and 3) Bill Jordan's (Chapter 21) use of the critical aspects of the humanities extended to ecological problem-solving. All of these reminded us that the human dimensions of ecosystems will unavoidably stretch the limits of solely scientific descriptions. Taken together, these studies revalidate the central problem of complex open systems: the inevitability of surprises; the necessity of research and planning methodologies that are broadly interdisciplinary; and the value of integrating formal academic expertise with local knowledge and interests.

In another vein, I noticed here—as on numerous other occasions—a problem with inchoate assumptions, cultural truisms, and teleological "ghosts." It appeared on the first day in discussions of nineteenth century science as a still-lingering uncomfortableness that lack of predictability is a sign of "softness," and that this is somehow a problem for ecology. I think it is long overdue that scientists, and the public as well, recognize that the nature of explanation for living systems is importantly different from nineteenth century conceptions of hard science (Drury 1991; Visuader 1986).

Living systems are alive. They are open to uncertainty and new arrangements; it is the study of this openness that makes an ecological perspective both interesting and necessary. Good ecological science invariably will have a large element of unpredictability. While ecosystem studies strive for thoroughness, they always will be subject to novelty and uniqueness. Indeed, it may be important to guard against false precision or the apparent need for it.

Another problem of systems thinking is a predilection toward implicit "intelligence" or "intentionality." As Jane Hall showed us (Chapter 5), Adam Smith's "invisible hand" still appears very much alive in neo-classical economic thinking and models. The approach as a whole is buoyed by buried assumptions, blind optimism, and unarticulated beliefs that growth and progress are natural and right.

A similar bias in popular thought and in formal policy was displayed again through Fred Wagner's and Charles Kay's analysis of the management of Yellowstone National Park (Chapter 20). Assumptions that "nature knows best"—that it is somehow goal-directed toward diversity or a "correct" ecology—are, and will continue to be, difficult obstacles. These beliefs lie at the core of many peoples' fundamental conceptions of the world. They are exceedingly difficult to examine openly or rationally (see Egerton, Chapter 2).

Comparable problems appear routinely in academic interdisciplinary forums as well, when narrow disciplinary descriptions of causation meet and clash at their boundaries. This is particularly true for explanations of human action—not only along classical scientific and humanistic lines, but also among the social sciences themselves.

In conclusion, there are three general points I would emphasize. First is the problem of institutional structures and the corresponding impediments to integrative study. Throughout the arena of education and research, we are confronted with a hard architecture of knowledge. Universities are physically constructed in ways that make it difficult for disciplines to work together or to find support for collaboration. Universities offer a universe of perspectives but they seldom provide a forum for the useful integration of these perspectives. Moreover, people cannot just come together briefly; they must have the opportunities and support to stay together if they are to work effectively on these complex issues. It is only through continuing collaboration that the strength of our individual disciplines can be woven into a broadened perspective. Alfred North Whitehead (1929:4) said the aim of education is "the acquisition of the art of the utilization of knowledge." On ecological and human ecological themes, that art is necessarily a collaborative one. To realize these possibilities, we must create new educational and research opportunities and find lasting support for them.

Second, for the areas of applied research, planning, and decision-making, much further effort must be extended to the development of new methodologies. In many cases, it is not that we lack knowledge, but that we have yet to devise means to coordinate or apply what is available. In order to identify and solve the real problems that confront us, there must be many more occasions like the conference from which this book grew. We might hope that a future conference could launch this agenda with a benchmark conference on comparative and combinational methodologies.

Third, and last, I would like to say something about the conceptual, theoretical, and philosophical dimensions underlying this meeting. The influential psychologist Kurt Lewin (Marrow 1969:ix) once said, "There is nothing so practical as a good theory." But theoretical work is often devalued and unsupported; this seems to be doubly true for interdisciplinary and ecological studies. Nonetheless, there is a tremendous need for substantive conceptual work and for safe harbors to develop integrative frameworks.

One of the principal goals of the meeting was to set the stage for this kind of general theory of correction—not in the prescriptive sense, but as a laying out of current understanding and confusions, and as a scaffold for conceptual rearrangement and extension.

In earlier times, this was the work of philosophy. Unfortunately, philosophy in the modern world has become as specialized and self-referential as most other fields. It is exceedingly difficult to find a philosopher knowledgeable and interested in the interdisciplinary problems of ecology and human society. Nonetheless, it is for just this kind of comprehension we all are grasping.

Near the end of his life, Alfred North Whitehead (1941:664) wrote a letter to his publisher and said about "the progress of philosophy" that "it does not involve reactions of agreement or dissent," he said, "It essentially consists in the enlargement of thought whereby contradictions are transformed into partial aspects of wider points of view." At this meeting, we have certainly taken some steps in that direction; I trust we will continue—individually and together—through the enriching contacts we have made here.

Recommended Readings

Drury, W.H. (1991). Humans in ecological perspective. In: S. Suzuki, R.J. Borden, and L. Hens, eds. *Human Ecology—Coming of Age: An International Overview*, pp. 25-246. VUB Press, Brussels.

Marrow, A.J. (1969). *The Practical Theorist: The Life and Work of Kurt Lewin*. Basic Books, New York.

Visauder, J. (1986). Philosophy and human ecology. In: R.J. Borden, J. Jacobs, and G.L. Young, eds. *Human Ecology: A Gathering of Perspectives*, pp. 117-127. Society for Human Ecology, College Park, Maryland.

Whitehead, A.N. (1929). *The Aims of Education and Other Essays*. Macmillan, New York.

Whitehead, A.N. (1941). Explanatory note to introduce "The Philosopher's Summary." In: P.A. Schilpp, ed. *The Philosophy of Alfred North Whitehead*, pp. 663-665. Tudor, New York.

23
Part II: A Human Ecologist's Perspective

Christine R. Padoch

I came to this conference as an anthropologist with a mission: to convince the ecologists in the group that people have influenced all those ecosystems you study; to persuade you that those influences are often subtle and counterintuitive and cannot be dismissed as "disturbance"; and to assure you that we have much to gain by collaborating in research. On the very first day of this conference I realized that ecologists have no need of outside missionaries. Among the natural scientists present are many whose work treats human influences with great sophistication. Now, at the very end of this conference, I find that I still want to make several comments. They are, however, but amplifications of points made by others, and I offer them not as a missionary from social to natural science, but rather as a colleague with a few comments and a few data to share.

During the conference, interest was repeatedly shown in correlations between the size and density of human populations on a global or large regional scale, and major environmental changes. Although several authors eloquently emphasized that global demographic measures are of limited utility and can be misleading in trying to explain or predict environmental change (Turner and Meyer, Chapter 4; Richerson, Chapter 11, this volume), I would like to support these warnings with my own observations. People are an extremely complex, variable, and changeable lot. Merely counting them, measuring how dense they are on the ground, and how quickly they are increasing or decreasing tells you something interesting but very limited about human populations and how resources are used.

My own research takes me to small villages in tropical forests where conservative traditions and geographical isolation might be thought to limit diversity, variability, and change in resource use. I invariably find the opposite to be true. In one village along Peru's Ucayali river, my colleagues and I found resource use to be both amazingly complex as well as mutable. The farmers of the small, isolated settlement practice twelve distinct kinds

303

of agriculture. In 1985, the 46 constituent households of that village opted for 39 different combinations of those twelve kinds of agriculture. Some cut far more forest than others, some cultivated monocultures, some managed wonderfully diverse forest-gardens and some did both. In a subsample of those 46 families, we determined that 75% significantly changed their repertoire of agricultural practices from 1985 to 1986.

The differences among families in a traditional Amazonian village are undetectable and perhaps unimportant on a global or even regional scale. I use that very small scale because that indeed is the scale on which I study and try to understand the world. Patterns of great variety and rapid change in resource management, such as those I detected, are more characteristic than not of human groups of whatever size in both the present and in the past. Throughout the world, women and men, rich and poor, urban, rural, and suburban, and different ethnic groups use resources and generate impacts on their environments in significantly different ways. And absolutely no group does it "exactly as their ancestors did." It is that variability—even on the minutest scale—that is the stuff of evolutionary change and, at times, the source of our best solutions for the future. The simple, highly-aggregated picture tells us little about how the people on the ground differ among themselves and over time in how they use resources. Such data, therefore, can tell us little about what important changes can be wrought or what changes are already in the offing.

My second example again illustrates the limitations of simple population data in understanding people as components of ecosystems. In addition, I hope it suggests how complex the study of "subtle" causes and effects of humans in ecosystems can be. The Amazonian village I have mentioned is a very small community at the edge of a very large forest, some of it a managed agroforest. The population density of the province where it lies has been estimated at only about 1 person·km^{-2}.

On the other side of the world in Kalimantan (Indonesian Borneo), I am studying another community of traditional shifting cultivators in a district with 54 times that density of humans. These latter folk also live surrounded by agroforests of high species diversity. Viewed from above, much of their district seems forested. These diverse woodlands—in one plot we found 44 tree species in 0.2 ha and some plots doubtless include more—have extremely different histories. Some are truly "created" forests. They were once cleared land that was planted with many useful species and enriched by managed "spontaneous" arrivals (a history similar to the agroforests of the sparsely-settled Peruvian Amazon). Other plots in the high-density Bornean district were never clear-cut, but have been managed over decades or more using more subtle techniques such as seasonal selective weeding or occasional interplanting. Both types are species diverse and have the appearance of forests; both are rather subtle transformations of a tropical forest system. Yet, some resulted from subtle management interventions, others from "unsubtle" clear-cutting, burning, planting, and weeding.

As this example shows, human population density is often a poor predictor of the extent of ecosystem transformation or of the level of biological diversity. The appearance of a landscape may also be a poor indicator of its history. Despite the frequently gross nature of human alterations of the earth, the remotely sensed image, like highly aggregated census data, can be not only limited but also misleading in telling us of past resource management and, therefore, also in helping us cope with the future.

These comments and examples, finally, are meant to bolster the point made frequently at the Conference that the data and insights social scientists offer are indeed valuable to those who would understand ecosystems and predict directions of change. Perhaps in my ignorance I suppose that 100 voles in one place today use more or less the same volume of resources if not the same resources as 100 voles will use somewhere else tomorrow. Such predictability is obviously not true of people. Despite one call to very limited collaboration heard here, social scientists have more to contribute to ecological inquiry than more accurate data on numbers of humans. In true collaboration we can indeed enlighten each others' research with data and insights about which the other didn't even know to ask. Such understanding cannot result from assigning each other a limited and predetermined role, but from joining together in planning, design, execution, and interpretation of research. This is clearly a difficult assignment given our disciplinary limitations. In the course of Conference discussions we've all seen that differences exist among us in what we hold to be significant, in how we think science is best done, and in what kinds of data we deem valuable. These differences, however, do not neatly follow disciplinary divisions between social scientists and natural scientists. There are in each camp some who are more able and eager to consider the insights and interests of the other. As we get down more frequently to specific research questions and tasks, and farther away from disciplinary wisdom and prejudices, I sincerely believe that misunderstandings are sure to diminish and scientific progress must result.

23

Part III: A Marine Ecologist's Perspective—Humans as Capstone-Species

Juan C. Castilla

The contributions to the Conference all converge toward a single view of humans in trophic dynamics of communities. Mankind is the "critical species" at the top of all trophic webs. According to my view, there are numerous reasons for that, the most important of which is humans' capacity to learn. Humans have the ability and potential to build or to destroy, to sustain or to over-exploit, to conserve, to exterminate, or to accomplish a sound balance between their extremes. No other creature in nature bears such ecological power. In addition, humans are not only a critical or keystone predator species (*sensu* Paine 1966), but also a top competitor. Man uses technology and has the capacity for direction and decision making. In my research I have unified all ecological potentials of humans, depicting them as the "capstone species" of ecological systems.

In spite of their capstone role, humans (as an ecologically important component of communities) are usually left out or ignored in most ecological studies. Indeed, critical human roles in communities or ecosystems (taking into account human peculiarities) have not been truly incorporated, for instance, in present ecological modeling. Further, in applied fields such as fisheries, there is still a need for the incorporation of human behavioral aspects (but see Walters 1986; Kitchell and Carpenter, Chapter 10, this volume).

The neglect of human behavioral characteristics is less remarkable given that even direct, indirect, or subtle historical effects of humans on communities are seldom addressed in the ecological literature. The same can be said regarding lagged ecological effects or unexpected actions occurring at a distance. There is a wealth of ways in which humans need to be accommodated in ecological studies.

23
Part IV: A Theoretical Ecologist's Perspective: Toward a Unified Paradigm for Subtle Human Effects and an Ecology of Populated Areas

Stephen W. Pacala

In going through my notes, I came up with four kinds of recommendations that collectively, through the talks and the discussion, we as a body made for how to integrate humans further into ecological paradigms. These are recommendations about how research could be done right now—proximal kinds of research, research that in many cases would be similar to that already being done.

The first recommendation is that we need to further document how ecological processes and variables depend in particular on things that humans do. Now, we saw some particularly good examples of this sort of research. These examples looked across temporal and spatial scales, and this is important, of course, because ecologists and social scientists work inevitably at different scales because they're usually working on different kinds of things. One example of the global scale is the by now famous relationship between the world's population and carbon dioxide (Turner and Meyer, Chapter 4). The examples that we saw during the Conference included Mark McDonnell's and his colleagues' (Chapter 15) study of the gradient from natural areas in the Bronx, New York to rural Connecticut. David Wigston, inspired by Charles Ives, called this the gradient from "Central Park in the Dark" to "Three Places in New England." John Aber (Chapter 14) talked about another, an east-west gradient of atmospheric deposition. Jonathan Cole et al. (Chapter 12) talked about a large-scale study of rivers that integrated many spatial scales. We also saw some studies that went through long temporal transects: David Foster's (Chapter 9), Michael William's (Chapter 3), and Peter Richerson's (Chapter 11). So, there are a lot of examples of studies of the influence of human activities on ecological processes, but we need a lot more. There are surprisingly few examples in the ecological literature, at least given the importance of the problem at hand, and given the amount that there is to be done on this kind of relationship. We need sound documentation of the phenomena so we have a firm basis for explanation.

The second recommendation is that we need to turn that relationship over and look at other the other half of the ecology—the ecosystems to humans feedback. We have to look at how human decision making is affected by the state of ecological systems. Pete Vayda showed us an interesting example of this in land use in Borneo (Chapter 6). I thought actually the most interesting thing at the Conference that was said on the topic was the discussion of the Biosphere 2 program (Nelson et al. Chapter 22) because this is likely to be the single largest experiment ever done in the history of humankind to test the effect of ecosystem structure on human behavior. In particular, I'd like to urge Linda Leigh (a member of the team that will live in Biosphere 2) to live in a savanna. If the social determinists are right, then savanna dwellers will become belligerent and sequester all the carbon and all that sort of thing after a couple of years.

Now, the third recommendation deals with what were the most exciting studies for me. These studies were the ones that looked at the full feedback with the effect of humans on ecosystems, and then how ecosystem states affect human behavior. The best example, I think, was James Kitchell's and Steve Carpenter's (Chapter 10). They know something of the decision-making of Wisconsin anglers and the interplay between that decision-making process and the foodweb, and the influence on the population dynamics of fishes in those lakes. The predictive power that results from an understanding of that feedback was enormously impressive to me. I know it's frustrating to them, but it's the envy of the rest of us. I wish we all had two states that we had to decide between. It's really a marvelous example. Eduardo Rapoport (Chapter 16) also had an excellent study that looked at an urban-to-rural gradient and spatial study in Mexico City. He also coupled that with detailed studies of decision-making in Mexican households.

Fourth and finally, I'd like to join Orie Loucks in a call for more ecological theory and modeling, being a modeler myself, that couples humans and ecosystems. We need two kinds of models. We need special models of systems that can be used in an applied context. We saw a marvelous example of this by Wolf Grossmann (Chapter 18) who really integrated descriptive studies of ecological interactions together with studies of human decision-making in modeling to solve some applied problems that are quite thorny. But we also need, I think, some very simple models. Minimal models of ideas, if you like. For example, John Lawton has called for the incorporation of human beings into models of food webs to find out how the special attributes of human beings—the rapid decision-making, the ability to communicate, the vertical functional responses, the lack of coupling to the dynamics of their prey—how all those things might affect food webs in principle. We need simple models to sensitize our imagination about what is possible here, so we can begin to formulate interesting hypotheses. Also Jane Hall (Chapter 5) had a particularly interesting insight to offer in this regard. She suggested that we should incorporate the assumptions behind the standard economic model—the model of explosive growth—into ecological models, as apparently some of the economists are doing now, because government policy is structured in such a way that it tends to make society adhere to the model whenever society deviates from it. This is a degree of determinism that might actually make that coupling fruitful and I think we ecologists ought to get into that game as well.

None of these four recommendations is by any means deep, and that's really the beauty of it. I think that for researchers the recommendations yield a very clear set of things that we have to go out and do. It's obvious what needs to be done, and I think that we will do it. I think it's already being done in studies that can serve as exemplars for further work. And if we do these things, and, moreover, if we train students to integrate these kinds of studies better than anybody here at this Conference can, then it's inevitable that human beings and ecosystems will become part of the same paradigm.

24
Humans as Components of Ecosystems: A Synthesis

Steward T.A. Pickett and Mark J. McDonnell

Introduction

The call for including humans as components of ecological research is not new. As we described in Chapter 1, the obvious negative impacts of human activities have been well documented, but the more subtle causes and effects of humans on ecosystems have not traditionally been the subject of study by North American ecologists. Similarly, human populated areas are typically not considered within the research domain of ecologists. Ironically, however, the very first paper published in the Ecological Society of America's journal, *Ecology*, was on the influence of weather on infection rates of pneumonia and influenza in the citizens of New York City and Boston (Huntington 1920). Since Huntington's paper, there have been relatively few studies that include humans as part of an ecological system published in the journal. The relative lack of ecological research that includes humans may reflect a long-standing tension in ecology (Kingsland 1985) between historical and instantaneous approaches.

One of the earliest calls for including humans as components of ecological study by a North American ecologist was by Charles C. Adams (1935, 1938). Some 50 years ago, Adams (1938:501) stated "the time is rapidly approaching when ecologists—or others more alert to the true ecological situation—will be called upon to assist in a fresh integration of the biological and the social sciences" in order to more effectively assess the influence of human activities on the environment. This call has been echoed by other ecologists including Curtis (1956), Holling and Orians (1971), Stearns and Montag (1974), Bornkamm et al. (1982), Ehrlich (1985), Brown and Roughgarden (1989), and McDonnell and Pickett (1990). The recent appearance of a new North American journal entitled *Ecological Applications* is yet another indication that ecologists are willing to broaden the scope of the field to explicitly include human activities. Unfortunately, in

310

order to effectively include humans as components of ecosystems, we need to re-evaluate how ecological systems are structured and how they function. Because ecologists have not been trained to study humans, we need to utilize theories and tools from other disciplines as well as develop new approaches for explicitly including humans in ecological research (see Hall, Chapter 5; Vayda, Chapter 6; Boyden, Chapter 7). In order to accomplish this task, we must begin more effectively to build bridges between ecologists, social scientists, physical scientists, geographers, historians, human ecologists, and managers.

This book has been designed to educate ecologists about the importance of incorporating humans and their effects in ecological concepts, models, and studies, and to highlight especially the neglected or non-obvious role of humans (i.e., subtle causes and effects). The message to social scientists, anthropologists, historians, geographers, and economists is the existence of an opportunity to integrate their approaches and ideas with those of ecologists. This chapter will synthesize the insights from the foregoing chapters: to illustrate the prominent role of humans in the functioning of the Earth's ecosystems; to explore the reasons for the historical neglect of humans as components of ecosystems by North American ecologists and, thereby, help overcome it; to flag the emergence of a new ecological paradigm that accommodates human activities into ecological subject matter; to examine methods for detecting and studying subtle causes and effects, and human-populated systems; and finally, to address what ecologists need to do in the future in order to effectively include humans as components of ecological systems.

Ubiquity of Human Effects

Over 120 years have passed since George Perkins Marsh published his 1864 treatise on the role of humans in the functioning of Earth's ecological systems (Marsh 1965; Williams, Chapter 3; Turner and Meyer Chapter 4). One of the major contributions of his book was the assessment of the cumulative impacts of humans over long time scales and large areas. We now know that the human impacts on Marsh's world of the mid 1800s provided only a small glimpse of the highly modified world we live in today (Turner et. al 1990a; Goudie 1990). As a result of a five-fold increase in the world's human population, coupled with technological advances in agriculture and industry, there are now numerous documented cases of human activities that have altered regional and global ecosystems (Thomas 1956; MacDonald and Sertorio 1989; Turner et al. 1990a; Schlesinger 1991).

We began with a definition of the range of human causes and effects on ecosystems (McDonnell and Pickett, Chapter 1) in order to draw attention to those that have been neglected by ecology. Russell (Chapter 8) improved on the conceptual framework by classifying various kinds of subtlety in cause and effect. The separation of subtlety into cause and effect is critical because subtle causes can have effects of large magnitude and vice versa. The chapters in Section I of this volume summarize key points from the social sciences and humanities concerning the nature and magnitude of human effects on environment, and show how social scientists and historians work (Williams, Chapter 3; Turner and Meyer, Chapter 4; Hall, Chapter 5; Vayda, Chapter 6; Boyden, Chapter 7). The entire book is replete with examples of human impacts on ecosystems, and some of the impacts are quite subtle or historically distant (Foster, Chapter 9; Aber, Chapter 14; Wagner and Kay, Chapter 20). Many areas that do not appear to be populated have in fact been, and so contain historic echoes, lagged effects, or are exposed to indirect effects or unsuspected effects from distant human actions (Castilla, Chapter 13; Aber, Chapter 14; Grossmann, Chapter 18; Cole et al., Chapter 12). One especially common scenario is the generation of subtle, yet significant, effects on lakes and aquatic systems by human modification of a linked terrestrial environment (Richerson, Chapter 11; Kitchell and Carpenter, Chapter 10). Combined, these examples demonstrate how ubiquitous the human modification of the planet has been. In addition to the contributors to this volume, there is a growing consensus among physical, social, and biological scientists that virtually every ecological system on Earth has, in some way, been affected by humans (Botkin 1990; Botkin et. al. 1989; Turner et al. 1990a; Likens 1992). Therefore, in order to effectively study Earth's ecosystems for the purpose of obtaining both basic knowledge and answers to environmental problems, we must develop the appropriate tools and conceptual frameworks to include humans as components of ecological systems (see Jutro, Chapter 19; Wagner and Kay, Chapter 20; Jordan Chapter 21).

Why Have Humans Been Neglected as Components of Ecosystems?

As described by Egerton (Chapter 2) and Williams (Chapter 3), the perceived role of humans on Earth has deep mythological, philosophical, and religious roots (cf. Oelschlaeger 1991). It is generally agreed that the modern view of humans was significantly influenced by the writings of George Perkins Marsh, especially his book entitled *Man and Nature* (Marsh 1965; Turner and Meyer, Chapter 4). Marsh not only perceived humans as separate from nature, he viewed natural systems as being in balance (i.e., in equilibrium) and self-regulating. This view promoted the assumption that natural disturbances such as fire or hurricanes, and thus, of course, human disturbances, had little or no lasting influence on the structure and function of ecological systems. This is the classical paradigm in ecology referred to as the "equilibrium paradigm" (Simberloff 1982; Pickett et al. 1992). The classical paradigm is encompassed in the cultural metaphor of the "balance of nature" (Pickett et al. 1992). From an operational point of view, the classical paradigm implied that ecologists, seeking to understand how ecological systems were structured and functioned, should work in areas free of human disturbance.

In the last 20 years, there has been a growing body of scientific evidence suggesting that the classical "equilibrium paradigm" is flawed (Botkin 1990; Pickett et al. 1992). In the field of plant ecology, researchers have, in general, failed to find steady-state or "climax" communities (Drury and Nisbet 1973; Pickett and McDonnell 1989). Instead, plant communities appear to have multiple persistent states and multiple pathways of vegetation change (Niering 1987; Pickett 1989; Pickett and McDonnell 1989; Luken 1990; Pickett et al. 1992). In addition, ecologists now acknowledge that disturbances such as fire, hurricanes, and floods can and do affect the structure of ecological systems (Pickett and White 1985). Thus, a new "nonequilibrium paradigm" has emerged that incorporates current knowledge of how ecological systems are structured and function (Pickett et. al. 1992). Although the new paradigm recognizes the existence of relatively stable states under certain circumstances, or the possibility of stable distributions of states in certain landscapes, it differs from the classical paradigm in several critical points. In the nonequilibrium paradigm, ecological systems are viewed as driven by process rather than end point, and as open systems potentially regulated by external forces (Pickett et al. 1992).

This contemporary paradigm encapsulates significant changes in the way ecologists practice their science; what is especially important to the subject of this book is that the nonequilibrium paradigm explicitly permits the inclusion of humans as components of ecological systems studied by ecologists (Egerton, Chapter 2; Pickett et al. 1992). If external forces potentially regulate the structure and function of some ecological systems, then human activities, both obvious and subtle, need to be considered as important drivers of change. As described in the previous section, human activities are now significantly affecting the flux of materials, energy, organisms, and water within Earth's ecological systems. In addition, human caused disturbances on ecosystems currently rival, in both magnitude and scale, natural disturbances such as floods and hurricanes (Turner et al. 1990a). The study of populated areas is an opportunity to address both basic ecological questions and environmental problems (see McDonnell et al., Chapter 15; Rapoport, Chapter 16). One basic question, for example, is: Does anthropogenic disturbance affect the structure and function of ecosystems differently than natural disturbance? Finally, the recognition of the important role of natural and anthropogenic forces acting on ecological systems has compelled ecologists to expand their typical spatial scale of study of 1 to 1000 m^2 to landscape and regional scales ranging from 1 to 1,000 km^2. This shift in spatial scales has resulted in the utilization of new tools, concepts, and models for ecology, some of which have been borrowed from other disciplines such as geography, while others are completely new (Gardner et al., Chapter 17; Grossmann, Chapter 18).

What Is a "Natural" System?

The problem of what is "natural" has recurred throughout the book. The definitions of natural range from those denoting a "pure state, unsullied by humans," to "unbuilt or uncultivated; primitive" (Random House 1966). Many of the uses of the term illustrate the biases that have excluded humans and their effects from the purview of ecology (Egerton, Chapter 2). The problem is of both theoretical and practical importance. Theory and management each require some reference state which simplifies the world and excludes at least some of the influences that might otherwise affect the system of interest (Sober 1984). As described above, traditionally, most ecologists have considered humans as separate from nature, and have tried to reconstruct or find systems from which humans or their influences were absent. The research and scholarship collected in this book clearly demonstrate that such a search is unrealistic. Models, experiments, and management must all accommodate the subtle effects of humans, and better understand the ecology of populated areas. Although the theoretical reference states used by ecololgists may explicitly, as a simplifying assumption, hold human effects constant or null, the models, experiments, and interpretations of real cases must often admit a significant, if subtle, human role.

Methodologies for Detecting Subtle Causes and Effects, and Human Populated Systems

Throughout the book, much concern has been paid to methods for detecting and dealing with subtle effects or human-populated systems (see Russell, Chapter 8; Richerson, Chapter 11; Aber, Chapter 14; McDonnell et al., Chapter 15). The challenge to ecology is to develop conceptual and methodological tools to discover subtle effects or to investigate the ecological effects of humans in the environments they inhabit or indirectly affect. The methodological concern is appropriate because most of ecology does not have much experience with these issues. The methods shown in this volume to expose the human component of ecosystems include: 1) hypothesis testing à la Popper (1965; Russell, Chapter 8; Cole et al., Chapter 12; Aber, Chapter 14); 2) meta-analysis and comparison (Rapoport, Chapter 16); 3) historical analysis (Williams, Chapter 3; Foster, Chapter 9; Kitchell and Carpenter, Chapter 10); 4) ecological restoration as experiment (Jordan, Chapter 21); 5) bounded system modeling (Nelson et al., Chapter 22); 6) progressive contextualization (Vayda, Chapter 6); 7) scaling and hierarchical analysis of cause (Gardner et al., Chapter 17; Grossmann, Chapter 18); and 8) multi-causal analysis and synthesis (Turner and Meyer, Chapter 4; Hall, Chapter 5; Boyden, Chapter 7).

Although several authors have promoted one approach over the others, it is clear that insights have been derived from all of the approaches, but specially those that accommodate multiple or interacting causes, and expose historical effects (Russell, Chapter 8; Richerson, Chapter 11). Ecology has a long tradition of using falsification methods for testing expectations (Strong 1982; Peters 1991), but these techniques are especially difficult to employ in multi-causal complex systems (Hilborn and Stearns 1982; Richerson, Chapter 11), such as those dominated by humans. For example, there is some debate within the scientific community about whether true replication, a key component of statistical hypothesis testing, can even be achieved on the landscape scale (Hargrove and Pickering 1992). The use of the broader method of confirmation, instead of the narrower method of falsification, is prominent in biology and ecology because it employs a variety of methods that can expose multiple and historical causation. Confirmation involves assessment of 1) the degree of fit between data and theory; 2) the independent support of assumptions of the theory; and 3) the variety of independent evidence (Lloyd 1988). Any successful ecological theory that includes humans as components of an ecological system would benefit from all these modes of confirmation. Because it uses a variety of approaches, confirmation is an especially robust approach for developing models and theories of complex, multi-causal systems (Pickett and Kolasa 1989), and it should prove to be a powerful tool for studying human dominated systems.

What Will It Take to Include Humans as Components of Ecosystems?

Worldwide, the call from ecologists, policy makers, and managers for programs focusing on populated areas has been getting louder (Lubchenco et al. 1991; Levin 1992; Likens 1992). In order to meet future needs of understanding humans as components of ecosystems, several things are required: 1) more integrated research of the sort illustrated throughout this book; 2) effective and sustained communication between social scientists, economists, historians, and ecologists; 3) joint training of students; and 4) institutional and financial support and encouragement of interdisciplinary studies involving humans as components of ecosystems. Constructing conceptual or quantitative models that integrate social, physical, and ecological processes is an important approach to meeting these goals. Such models are likely to be most useful when focused on a regional scale. Scientists and scholars who are concerned with human effects and the reciprocal interactions of humans and ecological systems, can come together around integrated regional models.

Ecologists who do not account for the effects of humans in their research and scholarship are likely to miss important factors structuring and organizing their systems. As described throughout this volume, humans have affected every ecosystem on Earth. The level at which humans need to be included as components of an ecological system depends on the questions being asked and the processes under study. Leaving humans out of ecology may result in erroneous explanations of phenomena in the natural world. Furthermore, integrating humans as components of ecological systems, especially at the levels of community, ecosystem, landscape, and the biosphere, presents an opportunity to advance ecology and to interact with other disciplines.

Conclusions

This book has attempted to integrate ideas, concepts, and tools from a variety of disciplines concerning the role of humans in ecological systems. The themes that have emerged from the book chapters as well as from other sources are as follows:

1) Human impact on the environment, some of it quite subtle or historically distant, is virtually ubiquitous;
2) A more refined understanding of the concept of subtle human causes and effects provides new insight into the role of humans in ecological systems;
3) A notion of populated systems and their ecology is beginning to emerge in terrestrial systems. The understanding of aquatic ecosystems as components of populated landscapes is more advanced;
4) The historic perspective and the comparative method have yielded much insight into the role of humans;
5) Natural systems are inconstant in the face of changing modes of human use, management, habitation and neglect;
6) Ecology needs new, integrated models and theories in order to effectively address both basic questions and environmental problems in human dominated ecosystems;
7) Methodological complexity required to study the role of humans can be dealt with hierarchically or by disarticulation and synthesis; and
8) The natural regulation, or classical, paradigm in ecology that underwrites traditional management schemes has failed.

All of these themes combine to yield a resounding answer to Marsh's great question of 1864: Humans are indeed part of nature, and must therefore become a part of the subject matter of modern ecology.

Bibliography

Abele, L.G., D.S. Simberloff, D.R. Strong, and A.B. Thistle. (1984). Preface. In: D.R. Strong, D. Simberloff, L.G. Abele, and A.B. Thistle, eds. *Ecological Communities*, pp. viii-x. Princeton University Press, Princeton, New Jersey.

Aber, J.D. and W.R. Jordan. (1985). Restoration ecology: an environmental middle ground. *BioScience* 35:399.

Aber, J.D. and J.M. Melillo. (1991). *Terrestrial Ecosystems*. Saunders College Publishing, Philadelphia.

Aber, J.D., J.M. Melillo, and C.A. Federer. (1982). Predicting the effects of rotation length, harvest intensity, and fertilization on fiber yield from northern hardwood forests in New England. *Forest Sci.* 28:31-45.

Aber, J.D., K.J. Nadelhoffer, P. Steudler, and J.M. Melillo. (1989). Nitrogen saturation in northern forest ecosystems. *BioScience* 39:378-386.

Aber, J.D., J.M. Melillo, and C.A. McClaugherty. (1990). Predicting long-term patterns of mass loss, nitrogen dynamics, and soil organic matter formation from initial fine litter chemistry in temperate forest ecosystems. *Can. J. Bot.* 68:2201-2208.

Adams, C.C. (1935). The relation of general ecology to human ecology. *Ecology* 16:316-335.

Adams, C.C. (1938). A note for social-minded ecologists and geographers. *Ecology* 19:500-502.

Adams, R. McC. (1990). Foreword: the relativity of time and transformation. In: B.L. Turner, II, W.C. Clark, R.W. Kates, J.F. Richards, J.T. Matthews, and W.B. Meyer, eds. *The Earth as Transformed by Human Action*, pp. vii-x. Cambridge University Press, Cambridge.

Adisoemarto and E.F. Brunig, eds. (1978). *Transactions of the Second International MAB-IUFRO Workshop on Tropical Rainforest Ecosystem Research, 21-25 October 1978*. (Special Report No 2) Chair of World Forestry, University of Hamburg.

Agee, J.K. and D.R. Johnson. (1988). A direction for ecosystem management. In: J.K. Agee and D.R. Johnson, eds. *Ecosystem Management for Parks and Wilderness*, pp. 226-232. University of Washington Press, Seattle.

Agren, G.I., and E. Bosatta. (1988). Nitrogen saturation of terrestrial ecosystems. *Environ. Pollut.* 54:185-197.

Ahern, G.P. (1929). *Deforested America—Statement of the Present Forest Situation in the United States*. (U.S. Senate Document No. 216) U.S. Government Printing Office, Washington, D.C.

Ahlgren, C.E. and I.F. Ahlgren. (1983). The human impact on northern forest ecosystems. In: S.L. Flader, ed. *The Great Lakes Forest: An Environmental and Social History*, pp. 33-51. University of Minnesota Press, Minneapolis, in association with Forest History Society, Santa Cruz.

Air Pollution. (1969). *Proceedings of the First European Congress on the Influence of Air Pollution on Plants and Animals*. Wageningen, April 22-27, 1968. Center for Agricultural Publishing and Documentation. Wageningen, Netherlands.

Airola, T. M. and K. Buchholz. (1984). Species structure and soil characteristics of five urban sites along the New Jersey Palisades. *Urban Ecol.* 8:149-164.

Alder, G.H. (1987). Influence of habitat structure on demography of two rodent species in eastern Massachusetts. *Can. J. Zool.* 65:903-912.

Alexander, M. (1977). *Soil Microbiology*, 2nd ed. J. Wiley and Sons, New York.

Alfaro, R., E. Bustamante, J. Torres, H. Treviño, and W. Wurtsbaugh. (1982). *La Pesqueria en el Lago Titicaca (Peru): Presente y Futuro*. United Nations Food and Agriculture Organization, Rome.

Alland, S. (1970). *Adaptation in Cultural Evolution: An Approach to Medical Anthropology*. Columbia University Press, New York.

Allen, J. (1991). *Biosphere 2: The Human Experiment*. Penguin Books, New York.

Allen, J. and M. Nelson. (1988). *Space Biospheres*, 2nd ed. Synergetic Press, Oracle, Arizona.

Allen, J.C. and D.F. Barnes. (1985). The causes of deforestation in developing countries. *Ann. Assoc. Am. Geog.* 75:163-184.

Allen, T.F.H. and T.B. Starr. (1982). *Hierarchy. Perspectives for Ecological Complexity*. University of Chicago Press, Chicago.

Alling, A., M. Nelson, L. Leigh, R. Frye, T. MacCallum, N. Alvarez-Romo, and J. Allen. (1993). Biosphere 2 test module experiments. In: R.J. Beyers and H.T. Odum, eds. *Ecological Microcosms*. Springer-Verlag, New York. (in press).

Altshuller, A.P. (1983). Measurements of the products of atmospheric photochemical reactions in laboratory studies and in ambient air-relationships between ozone and other products. *Atmosph. Environ.* 17:2383-2427.

Anders, M.B. (1986). Andean archaeology. *Inst. Archaeol. Monog. (UCLA)* 27:201-224.

Anderson, E. (1956). Man as a maker of new plants and new plant communities. In: W.L. Thomas, ed. *Man's Role in Changing the Face of the Earth*, pp. 763-777. University of Chicago Press, Chicago.

Angyal, A.J. (1983). *Loren Eiseley*. Twayne, Boston.

Anonymous. (1959). *Classification of Land Cover Types by Towns in Worcester County, Massachusetts*. Cooperative Extension Service, University of Massachusetts, Amherst.

Anonymous. (1967). *Administrative Policy for the Management of Ungulates*. Yellowstone National Park, United States Department of the Interior, National Park Service, Washington, D.C.

Anonymous. (1968). *Compilation of the Administrative Policies for the National Parks and National Monuments of Scientific Significance (Natural Area Category)*. United States Department of the Interior, National Park Service, Washington, D.C.

Anonymous. (1977). *Management Objectives for Northern Yellowstone Elk*. Yellowstone National Park Information Paper No. 10, United States Department of the Interior National Park Service, Washington, D.C.

Anonymous. (1988). *Management Policies*. United States Department of the Interior, National Park Service, Washington, D.C.

Ashdown, M. and J. Schaller. (1990). *Geographic Information Systems and their Application in MAB-Projects, Ecosystem Research, and Environmental Monitoring*. (MAB Mitteilungen Nr. 34) Deutsches Nationalkomitee, Bonn. (German and English).

Austin, M. P. (1985). Continuum concept, ordination methods, and niche theory. *Ann. Rev. Ecol. Syst.* 16:39-61.

Ausubel, J.H. and H.E. Sladovich, eds. (1989). *Technology and Environment*. National Academy Press, Washington, D.C.

Baath, E., B. Lundgren, and B. Soderstrom. (1981). Effects of nitrogen fertilization on the activity and biomass of fungi and bacteria in a podzolic soil. *Zentralblat. Bacteriol. Parasitenkd. Infektionskr.* Hyg. Abt. 1:orig. Reihe C.2:90-98.

Baath, E. (1989). Effects of heavy metals in soil on microbial processes and populations (a review). *Water Air Soil Pollut.* 47:335-379.

Backman, A. (1984). 1000-year record of fire-vegetation interactions in the northeastern United States: a comparison between coastal and inland regions. M.S. Thesis, University of Massachussets, Amherst, Massachusetts.

Bailey, R.G. (1986). The Zaïre River system. In: B.R. Davies and K.F. Walker, eds. *The Ecology of River Systems*, pp. 201-214. Dr. W. Junk, The Hague.

Bailey, G. and J. Parkington, eds. (1988). *The Archaeology of Prehistoric Coastlines*. Cambridge University Press, Cambridge.

Baker, A.H. and H.I. Patterson. (1986). *Farmer's Adaptations to Markets in Early Nineteenth-century Massachusetts*. Dublin Seminar, Dublin, Ireland.

Baksh, M. (1985). Faunal food as a "limiting factor" on Amazonian cultural behavior: a Machiguenga example. *Res. Econ. Anthropol.* 7:145-175.

Baldwin, H.I. (1942). *Forestry in New England*. (Publication No. 70) National Resources Planning Board, Boston.

Balée, W. (1989). The culture of Amazonian forests. *Adv. Econ. Bot.* 7:78-96.

Barkai, A. and C.D. McQuaid. (1988). Predator-prey role reversal in a marine benthic ecosystem. *Science* 242:62-64.

Barraclough, S.L. (1949). Forest land ownership in New England. Ph.D. Thesis, Harvard University, Cambridge.

Barraclough, S.L. and E.M. Gould. (1955). Economic analysis of farm forest operating units. *Harvard Forest Bull.* No. 26.

Barrett, S.W. and S.F. Arno. (1982). Indian fires as an ecological influence in the northern Rockies. *J. Forest.* 80:647-651.

Beatley, J.C. (1953). The primary forests of Jackson and Vinton counties, Ohio. Ph.D Thesis, Ohio State University, Columbus.

Bengtsson, G. and L. Tranvik. (1989). Critical metal concentrations for forest soil invertebrates. *Water Air Soil Pollut.* 47:381-417.

Bennekom, A.J. van and W. Salomons. (1981). Pathways of nutrients and organic matter from land to ocean through rivers. In: J.N. Martin, J. Barton, and D. Eisma, eds. *River Input to Ocean Systems*, pp. 33-51. UNESCO-UNEP, Rome, Italy, 1979.

Berner, E.K. and R.A. Berner. (1987). *The Global Water Cycle*. Prentice Hall, Englewood Cliffs, New Jersey.

Berry, B.J.L. (1990). Urbanization. In: B. L. Turner, II, W.C. Clark, R.W. Kates, J.F. Richards, J.T. Matthews, and W.B. Meyer, eds. *The Earth as Transformed by Human Action*, pp. 103-119. Cambridge University Press with Clark University, Cambridge.

Bertalanffy, L.V. (1969). *General Systems Theory: Essays on its Foundations and its Development*, rev. ed. Braziller, New York.

Betancourt, J.L. and T.R. Van Devender. (1981). Holocene vegetation in Chaco Canyon, New Mexico. *Science* 214:656-658.

Betancourt, J.L., J.S. Dean, and H.M. Hull. (1986). Prehistoric long-distance transport of construction beams, Chaco Canyon, New Mexico. *Am. Ant.* 51:370-375.

Bidwell, P. (1916). Rural economy of New England at the beginning of the 19th century. *Trans. Conn. Acad. Arts Sci.* 13:1-75.

Bienen, H. and H.J. Leonard. (1985). Environment, economic growth, and distribution in the Third World. In: H.J. Leonard, ed. *Divesting Nature's Capital: The Political Economy of Environmental Abuse in the Third World*, pp. 51-89. Holmes and Meier, New York.

Birks, H.H., H.J.B. Birks, P.E. Kaland, and D. Moe, eds. (1988). *The Cultural Land-*

scape—Past, Present, and Future. Cambridge University Press, Cambridge.

Black, J.D. and A. Brinser. (1952). *Planning One Town: Petersham, a Hill Town in Massachusetts*. Harvard University Press, Cambridge.

Black, J.D. and G.W. Westcott. (1959). *Rural Planning of One County. Worcester County, Massachusetts*. Harvard University Press, Cambridge.

Boardo, E.L. (1988). Incentive policies and forest use in the Philippines. In: R. Repetto and M. Gillis, eds. *Public Policies and the Misuse of Forest Resources*, pp. 165-203. Cambridge University Press, Cambridge.

Bocking, S.A. (1991). Environmental concerns and ecological research in Great Britain and the United States, 1950-1980. Ph.D. Thesis, University of Toronto, Toronto.

Bogart, E.L. (1948). *Peacham: The Story of a Vermont Hill Town*. Vermont Historical Society, Montpelier, Vermont.

Bohn, H.L. (1972). Soil adsorption of air pollutants. *J. Environ. Qual.* 1:372-377.

Bohn, H.L. and R.K. Bohn. (1986). Soil bed scrubbing of fugitive gas releases. *J. Environ. Sci. Health* A:21:561-569.

Bolger, D.T., A.C. Alberts, and M.E. Soulé. (1991). Occurrence patterns of bird species in habitat fragments: sampling, extinction, and nested species subsets. *Am. Nat.* 137:155-156.

Bonetto, A.A. (1986). The Paraná River system. In: B.R. Davies and K.F. Walker, eds. *The Ecology of River Systems*, pp. 541-556. Dr W. Junk, The Hague.

Bonnicksen, T.M. (1989a). Statement of Dr. Thomas M. Bonnicksen Department Head and Professor Department of Recreation and Parks Texas A&M University before the Committee on Interior and Insular Affairs Subcommittee on National Parks and Public Lands and the Committee on Agriculture Subcommittee on Forests, Family Farms, and Energy. United States House of Representatives, January 31, 1989, Washington, D.C.

Bonnicksen, T.M. (1989b). Standards of authenticity for restoring forest communities. Presented at the First Annual Conference of the Society for Ecological Restoration, Oakland, California, January 16-20, 1989. (unpublished).

Bonnicksen, T.M. and E.C. Stone. (1981). The giant sequoia-mixed conifer forest community characterized through pattern analysis as a mosaic of aggregations. *Forest Ecol. Manage.* 3:307-328.

Bonnicksen, T.M. and E.C. Stone. (1982a). Reconstruction of a presettlement giant sequoia-mixed conifer forest community using the aggregation approach. *Ecology* 63:1134-1148.

Bonnicksen, T.M. and E.C. Stone. (1982b). Managing vegetation within U.S. national parks: a policy analysis. *Envir. Manage.* 6:101-102.

Bonnicksen, T.M. and E.C. Stone. (1982c). Managing vegetation within U.S. national parks: a policy analysis. *Envir. Manage.* 6:109-122.

Bonnicksen, T.M. and E.C. Stone. (1985). Restoring naturalness to national parks. *Environ. Manage.* 6:109-122.

Borchert, G. and S. Kempe. (1985). A Zambezi aqueduct. In: E.T. Degens, S. Kempe, and R. Herrera, eds. Transport of Carbon and Minerals in Major World Rivers, Part 3. *Mitt. Geol.-Paläont. Inst. Univ. Hamburg, SCOPE/UNEP Sonderbd.* 58:443-458.

Bormann, F.H. (1982). The effects of air pollution on the New England landscape. *Ambio* 11:338-346.

Bormann, F.H. and G.E. Likens. (1979). *Pattern and Process in a Forested Ecosystem*. Springer-Verlag, New York.

Bornkamm, R., ed. (1982). *Urban Ecology*. Blackwell, Oxford.

Bornstein, R.D. (1968). Observations of the urban heat island effect in New York City. *J. Applied Meteorol.* 7:575-582.

Botkin, D.B. (1990). *Discordant Harmonies*. Oxford University Press, New York.

Botkin, D.B., M.F. Caswell, J.E. Estes, and A.A. Orio, eds. (1989). *Changing the Global Environment: Perspectives on Human Involvement*. Academic Press, Boston.

Botts, A.K. (1934). Northbridge, Massachusetts, a town that moved down hill. *J. Geog.* 33:249-260.

Boulding, K. (1966). The economics of the coming spaceship Earth. In: H. Jarret, ed. *Environmental Quality in a Growing Economy*, pp. 3-14. Johns Hopkins Press, Baltimore.

Bourne, L.S. and J.W. Simmons. (1982). Defining the area of interest: definition of the city, metropolitan areas, and extended urban regions. In L.S. Bourne, ed. *Internal Structure of the City*, pp. 57-72. Oxford University Press, New York.

Boyce, M.S. (1992). Wolf recovery for Yellowstone National Park. In: D.R. McCullough and R.H. Barrett, eds. *Wildlife 2001: Populations*, pp. 123-138. Elsevier Science Publishers Ltd., Essex.

Boyd, R. (1973). World dynamics: a note. *Science* 177:516-519.

Boyd, R., P. Gasper, and J.D. Trout, eds. (1991). *The Philosophy of Science*. MIT Press, Cambridge, Massachusets.

Boyden, S. (1987). *Western Civilization in Biological Perspective: Patterns in Biohistory.* Oxford University Press, Oxford.

Boyden, S. (1992a). *Biohistory: The Interplay Between Human Society and the Biosphere, Past and Present*. UNESCO, Paris, and Parthenon Publishing Group, Carnforth, United Kingdom.

Boyden, S. (1992b). The human aptitude for culture and its biological consequences. *Perspect. Human Biol.* (Centre for Human Biology, University of Western Australia.) 1:29-108.

Boyden, S., S. Dovers, and M. Shirlow. (1990). *Our Biosphere Under Threat: Ecological Realities and Australia's Opportunities*. Oxford University Press, Melbourne.

Boyle, J.R. and A.R. Ek. (1973). Whole tree harvesting: nutrient budget evaluation. *J. Forest.* 71:760-762.

Bradshaw, A.D. and T. McNeilly. (1981). *Evolution and Pollution*. E. Arnold, London.

Bramwell, A. (1989). *Ecology in the 20th Century: A History*. Yale University Press, New Haven, Connecticut.

Branch, G.M. (1975). Notes on the ecology of *Patella concolor* and *Cellana capensis*, and the effect of human consumption on limpet populations. *Zool. Afric.* 10:75-78.

Bratman, M.E. (1992). Practical reasoning and acceptance in a context. *Mind* 101:1-15.

Braun, C.E., D.R. Stevens, K.M. Giesen, and C.P. Melcher. (1991). Elk, white-tailed ptarmigan, and willow relationships: a management dilemma in Rocky Mountain National Park. *Trans. N. Am. Wildlife Nat. Res. Conf.* 56:74-85.

Bray, J. R. and J. T. Curtis. (1957). An ordination of the upland forest communities of southern Wisconsin. *Ecol. Monog.* 27:325-349.

Brock, T.D. (1985). *Lake Mendota: A Eutrophic Lake*. Springer-Verlag, New York.

Bronny, C. (1989). One-two punch: grazing history and the recovery potential of oak savannas. *Rest. Manage. Notes* 7:73-76.

Brookfield, H.C. and Y. Byron. (1990). Deforestation and timber extraction in Borneo and the Malay Peninsula: the record since 1965. *Global Environ. Change* 1:42-56.

Broughton, J.G., D.W. Fisher, Y.W. Isachsen, and L.V. Richard. (1966). *Geology of New York: A Short Account*. Leaflet No. 20. The University of the State of New York, The State Ed. Dept. and New York State Mus. and Sci. Ser. Educ., Albany, New York.

Browman, D.L. (1981). New light on Andean Tiwanaku. *Am. Sci.* 69:408-419.

Browman, D.L. (1974). Pastoral nomadism in the Andes: pre-Columbian considerations. *Curr. Anthropol.* 15:188-196.

Browman, D.L. (1991). The dynamics of the Chiripa polity. Paper presented at 47th International Congress of Americanists, July 11, 1991, New Orleans. (unpublished)

Brown, F. (1895). *Decline and Fall of Petersham, Worcester County*. Town of Petersham, Massachusetts.

Brown, A.A. and K.P. Davis. (1973). *Forest Fire: Control and Use*, 2nd ed. McGraw-Hill, New York.

Brown, J. and J. Roughgarden., (1989). US ecologists address global change. *Trends Ecol. Evol.* 4:255-256.

Brown, L., ed. (1991). *State of the World 1991: A World Watch Institute report on progress toward a sustainable society.* Norton, New York.

Browne, C.A. (1942). *A Source Book of Agricultural Chemistry.* Chronica Botanica, Waltham. Reprinted 1977, Arno Press, New York.

Brunskill, G.J. (1986). Environmental features of the Mackenzie system. In: B.R. Davies and K.F. Walker, eds. *The Ecology of River Systems,* pp. 435-472. Dr. W. Junk, The Hague.

Brush, G.S. and F.W. Davis. (1984). Stratigraphical evidence of human disturbance in an estuary. *Quat. Res.* 22:91-108.

Bulmer, R.N.H. (1982). Traditional conservation practices in Papua New Guinea. In: L. Morauta, J. Pernetta, and W. Heaney, eds. *Traditional Conservation in Papua New Guinea,* pp. 59-77. Institute of Applied Social and Economic Research, Boroko, Papua New Guinea.

Bunyard, P. and E. Goldsmith, eds. (1988). *Gaia: The Thesis, the Mechanisms and the Implications.* Wadebridge Ecological Centre, Camelford, United Kingdom.

Burgess, R.L. and D.M. Sharpe, eds. (1981). *Forest Island Dynamics in Man-Dominated Landscapes.* Springer-Verlag, New York.

Buss, L. W. (1986). Competition and community organization on hard surfaces. In: J. Diamond and T. J. Case, eds. *Community Ecology.* Harper and Row, New York.

Buzas, M.A. and S.J. Culver. (1991). Species diversity and dispersal of benthic foraminifera. *BioScience* 41:483-489.

Cairns, J., Jr. (1987). Disturbed ecosystems as opportunities for research in restoration ecology. In W.R. Jordan III, M.E. Gilpin, and J.D. Aber, eds. *Restoration Ecology: A Synthetic Approach to Ecological Research,* pp. 307-320. Cambridge University Press, Cambridge.

Cairns, J., Jr. (1988). Restoration ecology: the new frontier. In J. Cairns, Jr., ed. *Rehabilitating Damaged Ecosystems,* Vol I., pp. 2-11. CRC Press, Inc., Boca Raton, Florida.

Cambray, J.A., B.R. Davies, and P.J. Ashton. (1986). The Orange-Vaal River system. In: B.R. Davies and K.F. Walker, eds. *The Ecology of River Systems,* pp. 89-122. Dr W. Junk, The Hague.

Caraco, N.F., A. Tamse, O. Boutros, and I. Valiela. (1987). Nutrient limitation of phytoplankton growth in brackish coastal ponds. *Can. J. Fish. Aquat. Sci.* 44:473-476.

Carlisle, E.F. (1983). *Loren Eiseley: The Development of a Writer.* University of Illinois Press, Urbana.

Carlson, D.A. and C.P. Leiser. (1966). Soil beds for the control of sewage odors. *J. Water Pollut. Control Fed.* 38:829-840.

Carmouze, J.-P., C. Arze, and J. Quintanilla. (1977). La regulation hydrique des lacs Titicaca et Poopo. *Cah. ORSTOM Ser. Hydrobiol.* 11:269-283.

Carney, H.J., M.W. Binford, R.R. Marin, and C.R. Goldman. (1993). Nitrogen and phosphorus dynamics and retention in ecotones of Lake Titicaca, Bolivia/Peru. *Hydrobiologia.* (in press).

Carpenter, S.R., ed. (1988). *Complex Interactions in Lake Communities.* Springer-Verlag, New York.

Carpenter, S.R. and J.F. Kitchell. (1988). Consumer control of lake productivity. *BioScience* 38:764-769.

Carpenter, S.R., J.F. Kitchell, and J.R. Hodgson. (1985). Cascading trophic interactions and lake productivity. *BioScience* 35:634-639.

Carpenter, S.R., T.M. Frost, J.F. Kitchell, T.K. Kratz, D.W. Schindler, J. Shearer, W.G. Sprules, M.J. Vanni, and A.P. Zimmerman. (1991). Patterns of primary production and herbivory in 25 North American lake ecosystems. In: J. Cole, S. Findlay, and G. Lovett, eds. *Comparative Analyses of Ecosystems: Patterns, Mechanisms, and Theories,* pp. 67-96. Springer-Verlag, New York.

Carrier, J.G. (1982). Conservation and conceptions of the environment: a Manuş case study. In: L. Morauta, J. Pernetta, and W. Heaney, eds. *Traditional Conservation in*

Papua New Guinea, pp. 39-43. Institute of Applied Social and Economic Research, Boroko, Papua New Guinea.

Carrier, J.G. (1987). Marine tenure and conservation in Papua New Guinea. In: B.J. McCay and J.M. Acheson, eds. *The Question of the Commons: the Culture and Ecology of Communal Resources*, pp. 142-167. University of Arizona Press, Tucson.

Carson, R. (1970). *Silent Spring*. Fawcett Publications, Greenwich, Connecticut. Reprinted from 1962 edition.

Castilla, J.C. (1988). Ecosistemas intermareales y submareales de fondos duros en el Cono Sur de Sudamérica; una oportunidad única para estudios regionales integrales. *Inf. UNESCO Ciencias Mar* 47:115-123.

Castilla, J.C., and L.R. Durán. (1985). Human exclusion from the rocky intertidal zone of central Chile: the effects on *Concholepas concholepas* (Gastropoda). *Oikos* 45:391-399.

Castilla, J.C. and R.T. Paine. (1987). Predation and community organization on eastern Pacific, temperate zone, rocky intertidal shores. *Rev. Chil. Hist. Nat.* 60:131-151.

Central Intelligence Agency. (1990). *The World Factbook*. Central Intelligence Agency, Washington, D.C.

Chadwick, E. (1843). Report on the sanitary conditions of the labouring population of Great Britain. W. Clowes and Sons, London, England.

Chagnon, N.A. and R. Hames. (1979). Protein deficiency and tribal warfare in Amazonia: new data. *Science* 203:10-15.

Chaloupka, M.Y. and S.B. Domm. (1986). Role of anthropochory in the invasion of coral cays by alien flora. *Ecology* 67:1536-1547.

Chapman, D. and T. Drennan. (1990). Equity and effectiveness of possible CO_2 treaty proposals. *Contemp. Policy Iss.* 8:29-42.

Chapman, J., P.A. Delcourt, P.A. Cridlebaugh, A.B. Shea, and H.R. Delcourt. (1982). Man-land interaction: 10,000 years of American impact on native ecosystems in the lower Little Tennessee Valley. *Southeast. Archaeol.* 1:115-21.

Chase, A. (1986). *Playing God in Yellowstone: The Destruction of America's First National Park*. The Atlantic Monthly Press, Boston.

Chase, F.H. (1890). Is agriculture declining in New England? *New Eng. Mag.* 2:449-452.

Chisholm, A. and Dumsday, R. (1987). *Land Degradation: Problems and Policies*. Cambridge University Press, Cambridge.

Churcher, P.B. and J.H. Lawton. (1987). Predation by domestic cats in an English village. *J. Zool.* 212:439-455.

Clark, C.E. (1970). *The Settlement of Northern New England 1610-1765*. Knopf, New York.

Clark, W.C., coordinator. (1988). The human dimensions of global change. In: Committee on Global Change. *Toward an Understanding of Global Change*, pp. 134-200. National Academy Press, Washington, D.C.

Clements, F.E. (1904). *The Development and Structure of Vegetation*. University of Nebraska Press, Lincoln, Nebraska.

Cline, A.C., N.L. Munster, R.J. Lutz, and M.E. Raymond. (1938). *The Farm Woodlands of the Town of Hardwick, Massachusetts*. Worcester County Land Use Planning Project. Worcester, Massachusetts.

Cole, G.F. (1971). An ecological rationale for the natural or artificial regulation of ungulates in parks. *Trans. N. Am. Wildlife Nat. Res. Conf.* 36:417-425.

Cole, J.C., G. Lovett, and S. Findlay, eds. (1991). *Comparative Analyses of Ecosystems: Patterns, Mechanisms and Theories*. Springer-Verlag, New York.

Colinvaux, P.A. (1976). Review of *Hum. Ecol.* F. Sargent, II, ed. *Hum. Ecol.* 4:263-266.

Collins, B.S., K.P. Dunne, and S.T.A. Pickett. (1985). Responses of forest herbs to canopy gaps. In: S.T.A. Pickett and P.S. White, eds. *The Ecology of Natural Disturbance and Patch Dynamics*, pp. 217-234. Academic Press, New York.

Collot, D. (1980). *Les macrophytes de quelques lacs andins*. Convenio UMSA-ORSTOM. La Paz and Paris.

Committee on Global Change. (1988). *Toward an Understanding of Global Change: Initial*

Priorities for U.S. Contributions to the International Geosphere-Biosphere Program. National Academy Press, Washington, D.C.

Committee on the Applications of Ecological Theory to Environmental Problems. (1986). *Ecological Knowledge and Environmental Problem-Solving*. National Academy Press, Washington, D.C.

Committee on Earth Sciences. (1990). *Our Changing Planet: The FY 1991 U.S. Global Change Research Program*. Federal Coordinating Council for Science, Engineering, and Technology. Office of Science and Technology Policy, Washington, D.C.

Commoner, B. (1972). *The Closing Circle*. Knopf, New York.

Conrad, G.W. and A.A. Demarest. (1984). *Religion and Empire: The Dynamics of Aztec and Inca Expansionism*. Cambridge University Press, Cambridge.

Conservation International (CI). (1991). *Conservation International 1990 Annual Report. Tropicus*. Conservation International, Washington, D.C.

Cook, H.O. (1917). *The Forests of Worcester County. The Results of a Forest Survey of the Fifty-Nine Towns in the County and a Study of Their Lumber Industry*. Potter Printing, Boston.

Cook, H.O. (1961). *Fifty Years a Forester*. Massachusetts Forests and Park Association. Reynolds Printing, New Bedford, Massachusetts.

Cook, S.F. (1973). The significance of disease in the extinction of New England Indians. *Hum. Biol.* 45:485-508.

Cook, S.F. and W. Borah. (1972). *Essays in Population History: Mexico and the Caribbean*. Vol. 1. University of California Press, Berkeley, California.

Coolidge, M.C. (1948). *The History of Petersham, Massachusetts*. Powell Press, Hudson, Massachusetts.

Costanza, R. (1989). What is ecological economics. *Ecol. Econ.* 1:1-7.

Costanza, R., ed. (1991). *Ecological Economics: the Science and Management of Sustainability*. Columbia University Press, New York.

Council on Environmental Quality (CEQ). (1991). *Environmental Quality: 21st Annual Report*, pp. 135-187. The Council on Environmental Quality, Washington, D.C.

Cowles, H.C. (1899). The ecological relations of the vegetation on the sand dunes of Lake Michigan. *Bot. Gaz.* 27: 95-1117, 167-202, 281-308, 361-391.

Craig, B.W. and A.J. Friedland. (1991). Spatial patterns in forest composition and standing dead red spruce in montane forests of the Adirondacks and northern Appalachians. *Environ. Monit. Assess.* 18:129-144.

Cramp, S. (1963). Toxic chemicals and birds of prey. *Brit. Birds* 56:124-139.

Cronon, W. (1983). *Changes in the Land: Indians, Colonists, and the Ecology of New England*. Hill and Wang, New York.

Cropper, M. and W. Oates. (1992). Environmental economics: a survey. *J. Econ. Lit.* 30:675-740.

Crosby, A.F. (1986). *Ecological Imperialism: The Biological Expansion of Europe, 900-1900*. Cambridge University Press, Cambridge.

Crosby, A.W. (1972). *The Columbian Exchange: Biological and Cultural Consequences of 1492*. Greenwood Press, Westport, Connecticut.

Cullen, P. (1990). The turbulent boundary between water science and water management. *Freshwater Biol.* 24:201-209.

Currier, A.N. (1891). The decline of rural New England. *Pop. Sci. Monthly* 38:384-389.

Curtis, J.T. (1956). The modification of mid-latitude grasslands and forests by man. In: W. Thomas, ed. *Man's Role in Changing the Face of the Earth*, pp. 721-736. University of Chicago Press, Chicago.

Curtis, J.T. (1959). *The Vegetation of Wisconsin: An Ordination of Plant Communities*. University of Wisconsin Press, Madison, Wisconsin.

Curtis, J.T. and R.P. McIntosh. (1951). An upland forest continuum in the prairie-forest border region of Wisconsin. *Ecology* 32:476-496.

Curtis, J.T. and M.L. Partch. (1948). Effect of fire on the competition between bluegrass and certain prairie plants. *Am. Midl. Nat.* 39:437-443.

Cwikowa, A., A. Grabowski, A.J. Lesinski, and S. Myxzkowski. (1984). Flora and vegetation of the Nieplomice Forest. In: W. Grodzinski, J. Weiner, and P.F. Maycock, eds. *Forest Ecosystems in Industrial Regions*, pp. 1-11. Springer-Verlag, New York.

D'Elia, C.F. 1987. Nutrient enrichment in the Chesapeake Bay: too much of a good thing. *Environment* 29:6-11, 30-33.

Dale, V.H. and T.W. Doyle. (1987). The role of stand history in assessing forest impacts. *Environ. Manage.* 11:351-357.

Daly, H. (1991). *Steady State Economics*. Island Press, Washington, D.C.

Daly, H. (1984). Alternative strategies for integrating economics and ecology. In: A.-M. Jansson, ed. *Integration of Economy and Ecology: An Outlook for the Eighties*, pp. 19-29. Proceedings from the Wallenberg Symposium, 1982. Marcus Wallenberg Foundation for International Cooperation in Science. Stockholm, Sweden.

Darwin, C. (1859). *The Origin of Species By Means of Natural Selection*. Penguin Books edition 1983, with introduction by J.W. Burrow. Penguin Books, Harmondsworth, U.K.

Darwin, C.R. (1987). *Charles Darwin's Notebooks, 1836-1844*. P.H. Barnett, P.J. Gautrey, S. Herbert, D. Kohn, S. Smith, eds. Cornell University Press, Ithaca, New York.

Davidson, D. (1980). *Essays on Actions and Events*. Clarendon Press, Oxford.

Davies, B.R. (1986). The Zambezi River system. In: B.R. Davies and K.F. Walker, eds. *The Ecology of River Systems*, pp. 225-268. Dr W. Junk, The Hague.

Davis, I.G. (1933). Agricultural production in New England. *Am. Geog. Soc. Spec. Publ.* 16:137-167.

Davis, M.B. (1965). Phytogeography and palynology of northeastern United States. In: H.E. Wright and D.G. Frey, eds. *The Quaternary of the United States*, pp. 377-401. Princeton University Press, Princeton, New Jersey.

Davis, M.B. (1989). Insights from paleoecology on global change. *Ecol. Soc. Am. Bull.* 70:222-228.

Davis, R.B. (1987). Palaeolimnological diatom studies of acidification of lakes by acid rain: an application of Quaternary science. *Quat. Sci. Rev.* 6:147-163.

Davis, R.B. and P.M. Stokes. (1986). Overview of historical and paleoecological studies of acidic air pollution and its effects. *Water Air Soil Pollut.* 30:311-318.

Davis, R.B., D. Anderson, and F. Berge. (1985). Palaeolimnological evidence that lake acidification is accompanied by loss of organic matter. *Nature* 316:436-438.

Day, F.P. (1983). Effects of flooding on leaf litter decomposition in microcosms. *Oecologia* 56:180-184

Day, G.M. (1953). The Indians as an ecological factor in the Northeastern forest. *Ecology* 34:329-346.

Day, J.A. and B.R. Davies. (1986). The Amazon River system. In: B.R. Davies and K.F. Walker, eds. *The Ecology of River Systems*, pp. 289-319. Dr W. Junk, The Hague.

de Candolle, A.P. (1809). Geographic agricole et botanique. *Nouveau cours complet d'agriculture, ou dictionnaire raisonne et universel d'agriculture* 6:355-373.

de Candolle, A.P. (1820). Geographie botanique. *Diction. sci. nat.* 18:359-422.

DeAngelis, D. (1991). *Dynamics of Nutrient Cycling and Food Webs*. Chapman and Hall, New York.

DeAngelis, D.L., P.J. Mulholland, A.V. Palumbo, A.D. Steinman, M.A. Huston, and J.W. Elwood. (1989). Nutrient dynamics and food web stability. *Ann. Rev. Ecol. Syst.* 20:71-95.

Dejoux, C. and A. Iltis. (1992). *Lake Titicaca: A Synthesis of Limnological Knowledge*. Kluwer Academic Publishers, Dordrecht, Amsterdam.

Delcourt, H.R. (1987). The impact of prehistoric agriculture and land occupation on natural vegetation. *Ecology* 34:341-46.

Delcourt, H.R., P.A. Delcourt, and T. Webb III. (1983). Dynamic plant ecology: the spectrum of vegetation change in space and time. *Quat. Sci. Rev.* 1:153-175.

Dempster, W.B. (1988). Biosphere II: design of a closed manned terrestrial ecosystem,

Society of Automotive Engineers , 18th Intersociety Conference on Environmental Systems, Technical Paper Series #881096. SAE, Warrendale, Pennsylvania.

Dempster, W.B. (1989). Biosphere II: technical overview of a manned closed ecological system. Society of Automotive Engineers, 19th Intersociety Conference on Environmental Systems, Technical Paper Series #181599, SAE. Warrendale, Pennsylvania.

Dempster, W.F. (1992). *Methods for Measurement and Control of Leakage in CELSS and its Application and Performance in the Biosphere 2 Facility.* COSPAR/IAF Space Congress, Washington, 1992, workshop F4.4-M.1.10X. (unpublished).

Denevan, W.M. (1970). Aboriginal drained-field cultivation in the Americas. *Science* 169:-647-654.

Denevan, W.M. (1976). *The Native Population of the Americas in 1492.* University of Wisconsin Press, Madison, Wisconsin.

Descy, J.-P. and A. Empain. (1984). Meuse. In: B.A. Whitton, ed. *Ecology of European Rivers*, p. 1-23. Blackwell, Oxford.

Di Persia, D.H. and J.J. Neiff. (1986). The Uruguay River system. In: B.R. Davies and K.F. Walker, eds. *The Ecology of River Systems*, pp. 599-622. Dr. W. Junk, The Hague.

Diamond, J.M. (1986). The environmentalist myth. *Nature* 324:19-21.

Diamond, J. (1988). The golden age that never was. *Discover* December 1988:71-79.

Díaz-Betancourt, M., I. López-Moreno, and E. H. Rapoport. (1987). Vegetación y ambiente urbano en la Ciudad de México. Las plantas de los jardines privados. In: E.H. Rapoport and I. López-Moreno, eds. *Aportes a la Ecologia Urbana de la Ciudad de México*, pp. 13-72. Instituto de Ecologia and MAB-UNESCO, Limusa, México, D.F.

Dickinson, R.E. (1966). The process of urbanization. In F.F. Darling and J.P. Milton, eds. *Future Environments of North America*, pp. 463-478. Natural History Press, Garden City, New York.

Dickson, D.R. and C.L. McAfee. (1988). *Forest statistics for Massachusetts—1972 and 1985.* Northeastern Forest Experiment Station Resource Bulletin NE-106. USDA Forest Service, Bromall, Pennsylvania.

Dillehay, T.D. (1984). A late Ice-Age settlement in southern Chile. *Sci. Am.* 251:100-109.

Dimitri, M.J. (1972). *La Región de los Bosques Andino-Patagónicos. Sinopsis General.* Instituto Nacional de Tecnologia Agropecuaria, Buenos Aires, Argentina.

Dobyns, H.F. (1966). Estimating aboriginal American population: an appraisal of techniques with a new hemispheric estimate. *Curr. Anthropol.* 7:395-416.

Dobyns, H.F. (1983). *Their Number Become Thinned: Native American Population Dynamics in Eastern North America.* University of Tennessee Press, Knoxville, Tennessee.

Doherty, R. (1977). *Society and Power: Five New England Towns, 1800-1860.* University of Massachusetts Press, Amherst, Massachusetts.

Donahue, B. (1983). The forests and fields of Concord: an ecological history. In: D.H. Fischer, ed. *Concord: the Social History of a New England Town 1750-1850*, pp. 14-63. Brandeis University, Waltham, Massachusetts. pp. 14-63.

Dony, J.G. (1967). *Flora of Hertfordshire.* Hitchin Museum, England.

Dovers, S. (1993). The history of natural resource use in rural Australia: practicalities and ideologies. In: G. Lawrence, F. Vanclay, and B. Furze, eds. *Agriculture, Environment, and Society: the Australian Experience.* Macmillan, Sydney. (in press).

Driscoll, C.T., C.P. Yatsko, and F.J. Unangst. (1987). Longitudinal and temporal trends in the water chemistry of the North Branch of the Moose River. *Biogeochemistry* 3:37-62.

Drury, W.H. and I.C.T. Nisbet. (1973). Succession. *J. Arnold Arbor.* 54:331-368.

Drury, W.H. (1991). Humans in ecological perspective. In: S. Suzuki, R.J. Borden, and L. Hens, eds. *Human Ecology—Coming of Age: An International Overview*, pp. 25-246. VUB Press, Brussels.

Dueser, R.D. and H.H. Shugart, Jr. (1978). Microhabitats in a forest floor small mammal fauna. *Ecology* 59:89-98.

Dumas, J.B.A. (1841). On the chemical statistics of organized beings. *Philos. Mag.* 19:337-347, 456-469.

Dunlap, T.R. (1988). *Saving America's Wildlife*. Princeton University Press, Princeton, New Jersey.

Durán, L.R. and J.C. Castilla. (1989). Variation and persistence of the middle rocky intertidal community of central Chile, with and without human harvesting. *Mar. Biol.* 103:555-562.

Durán, L.R., J.C. Castilla, and D. Olivia. (1987). Intensity of human predation on rocky shores at Las Cruces in central Chile. *Environ. Conserv.* 14:143-149.

Dwight, T. (1821-22). *Travels in New-England and New York in 1821*. 4 Vols. T. Dwight, New Haven, Connecticut.

Eck, G.W. and L. Wells. (1987). Recent changes in Lake Michigan's fish community and their probable causes, with emphasis on the role of the alewife (*Alosa pseudoharengus*). *Can. J. Fish. Aquat. Sci.* 44:(Suppl. 2):53-60.

Edmond, J.M., A. Spivack, B.C. Grant, M. Hu, Z. Chen, S. Chen, and X. Zeng. (1985). Chemical dynamics of the Changjiang estuary. *Cont. Shelf Res.* 4:17-36.

Egerton, F.N. (1967). Studies of animal populations from Lamarck to Darwin. *J. Hist. Biol.* 1:225-259.

Egerton, F.N. (1973). Changing concepts of the balance of nature. *Quart. Rev. Biol.* 48:322-350.

Egerton, F.N. (1983). The history of ecology: achievements and opportunities, part one, *J. Hist. Biol.* 16:259-311.

Egerton, F.N. (1985a). The history of ecology: achievements and opportunities, part two. *J. Hist. Biol.* 18:103-143.

Egerton, F.N. (1985b). *Overfishing or Pollution: Case History of a Controversy on the Great Lakes*. Great Lakes Fishery Commission, Ann Arbor, Michigan.

Egerton, F.N. (1987). Pollution and aquatic life in Lake Erie: early scientific studies. *Environ. Rev.* 11:189-206.

Egerton, F.N. (1989). Missed opportunities: U.S. fishery biologists and productivity of fish in Green Bay, Saginaw Bay, and western Lake Erie. *Environ. Rev.* 13:33-63.

Egerton, F.N. and W.J. Christie. (1993). *The Role of Sea Lamprey Control in Restoration of Fisheries in the Great Lakes*. Great Lakes Fishery Commission, Ann Arbor, Michigan. (in press).

Ehrlich, P. (1981). Environmental disruption: implications for the social sciences. *Social Sci. Quart.* 62:7-2.

Ehrlich, P.R. (1985). Human ecology for introductory biology courses: an overview. *Am. Zool.* 25:379-394.

Ehrlich, P.R. and A.H. Ehrlich. (1990). *The Population Explosion*. Simon and Schuster, New York.

Ehrlich, P.R. and P.H. Raven. (1964). Butterflies and plants: a study in coevolution. *Evolution* 18:586-608.

Eiseley, L. (1958). *Darwin's Century: Evolution and the Men Who Discovered It*. Doubleday, New York.

Eiseley, L. (1964). *The Unexpected Universe*. Harcourt, Brace & World, New York.

Eiseley, L. (1973). *The Man Who Saw Through Time*. Charles Scribner's Sons, New York.

Eiseley, L. (1979). *Darwin and the Mysterious Mr. X: New Light on the Evolutionists*. Dutton, New York.

Ellen, R. (1982). *Environment, Subsistence, and System*. Cambridge University Press, Cambridge.

Elton, C. (1930). *Animal Ecology and Evolution*. Clarendon Press, Oxford.

Engel, J.R. (1983). *Sacred Sands: The Struggle for Community in the Indiana Dunes*. Wesleyan University Press, Middletown, Connecticut.

Erickson, G.L. (1981). The northern Yellowstone elk herd—a conflict of policies. *Proc. 61st Ann. Conf. West. Assoc. Fish Wildlife Agencies*, pp. 92-108.

Erickson, C.L. (1988). An archaeological investigation of raised field agriculture in the Lake Titicaca Basin, Peru. Ph.D. Thesis, University of Illinois, Champaign-Urbana, Illinois.

Erickson, C.L. and K.L. Candler. (1989). Raised fields and sustainable agriculture in the lake Titicaca basin of Peru. In: J.O. Browder, ed. *Fragile Lands of Latin America*, pp. 230-248. Westview Press, Boulder, Colorado.

Estes, J.A. and J.F. Palmisano. (1974). Sea otters: their role in structuring nearshore communities. *Science* 185:1058-1060.

Facelli, J. and S.T.A. Pickett. (1990). The dynamics of litter. *Bot. Rev.* 57:2-32.

Faliński, J.B., ed. (1968). *Synantropizacja Szaty Roslinnej (Synantropization of Vegetation Cover)*. I. Neofityzm i Apofitysm, No. 25; II. Flora i Roslinnosc Synantropijna Miast. No. 27. Materialy Zakladu Fitosocjologii Stosowanej Uniwersytet Warszawski (Materials of the Department of Applied Phytosociology of Warsaw University).

Feder, J. (1988). *Fractals*. Plenum Press, New York.

Fedoseyev, I.A. (1976). Vladimir Ivanovich Vernadsky (1983-1945). *Diction. Sci. Biog.* 13:616-620.

Feely, J.M. (1980). Did Iron Age man have a role in the history of Zululand wilderness landscapes? *S. Afric. J. Sci.* 76:150-152.

Fegeas, R.G., R.W. Claire, S.C. Gupthill, K.G. Anderson, and C.H. Hallam. (1983). Land use and land cover digital data. Geological Survey Circular 895-E. U.S. Geological Survey, Reston, Virginia, and Washington, D.C.

Firor, J. (1990). *The Changing Atmosphere. A Global Challenge*. Yale University Press, New Haven, Connecticut.

Fischman, R.L. (1991) *Biological Diversity and Environmental Protection: Authorities to Reduce Risk*. Environmental Law Institute, Washington, D.C.

Fisher, A. and M. Hanneman. (1986). Option value and the extinction of species. In: K. Smith, ed. *Advances in Applied Microeconomics*. JAI Press, Greenwich, Connecticut.

Fisher, D.C. and M. Oppenheimer. (1991). Atmospheric nitrogen deposition and the Chesapeake Bay Estuary. *Ambio* 20:102-108.

Fisher, R.A. (1958). *The Genetical Theory of Natural Selection*. Dover, New York.

Fisher, R.T. (1925). Descent of the white pine woodlot. *Harvard Forest Bull.* 8:7-14.

Fisher, R.T. (1928). Soil changes and silviculture on the Harvard Forest. *Ecology* 9:6-11.

Fisher, R.T. (1931). The Harvard Forest as a demonstration tract. *Quart. J. Forest.* 25:1-11.

Fisher, R.T. (1933). New England's forests: biological factors. New England's prospect. *Am. Geog. Soc. Spec. Publ.* 16:213-223.

Fisher, R.T. and E.I. Terry. (1920). The management of second growth white pine in central New England. *J. Forest.* 18:1-9.

Flannery, K.V., J. Marcus, and R.G. Reynolds. (1989). *The Flocks of the Wamani: A Study of Llama Herders on the Punas of Ayachucho, Peru*. Academic Press, San Diego, California.

Flenly, J.R. and S.M. King. (1984). Late Quaternary pollen records from Easter Island. *Nature* 307:47-50.

Flohn, H. (1979). On the time scales and causes of abrupt paleoclimatic events. *Quat. Res.* 12:135-149.

Flores, J.A. (1987). Cultivation of the *Qocha* of the south Andean puna. In: D.L. Browman, ed. *Arid Land Use Strategies and Risk Management in the Andes*, pp. 271-296. Westview, Boulder, Colorado.

Food and Agriculture Organization of the United Nations (FAO). (1990). *Outline for a Draft International Convention on the Conservation and Sustainable Use of Biological Diversity*. FAO, Rome.

Forbes, S.A. (1887). The lake as a microcosm. *Peoria Sci. Assoc. Bull.* 77-87. (Reprinted in Forbes, S.A. (1977). *Ecological Investigations*. Arno Press, New York.)

Forman, R.T.T. (1989). The ethics of isolation, the spread of disturbance, and landscape ecology. In: M.G. Turner, ed. *Landscape Heterogeneity and Disturbance*, pp. 213-229. Springer-Verlag, New York.

Forman, R.T. and M. Godron. (1986). *Landscape Ecology*. John Wiley and Sons, New York.

Forrester, J.W. (1968). *Urban Dynamics*. MIT-Press, Cambridge.

Foster, D.R. (1988). Species and stand response to catastrophic wind in central New England, USA. *J. Ecol.* 76:135-151.

Foster, D.R. (1992). Land-use and vegetation dynamics in central New England: an historical perspective. *J. Ecol.* 80:753-772.

Foster, D.R. and E. M. Boose. (1992). Patterns of forest damage resulting from catastrophic wind in central New England, U.S.A. *J. Ecol.* 80:79-98.

Foster, D.R. and D.R. Smith (1991). *Harvard Forest Long Term Ecological Research Program—Abstracts of the Second Annual Harvard Forest Ecology Symposium*. Harvard Forest, Harvard University, Petersham, Massachusetts.

Foster, D.R. and T.M. Zebryk. (1993). Long-term vegetation dynamics and disturbance history of a *Tsuga* dominated forest in central New England. *Ecology*. (in press).

Foster, D.R., T. Zebryk, P.K. Schoonmaker, and L. Lezberg. (1992). Land-use history and vegetation dynamics of a hemlock forest in central New England. *J. Ecol.* 80:773-786.

Foster, J.R. and W.A. Reiners. (1983). Vegetation patterns in a virgin subalpine forest at Crawford Notch, White Mountains, New Hampshire. *Bull. Torrey Bot. Club* 110:141-153.

Frankfort, H., H.A. Frankfort, J.A. Wilson, and T. Jacobsen. (1946). *The Intellectual Adventure of Ancient Man*. University of Chicago Press, Chicago.

Franklin, J.F and R.T.T. Forman. (1987). Creating patterns by cutting: ecological consequences and principles. *Landscape Ecol.* 1:5-18.

Freedman, B. (1989). *Environmental Ecology: The Impacts of Pollution and Other Stresses on Ecosystem Structure and Function*. Academic Press, San Diego, California.

Frenkel, R.E. 1974. Floristic changes along Everitt Memorial Highway, Mount Shasta, California. Wassman J. Bio. 32:105-136.

Frenkel, R.E. (1970). *Ruderal Vegetation Along Some California Roadsides*. 2nd ed. 1977. University of California Press, Berkeley, California.

Friedland, A.J., R.A. Gregory, L. Karenlampi, and A.H. Johnson. (1984). Winter damage as a factor in red spruce decline. *Can. J. Forest Res.* 14:963-965.

Friedrich, G. and D. Müller. 1984. Rhine. In: B.A. Whitton, ed. *Ecology of European Rivers*, pp. 265-315. Blackwell, Oxford.

Frothingham, E.H. (1912). *Second-growth Hardwoods in Connecticut*. USDA Forest Serv. Bull. 96. USDA Forest Service, Washington, D.C.

Fryer, G. and T.D. Iles. (1972). *The Cichlid Fishes of the Great Lakes of Africa*. Oliver and Boyd, Edinburgh.

Furse, M. T., D. Moss, and J.F. Wright. (1984). The influence of seasonal and taxonomic factors on the ordination and classification of running-water sites in Great Britain and on the predation of their macro-invertebrate communities. *Freshwater Biol.* 14:257-280.

Gale, R.S. (1987). Learning from the past, preparing for the future. *Greater Yellowstone Rep.* 4:12-14.

Galloway, J.N., D. Zhao, J. Xiong, and G.E. Likens. (1987). Acid rain: China, United States, and a remote area. *Science* 236:1559-1562.

Garden, A. and R.W. Davies. (1988). The effects of a simulated acid precipitation on leaf litter quality and the growth of a detritivore in a buffered lotic system. *Environ. Pollut.* 52:303-313.

Gardner, R.H. and R.V. O'Neill. (1990). Pattern, process and predictability: the use of neutral model for landscape analysis. In: M.G. Turner and R.H. Gardner, eds. *Quantitative Methods in Landscape Ecology. The Analysis and Interpretation of Landscape Heterogeneity*, pp. 289-307. Springer-Verlag, New York.

Gardner, R.H., B.T. Milne, M.G. Turner, R.V. O'Neill. (1987). Neutral models for the analysis of broad-scale landscape pattern. *Landscape Ecol.* 1:19-28.

Gardner, R.H., V.H. Dale, R.V. O'Neill, and M.G. Turner. (1992). A percolation model of ecological flows. In: A.J. Hansen and F. di Castri, eds. *Landscape Boundaries: Consequences for Biotic Diversity and Ecological Flows*, pp. 259-269. Springer-Verlag, New York.

Garrison, J.R. (1987). Farm dynamics and regional exchange: the Connecticut Valley beef trade, 1670-1850. *Agric. Hist.* 61:11-17.

Gast, P.R. (1937). Contrasts between the soil profiles developed under pines and hardwoods. *J. Forest.* 35:11-16.

Gates, P.W. (1978). Two hundred years of farming in Gilsum. *Hist. New Hampshire* 23:1-24.

Gauch, H. G., Jr. (1982). *Multivariate Analysis in Community Ecology*. Cambridge University Press, Cambridge.

Geertz, C. (1984). Culture and social change: the Indonesian case. *Man* 19:511-532.

Ghazanshahi, J., T.D. Huchel, and J.S. Devinny. (1983). Alteration of southern California rocky shore ecosystems by public recreational use. *J. Environ. Manage.* 16:379-394.

Gibbs, R.J. (1972). Water chemistry of the Amazon River. *Geochim. Cosmochim. Acta* 36: 1061-1066.

Gifford, B. and Editors of *Outside*. (1990). Inside the environmental groups. *Outside* 15:69-84.

Giles, R.J., Jr. (1978). *Wildlife Management*. W.H. Freeman, San Francisco.

Gillis, M. (1988a). Indonesia: public policies, resource management, and the tropical forest. In: R. Repetto and M. Gillis, eds. *Public Policies and the Misuse of Forest Resources*, pp. 43-113. Cambridge University Press, Cambridge.

Gillis, M. (1988b). Malaysia: public policies and the tropical forest. In: R. Repetto and M. Gillis, eds. *Public Policies and the Misuse of Forest Resources*, pp. 115-164. Cambridge University Press, Cambridge.

Gillispie, C.C. (1968). Remarks on social selection as a factor in the progressivism of science. *Am. Sci.* 56:439-450.

Gilpin, M.E. (1987). Another view. *Rest. Manage. Notes* 5:21-22.

Glacken, C.J. (1967). *Traces on the Rhodian Shore: Nature and Culture in Western Thought from Ancient Times to the End of the Eighteenth Century*. University of California Press, Berkeley, California.

Gleason, H. A. 1917. The structure and development of the plant association. Bull. Torrey Bot. Club 43:463-481.

Gleick, J. (1988). *Chaos*. Bantam Books, New York.

Glenn, E., C. Clement, P. Brannon, and L. Leigh. (1990). Sustainable food production for a complete diet. *HortScience* 25:1507-1512.

Glitzenstein, J.S., C.D. Canham, M.J. McDonnell, and D.R. Streng. (1990). Effects of environment and land-use history on upland forests of the Cary Arboretum, Hudson Valley, New York. *Bull. Torrey Bot. Club* 117:106-122.

Godoy, C. and C.A. Moreno. (1989). Indirect effects of human exclusion from the rocky intertidal in southern Chile: a case of cross-linkage between herbivores. *Oikos* 54:101-106.

Good, N. (1965). A study of natural replacement of chestnut in New Jersey. Ph.D. Dissertation, Rutgers University, New Brunswick, New Jersey.

Gordon, J.C., S. Bishop, W. Burch, R. Cahn, R. Cahill, T. Clark, R. Dean, J. Franklin, G. Gumerman, B.J. Howe, B. Howell, R.W.E. Jones, D. Latimer, S.P. Leatherman, H. Mooney, V. Wyatt, and E. Zube. (1989). *National Parks: From Vignettes to a Global View*. National Parks and Conservation Association, Washington, D.C.

Goudie, A.S. (1986). *The Human Impact: Man's Role in Environmental Change*. Basil Blackwell, Oxford.

Goudie, A.S. (1990). *The Human Impact: Man's Role in Environmental Change*. 3rd edition. MIT Press, Cambridge, Massachusetts.

Gould, E.M. (1942). A budgetary analysis method for farm woodlands. M.F.S. Thesis, Harvard University, Cambridge.

Goulding, K.W.T. and A.E. Johnston. (1989). Long-term atmospheric deposition and soil acidification at Rothamsted Experimental Station, England. In: J.W.S. Longhurst, ed.

Acid Deposition: Sources, Effects, and Controls, pp. 213-228. Air Science Co., Corning, New York.

Graedel, T. E. and P. J. Crutzen. (1989). The changing atmosphere. *Sci. Am.* 261:58-68.

Graham, F.J. (1970). *Since Silent Spring*. Houghton-Mifflin Co., Boston.

Grassle, J.F. Deep-sea benthic diversity. (1991). *BioScience* 41:464-469.

Greenlee, J.M. and J.H. Langenheim. (1990). Historic fire regimes and their relation to vegetation patterns in the Monterey Bay area of California. *Am. Midl. Nat.* 124:239-253.

Greig-Smith, P. (1983). *Quantitative Plant Ecology*. University of California Press. Berkeley, California.

Griffith, B.G., E.W. Hartwell, and T.E. Shaw. (1930). The evolution of soils as affected by the old field white pine-mixed hardwood succession in central New England. *Harvard Forest Bull.* 15: 82 p.

Grinevald, J. (1988). A history of the idea of the biosphere. In: P. Bunyard and E. Goldsmith, eds. (1988). *Gaia: The Thesis, the Mechanisms and the Implications*, pp. 1-34. Wadebridge Ecological Centre, Camelford, United Kingdom.

Grossmann, W.D. (1983). Systems approaches towards complex systems. In: P. Messerli and E. Stucki, eds. *Fachbeiträge der schweizerischen MAB-Information*. Vol. 19. Bundesamt für Umweltschutz, Bern.

Grossmann, W.D. (1990). *Erfahrungen mit der hierarchischen Systemmethodik und dem Einsatz dynamischer Modelle im MAB6-Projekt Berchtesgaden*. (Experiences with the hierarchical systems method and with the application of dynamic models in the MAB6 Project Berchtesgaden). Institution for Ecosystems and Environmental Studies. Austrian Academy of Sciences, Vienna.

Grossmann, W.D. (1991a). Model- and Strategy-Driven Geographical Maps for Ecological Research and Management. In: P. G. Risser and J. Mellilo, eds. *Long Term Ecological Research: An International Perspective*, pp. 241-256. SCOPE 47. John Wiley and Sons, Chichester, England.

Grossmann, W.D. (1991b). Einsatz von Risikokarten in der Waldschadensproblematik: Konzept, Probleme, und Ergebnisse im Projekt Lehrforst Rosalia (Assessing hypotheses on forest decline with geographical maps of risk: specifications and problems). *Centralblatt ges. Forstwesen* 108:3-13.

Grossmann, W.D. (1991c). Einsatz dynamischer Modelle in der Waldschadensproblematik: Anwendungsfelder, Probleme, und Ergebnisse im Projekt Lehrforst Rosalia. (Application of dynamic models to the problem of forest damage: possibilities, problems and results). *Centralblatt ges. Forstwesen* 108:15-35.

Grossmann, W.D. (1991d). Ergebnisse der Anwendung von dynamischen Karten ("Zeitkarten") im Lehrforst Rosalia. (Results of the application of dynamic maps in the demonstration forest Rosalia). *Centralblatt ges. Forstwesen* 108:215-235.

Grossmann, W.D. and B. Clemens-Schwartz. (1985). OLIMP (Olympic Impacts)- Ein Modell über die Auswirkungen geplanter Olympischer Winterspiele in Berchtesgaden 1992. (OLIMP- Olympic Impacts- A Model on the Effects of intended Olympic Winter Games in Berchtesgaden 1992). *MAB* 21:225-242.

Grossmann, W.D., and B. Grossmann. (1985). *Beratung des Bundesministeriums für Wissenschaft und Forschung in der Waldschadensproblematik*. Forschungsprojekt Schlussbericht. (Advice to the Federal Ministry for Science and Research regarding Forest Die-Back). Federal Ministry for Science and Research, Vienna.

Grossmann, W.D. and J. Schaller. (1986). Geographical maps on forest die-off, driven by dynamic models. *Ecol. Model.* 31:341-353.

Grossmann, W.D. and K.E.F. Watt (1992). Viability and sustainability of civilizations, corporations, institutions, and ecological systems. *Syst. Res.* 1:3-41.

Grossmann, W.D., W. Haber, F. Kerner, U. Richter, J. Schaller, and M. Sittard. (1983). Ziele, Fragestellungen und Methoden. Ökosystemforschung Berchtesgaden. MAB- Mitteilungen Nr. 16. Deutsches Nationalkomitee MAB, Bonn.

Grossmann, W.D., J. Schaller, and M. Sittard. (1984). *"Zeitkarten": eine neue Methode zum*

Test von Hypothesen und Gegenmassnahmen beim Waldsterben. (Dynamic maps: a new method to test hypotheses and counter policies in the problem area of forest die-back). Allgemeine Forstzeitschrift, Munich.

Guirand, F., ed. (1959). *Larousse Encyclopedia of Mythology*. Prometheus Press, New York.

Gulati, R.D., E.H.R.R. Lammens, M.-L. Meijer, and E. van Donk, eds. (1990). *Biomanipulation—Tool for Water Management*. Kluwer, Amsterdam.

Guppy, N. (1984). Tropical deforestation: a global view. *Foreign Affairs* 62:918-965.

Guthrie, D.A. (1974). Suburban bird population in southern California. *Am. Midl. Nat.* 92:461-466.

Haber, W. (ed.) (1985). *Mögliche Auswirkungen der geplanten Olympischen Winterspiele 1992 auf das Regionale System Berchtesgaden*. (Possible effects of intended Olympic Winter Games 1992 on the regional stystem of Berchtesgaden). Technical University, Freising.

Haber, W., W.D. Grossmann, and J. Schaller. (1984). Integrated evaluation and synthesis of data by connecting dynamic feedback models with a geographic information system. In: J. Brandt and P. Agger, eds. *Methodology in Landscape Ecological Research and Planning*. Proc. 1st International Seminar. Int. Assoc. of Landscape Ecol. Roskilde University Center, Roskilde.

Haeckel, E.H.P.A. (1866). *Generelle Morphologie der Organismen*, vol. 1 and 2. G. Reimer, Berlin.

Hagen, J.B. (1990). Atomic energy and ecosystem ecology: the emergence of a scientific specialty. *Ecol. Soc. Am. Bull.* 71:177-178.

Haines, T.A. (1986). Fish population trends in response to surface water acidification. In: Committee on Monitoring and Assessment of Trends in Acid Deposition, National Research Council. *Acid Deposition. Long-Term Trends*, pp. 300-334. National Academy Press, Washington, D.C.

Haken, H. (1978). *Synergetics*, (2nd edition). Springer-Verlag, Berlin.

Halffter, G., J. Sarukhán, and P. Reyes. (1977). ¿Puede ser la ciudad de México en ecosistema en equilibrio? *Primer Simposio de Contaminación Ambiental*, 7-9 Sept. 1977, México (mimeo), 58 pp. (unpublished).

Hall, A.R. (1983). *The Revolution in Science, 1500-1750*. Longman, New York.

Hall, D. (1990). Introduction to social and private costs of alternative energy technologies. *Contemp. Policy Iss.* 8:240-254.

Hall, D. and J. Hall. (1984). Concepts and measures of natural resource scarcity with a summary of recent trends. *J. Environ. Econ. Manage.* 11:363-379.

Hall, J., A. Winer, M. Kleinman, F. Lurmann, V. Brajer, and S. Colome. (1992). Valuing the health benefits of clean air. *Science* 255:812-817.

Hall, M. (1984). Prehistoric farming in the Mfolozi and Hluhluwe valleys of southeast Africa: an archaeo-botanical survey. *J. Archaeol. Sci.* 11:223-235.

Hall, T.S. (1969). *Ideas of Life and Matter: Studies in the History of General Physiology*, vol. 1 and 2. University of Chicago Press, Chicago.

Hames, R.B. and W.T. Vichas. (1982). Optimal diet breadth as a model to explain variability in Amazonian hunting. *Am. Ethol.* 9:258-278.

Hammer, D., ed. (1989). *Constructed Wetlands for Wastewater Treatment: Municipal, Industrial, and Agricultural*. Lewis Publishers, Boca Raton, Florida.

Hanemann, W.M. (1988). Economics and the preservation of biodiversity. In: E.O. Wilson, ed. *Biodiversity*, pp. 1193-199. National Academy Press, Washington, D.C.

Hanski, I. (1982). Distributional ecology of anthropochorous plants in villages surrounded by forest. *Ann. Bot. Fennici* 19:1-15.

Hardin, G. (1968). The tragedy of the Commons. *Science* 162:1243-1248.

Hargrove, W.W. and J. Pickering (1992). Pseudoreplication: a *sine qua non* for regional ecology. *Landscape Ecol.* 6:251-258.

Harris, G.P. (1986). *Phytoplankton Ecology: Structure, Function, and Fluctuation*. Chapman and Hall, New York.

Harris, M. (1987). Cultural materialism: alarums and excursions. In: K. Moore, ed. *Way-*

marks: *The Notre Dame Inaugural Lectures in Anthropology*, pp. 107-126. University of Notre Dame Press, Notre Dame, Indiana.

Hart, J.F. (1968). Loss and abandonment of cleared farmland in the eastern United States. *Ann. Assoc. Am. Geog.* 58:417-40.

Hart, R.C. (1982). The Orange River: Preliminary results. In: E.T. Degens, ed. *Transport of Carbon and Minerals in Major World Rivers, Part 1. Mitt. Geol.-Paläont. Inst. Univ. Hamburg, SCOPE/UNEP Sonderbd.* 52:435-436.

Hartig, J.H., J.F. Kitchell, D. Scavia, and S.B. Brandt. (1991). Rehabilitation of Lake Ontario: the role of nutrient reduction and food web dynamics. *Can. J. Fish. Aquat. Sci.* 48:1574-1580.

Hasler, A.D. (1947). Eutrophication of lakes by domestic drainage. *Ecology* 28:383-395.

Haycock, N. and T. Burt. (1990). Handling excess nitrates. *Nature* 348:291.

Hayek, F.A. (1973). Scientism and the study of society. In: J. O'Neill, ed. *Modes of Individualism and Collectivism*, pp. 27-67. Heinemann, London.

Hecht, S. and A. Cockburn. (1989). *The Fate of the Forest*. Verso, London.

Heidenreich, C. (1977). *Huronia: A History and Geography of the Huron Indians, 1600-1650*. McCelland and Stewart, Ontario.

Hendee, J.C. (1974). A multiple-satisfactions approach to game management. *Wildlife Soc. Bull.* 2:104-113.

Hendrix, P.F., R.W. Parmelee, D.A. Crossley, Jr., D.C. Coleman, E.P. Odum, and P.M. Groffman. (1986). Detritus food webs in conventional and no-tillage agroecosystems. *BioScience* 36:374-380.

Hennigan, R.D. (1985). Water-changing demands for a changing resource. In: L. Dworskyu, P. Otis, G.E. Galloway, and W.J. Reynolds, eds. *The Hudson River, Yesterday, Today, Tomorrow, Sixth Symp. on Hudson River Ecology*, pp. 36-42. Hudson River Environmental Society, Inc. New Paltz, New York.

Heuff, H. and K. Horkan. (1984). Caragh. In: B.A. Whitton, ed. *Ecology of European Rivers*, pp. 363-384. Blackwell, Oxford.

Hilborn, R. (1990). Marine biota. In: B.L. Turner, II, W.C. Clark, R.W. Kates, J.F. Richards, J.T. Matthews, and W.B. Meyer, eds. *The Earth as Transformed by Human Action*, pp. 371-385. Cambridge University Press, Cambridge.

Hilborn R. and S.C. Stearns. (1982). On inference in ecology and evolutionary biology: the problem of multiple causes. *Acta Biotheor.* 31:145-164

Hill, D.E., E.H. Sauter, and W.N. Gonick. (1980). *Soils of Connecticut*. Connecticut Agric. Exp. Station Bull. No. 787. Connecticut Agricultural Experiment Station, New Haven, Connecticut.

Hockey, P.A.R. (1987). The influence of coastal utilization by man on the presumed extinction of the Canadian black oystercatcher *Haematopus meadewaldoi* Bannerman. *Biol. Conserv.* 39:49-62.

Hockey, P.A.R. and A.L. Bosman. (1986). Man as intertidal predator in Transkei: disturbance, community convergence and management of a natural food resource. *Oikos* 46:3-14.

Hodges, C. and R. Frye. (1990). Soil bed reactor work of the environmental research laboratory of the University of Arizona in support of the research and development of Biosphere 2. In: M. Nelson and G.A. Soffen, eds. *Biological Life Support Systems*, pp. 33-40. Synergetic Press, Oracle, Arizona and National Technical Information Service publication, NASACP-3094, Washington, D.C.

Holling, C.S. and G. Orians. (1971). Toward an urban ecology. *Bull. Ecol. Soc. Am.* 52:2-6.

Holling, C.S. (1978). Myths of ecological stability. In: C.F. Smart and W.T. Stansbury, eds. *Studies in Crisis Management*. Butterworth, Toronto.

Holling, C.S. (1986). The resilience of terrestrial ecosystems: local surprise and global change. In: W.C. Clark and R.E. Munn, eds. *Sustainable Development of the Biosphere*, pp. 292-317. Cambridge University Press, Cambridge.

Horkan, K. (1984). Shannon. In: B.A. Whitton, ed. *Ecology of European Rivers*, pp. 345-362. Blackwell, Oxford.

Hough, F.B. (1882). *Report on Forestry*. USDA. U.S. Government Printing Office, Washington, D.C.

Houghton, J.T., G.J. Jenkins, and J.J. Ephraums, eds. (1990). Climate Change: the IPCC Scientific Assessment. Cambridge University Press, Cambridge.

Houghton Mifflin Co. (1989). *Information Please Almanac 1990*. Houghton Mifflin Co., New York.

Houston, D.B. (1982). *The Northern Yellowstone Elk: Ecology and Management*. Macmillan Publishing, New York.

Howarth, R.A. (1988). Nutrient limitation of net primary production in marine ecosystems. *Ann. Rev. Ecol. Syst.* 19:89-110.

Huntington, E. (1920). The control of pneumonia and influenza by weather. *Ecology* 1:1-23.

Hutchinson, G.E., E. Bonatti, U.M. Cowgill, C.E. Goulden, E.A. Ceventhal, M.E. Mallott, F. Margaritoria, R. Patrick, A. Racek, S.A. Roback, E. Stella, J.B. Ward-Perkins, and T.R. Wellman. (1970). Ianula: an account of the history and development of lago di Monterosi, Latium, Italy. *Trans. Amer. Phil. Soc.* N.S. 60:1-178.

Instituto Nacional de Estadistica (Peru) (INE). (1990-1). *Compendio Estadistico*. INE, Lima.

International Union for the Conservation of Nature and Natural Resources, United Nations Environmental Programme, World Resources Institute (IUCN, UNEP, WRI). (1990). *Biodiversity Conservation Strategy Programme*. World Resources Institute, Washington, D.C.

Ives, J.D. and B. Messerli. (1989). *The Himalayan Dilemma: Reconciling Development and Conservation*. Routledge, London.

Jackson, J.B.C. (1991). Adaptation and diversity of coral reefs. *BioScience* 41:475-482.

Jacobs, J. (1985). Diversity, maturity, and stability in ecosystems influenced by human activities. In: W. Van Dobbin and R.H. Lowe-McConnell, eds. *Unifying Concepts in Ecology*, pp. 263-274. Dr. W. Junk, The Hague.

Jacobsen, T. and R.M. Adams. (1958). Salt and silt in ancient Mesopotamian agriculture. *Science* 128:1251-1258.

Jeffers, J.N.R. (1981). *Preparatory statement for the meeting of the "Reconvened Expert Panel on Systems Analysis" of UNESCO's Man and Biosphere Programme MAB*. Institute of Terrestrial Ecology, Grange-over-Sands, United Kingdom.

Jerardino, A., J.C. Castilla, J.M. Ramírez, and N. Hermosilla. (1992). Early coastal subsistence patterns in central Chile: a systematic study of the maritime-invertebrate fauna from the site of Curaumilla-1. *Latin Am. Ant.* 3:43-62.

Jochim, M.A. (1981). *Strategies for Survival*. Academic Press, New York.

Johnson, B.M. and M.D. Staggs. (1992). The fishery. In: J.F. Kitchell, ed. *Food Web Management: A Case Study of Lake Mendota*, pp. 353-375. Springer-Verlag, New York.

Johnson, D.R. and J.K. Agee. (1988). Introduction to ecosystem management. In: J.K. Agee and D.R. Johnson, eds. *Ecosystem Management for Parks and Wilderness*, pp. 3-14. University of Washington Press, Seattle.

Jones, N. (1991). An ecological history of agricultural land use in two Massachusetts towns. Senior Thesis, Hampshire College, Amherst, Massachusetts.

Jones, R. (1969). Fire-stick farming. *Austr. Nat. Hist.* 16:224-228.

Jordan, W.R., III, M.E. Gilpin, and J.D. Aber. (1987). *Restoration Ecology: A Synthetic Approach to Ecological Research*. Cambridge University Press, Cambridge.

Joseph, L.E. (1990). *Gaia: The Growth of an Idea*. St. Martin's Press, New York.

Jutro, P.R. (1991). Biological diversity, ecology, and global climate change. *Environ. Health Perspect.* 96:167-170.

Kalkhoven, J. and P. Opdam. (1984). Classification and ordination of breeding bird data and landscape attributes. In: J. Brandt and P. Agger, eds. *Methodologies in Landscape Ecological Research and Planning*, Vol. 3, Theme 3, pp. 15-26. Roskilde. Roskilde Universitetsforlag GeoRue, Denmark.

Kamshilov, M.M. (1976). *The Evolution of the Biosphere*. Mir Publishers, Moscow.

Kania, G.S. and M.R. Conover. (1991). How government wildlife agencies should respond to local governments that pass antihunting legislation—a response. *Wildlife Soc. Bull.* 19:224-225.

Karl, T.R., R.R. Heim, Jr., and R.G. Quayle. (1991). The greenhouse effect in central North America: if not now, when? *Science* 251:1058-1061.

Kartawinata, K., H. Soedjito, T. Jessup, A.P. Vayda, and C.J.P. Colfer. (1984). The impact of development on interactions between people and forests in East Kalimantan: a comparison of two areas of Kenyah Dayak settlement. In: J. Hanks, ed. *Traditional Life-styles, Conservation, and Rural Development*, pp. 87-95. International Union for Conservation of Nature and Natural Resources (Commission on Ecology Papers No. 7), Gland, Switzerland.

Kartawinata, K., T.C. Jessup, and A.P. Vayda. (1989). Exploitation in Southeast Asia. In: H. Lieth and M.J.A. Werger, eds. *Tropical Rain Forest Ecosystems, Part B*, pp. 591-610. Elsevier, Amsterdam.

Kates, R.W., B. L. Turner II, and W. C. Clark. (1990). The great transformation. In: B.L. Turner, II, W.C. Clark, R.W. Kates, J.F. Richards, J.T. Matthews, and W.B. Meyer, eds. *The Earth as Transformed by Human Action*, pp. 1-17. Cambridge University Press, Cambridge.

Kay, C.E. (1988). The conservation of ungulate populations by Native Americans in the intermountain West: an evaluation of unavailable evidence. Unpub. ms., 95 pp.

Kay, C.E. (1990a). Yellowstone's northern elk herd: a critical evaluation of the "natural regulation" paradigm. Ph.D. Thesis, Utah State University, Logan, Utah.

Kay, C.E. (1990b). The role of Native American predation in structuring large mammal communities. Presented at the Sixth International Conference of the International Council of Archaeozoology. Smithsonian Institution, Washington, D.C. (unpublished).

Keddy, P.A. (1989). *Competition*. Chapman and Hall, New York.

Keegan, W. (1986). The optimal foraging analysis of horticultural production. The optimal foraging analysis of horticultural production. *Am. Anthropol.* 88:92-107.

Kempe, S. (1982). Long term records of the CO_2 pressure fluctuations in fresh water. In: E.T. Degens, ed. Transport of Carbon and Minerals in Major World Rivers, Part 1. *Mitt. Geol.-Paläont. Inst. Univ. Hamburg, SCOPE/UNEP Sonderbd.* 52:91-332.

Kennedy, J.J. (1985). Conceiving forest management as providing for current and future social value. *Forest Ecol. Manage.* 13:121-132.

Kent, J.D. (1987). Periodic aridity and prehispanic Titicaca basin settlement patterns. In: D.L. Browman, ed. *Arid Land Use Strategies and Risk Management in the Andes*, pp. 297-314. Westview, Boulder, Colorado.

Kerfoot, W.C. (1974). Net accumulation rates and the history of cladoceran communities. *Ecology* 55:51-61.

Kerfoot, W.C., and A. Sih, eds. (1987). *Predation: Direct and Indirect Impacts on Aquatic Communities*. University Press of New England, Hanover, New Hampshire.

Kerner, F. and L. Spandau, ed. (1990). *Sozioökonomische Auswirkungen und Szenarien der Berglandwirtschaft im Alpenpark Berchtesgaden* (Report on the consequences of different agricultural policies in Berchtesgaden). Rep. No. 10104040/5. Federal Environmental Office, Berlin.

Keystone Center. (1991). *Biological Diversity on Federal Lands: Report of a Keystone Policy Dialogue*. Keystone Center, Keystone, Colorado.

Kilgore, B.M. (1976). From fire control to fire management: an ecological basis for policies, 477-493. *Trans. 41st N. Am. Wildlife Nat. Res. Conf.* Wildlife Management Institute, Washington D.C.

Kimenker, J. (1983). The Concord farmer: an economic history. In: D.H. Fischer, ed. *Concord: the Social History of a New England Town 1750-1850*, pp. 139-197. Brandeis University, Waltham, Massachusetts.

Kingsland, S.E. (1985). *Modeling Nature: Episodes in the History of Population Ecology*. University of Chicago Press, Chicago.

Kirch, P.V. (1984). *The Evolution of Polynesian Chiefdoms*. Cambridge University Press, Cambridge.

Kirk, G.S., J.E. Raven, and M. Schofield. (1983). *The Presocratic Philosophers*, 2nd ed. Cambridge University Press, Cambridge.

Kitchell, J.A. and J.F. Kitchell. (1980). Size selective predation, light transmission, and oxygen stratification: evidence from the recent sediments of manipulated lakes. *Limnol. and Oceanog*. 25:389-402.

Kitchell, J.F., ed. (1992). *Food Web Management: A Case Study of Lake Mendota*. Springer-Verlag, New York.

Kitchell, J.F., and S.R. Carpenter. (1987). Piscivores, planktivores, fossils, and phorbins. In: W.C. Kerfoot and A. Sih, eds. *Predation: Direct and Indirect Impacts on Aquatic Communities*, pp. 132-146. University Press of New England, Hanover, New Hampshire.

Kitchell, J.F. and S.R. Carpenter. (1992). Summary: accomplishments and new directions of food web management in Lake Mendota. In: J.F. Kitchell, ed. *Food Web Management: A Case Study of Lake Mendota*, pp. 539-544. Springer-Verlag, New York.

Kitchell, J.F., and L.B. Crowder. (1986). Predator-prey interactions in Lake Michigan: model predictions and recent dynamics. *Environ. Biol. Fish*. 16:205-211.

Kitchell, J.F., and P.R. Sanford. (1992). Paleolimnological evidence of food web structure. In: J.F. Kitchell, ed. *Food Web Management: A Case Study of Lake Mendota*, pp. 31-47. Springer-Verlag, New York.

Kitchell, J.F., M.S. Evans, D. Scavia, and L.B. Crowder. (1988a). Food web regulation of water quality in Lake Michigan: report of a workshop. *J. Great Lakes Res*. 14:109-114.

Kitchell, J.F., S.M. Bartell, S.R. Carpenter, D.J. Hall, D.J. McQueen, W.E. Neill, D. Scavia, and E.E. Werner. (1988b). Epistemology, experimentation, and pragmatism. In: S.R. Carpenter, ed. *Complex Interactions in Lake Communities*, pp. 263-280. Springer-Verlag, New York.

Klein, R.G. (1979). Stone Age exploitation of animals in southern Africa. *Am. Sci*. 67:151-160.

Kohl, D.H., G.B. Shearer, and B. Commoner. (1971). Fertilizer nitrogen: contribution to nitrate in surface water in a corn-belt watershed. *Science* 174:1331-1334.

Koike, H. and N. Ohtaishi. (1985). Prehistoric hunting pressure estimated by the age composition of excavated Sika deer (*Cervus nippon*) using the animal layer of tooth cement. *J. Archaeol. Sci*. 12:443-456.

Kolata, A.L. (1986). The agricultural foundations of the Tiwanaku state: a view from the heartland. *Am. Ant*. 51:748-762.

Kowarik, I. (1990). Some response of flora and vegetation to urbanization in Central Europe. In: H. Sukopp and S. Hejny, eds. *Urban Ecology*, pp. 45-74. SBP Academic Publishing bv, The Hague.

Kozlowski, T.T. and C.F. Ahlgren, eds. (1974). *Fire and Ecosystems*. Academic Press, New York.

Kroeber, A.L. (1939). *Cultural and Natural Areas of Native North America*. University of California, Berkeley, California.

Krummel, J.R., R.H. Gardner, G. Sugihara, R.V. O'Neill, and P.R. Coleman. (1987). Landscape patterns in a disturbed environment. *Oikos* 48:321-324.

Kuhn, T.S. (1962). *The Structure of Scientific Revolutions*, 1st ed. University of Chicago Press, Chicago.

Kuhn, T.S. (1970). *The Structure of Scientific Revolutions*, 2nd ed. Enlarged. University of Chicago Press, Chicago.

Kummer, D. (1992). *Deforestation in the Postwar Philippines*. Department of Geography Research Paper #234. University of Chicago Press, Chicago.

Kurian G.T., ed. (1989). *Geo-Data, the World Geographical Encyclopedia*. Gale Research Co., Detroit, Michigan.

Lamarck, J.B. (1809). *Philosophie zoologique*, 2 v. Dentu, Paris, v. 1, p. 101 (translation

Zoological Philosophy, University of Chicago Press, Chicago, Chicago, 1984, p. 55).

Lamarck, J.B. (1801). *Système des animaux sans vertèbres*. Author, Paris (translation *Zoological Philosophy*, University of Chicago Press, 1984).

Landsberg, H.E. (1981). *The Urban Climate*. Academic Press, New York.

Lathrop, R.C. (1992). Nutrient loadings, lake nutrients, and water clarity. In: J.F. Kitchell, ed. *Food Web Management: A Case Study of Lake Mendota*, pp. 69-96. Springer-Verlag, New York.

Lathrop, R.C. and S.R. Carpenter. (1992). Phytoplankton and their relationship to nutrients. In: J.F. Kitchell, ed. *Food Web Management: A Case Study of Lake Mendota*, pp. 97-126. Springer-Verlag, New York.

Lauzanne, L. (1982). Les *Orestias* (Pices, Cyprinodontidae) du petit lac Titicaca. *Rev. Hydrobiol. Trop.* 15:39-70.

Lawrence, J.M. (1975). On the relationship between marine plants and sea urchins. *Oceanog. Mar. Biology Ann. Rev.* 13:213-286.

Leavitt, P.R., S.R. Carpenter, and J.F. Kitchell. (1989). Whole lake experiments: the annual record of fossil pigments and zooplankton. *Limnol. Oceanog.* 34:700-717.

Lee, D.B., Jr. (1973). Requiem for large-scale models. *Am. J. Planners* 39:163-178.

Lee, D.S. and J.W.S. Longhurst. (1989). Urban acidic deposition: a case study of greater Manchester. In: J.W.S. Longhurst, ed. *Acid Deposition: Sources, Effects, and Controls*, pp. 67-106. Air Science Co., Corning, New York.

Lee, K.E. (1985). *Earthworms: Their Ecology and Relationships With Soils and Land Use*. Academic Press, Sydney.

Legge, A.J. and P.A. Rowley-Conwy. (1987). Gazelle killing in Stone Age Syria. *Sci. Am.* 255:88-95.

Leigh, L., K. Fitzsimmons, M. Norem, and D. Stumpf. (1987). An introduction to the intensive agriculture biome of Biosphere II. In: B. Faughnan and G. Maryniak, eds. *Space Manufacturing 6*, pp. 76-81. American Institute of Aeronautics and Astronautics, Washington, D.C.

Leopold, A.S., S.A. Cain, C.M. Cottam, I.N. Gabrielson, and T.L. Kimball. (1963). Wildlife management in the national parks. *Trans. N. Am. Wildlife Nat. Res. Conf.* 28:28-45.

Levi-Strauss, C. (1978). Science: forever incomplete. *Johns Hopkins Mag.* 30:30-31.

Levieil, D. and B. Orlove. (1991). Importancia socio-ecónomica de la macrofitas. In: C. Dejoux and A. Iltis, eds. *El Lago Titicaca: Sintesis del conocimento Limnológico Actual*, pp. 509-516. ORSTOM/HISBOL, La Paz.

Levieil, D.P. (1987). Territorial use rights in fishing (TURFs) and the management of small-scale fisheries: the case of Lake Titicaca (Peru). Ph.D. Thesis, University of British Columbia, Vancouver.

Levin, S.A. (1992). Sustaining ecological research. *Bull. Ecol. Soc. Am.* 73:213-218.

Lewis, H.T. (1982). Fine technology and resource management in aboriginal North America and Australia. In: N.M. Williams and E.S. Hunn, eds. *Resource Managers: North American Hunter-Gatherers*, pp. 46-67. American Association for the Advancement of Science Selected Symposium Number 67. Westview Press, Boulder, Colorado.

Liddle, M.J. (1975). A selective review of the ecological effects of human trampling on natural ecosystems. *Biol. Conserv.* 7:17-36.

Liepolt, R. (1972). Uses of the Danube River. In: R.T. Oglesby, C.A. Carlson, and J.A. McCann, eds. *River Ecology and Man*, pp. 233-250. Academic Press, New York.

Likens, G.E., ed. (1972). *Nutrients and Eutrophication: The Limiting Nutrient Controversy*. Special Symposium, Vol. 1. Amer. Soc. Limnol. Oceanogr. Allen Press, Lawrence, Kansas.

Likens, G. 1989. (ed.). *Long-term Studies in Ecology: Approaches and Alternatives*. Springer-Verlag, New York.

Likens, G.E. (1991). Human-accelerated environmental change. *BioScience* 41:130.

Likens, G.E. (1992). *The Ecosystem Approach: Its Use and Abuse. Excellence in Ecology*. Vol. 3. Ecology Institute, Oldenforf/Luhe, Germany.

Likens, G.E., F.H. Bormann, N.M. Johnson, D.W. Fisher, and R.S. Pierce. (1970). Effects of forest cutting and herbicide treatment on nutrient budgets in the Hubbard Brook watershed-ecosystem. *Ecol. Monog.* 40:23-47.

Likens, G.E., F.H. Bormann, and N.M. Johnson. (1972). Acid rain. *Environment* 14:33-44.

Likens, G.E., F.H. Bormann, R.S. Pierce, and N.M. Johnson. (1977). *Biogeochemistry of a Forested Ecosystem.* Springer-Verlag, New York.

Limberg, K.E., M.A. Moran, and W.H. McDowell. (1986). *The Hudson River Ecosystem.* Springer-Verlag, New York.

Linnaeus, C. (1749). *Oeconimia Naturae.* Upsala, Sweden.

Linnaeus, C. (1760). *Polita Naturae.* Upsala, Sweden.

Linnaeus, C. (1775). *Miscellaneous Tracts Relating to Natural History, Husbandry, and Physick.* 3rd ed. Dodsley, Baker & Leigh, London. Reprinted 1977, Arno Press, New York.

Linnaeus, C. (1781). *Select Dissertations.* G. Robinson and J. Robson, London. Reprinted 1977, Arno Press, New York, pp. 129-166.

Llagosteras, A. (1979). 9,700 years of maritime subsistence on the Pacific, an analysis by means of bio-indicators in the north of Chile. *Am. Ant.* 44:309-324.

Lloyd, E.A. (1988). *The Structure and Confirmation of Evolutionary Theory.* Greenwood Press, New York.

Lloyd, G.E.R. (1970). *Early Greek Science: Thales to Aristotle.* Chatto & Windus, London.

Loeb, R. E. (1987). The tragedy of the commons. *J. Forest.* 85:28-23.

Long, E. (1889). Forest fires in Southern pines. *Forest Leaves* 2:37-41.

Long, P., ed. (1969). *The New Left: A Collection of Essays.* Porter Sargent, Boston.

Lord, N.W. (1884). Iron manufacture in Ohio. Report of the Geological Survey. *Ohio Econ. Geol.* 5:438-554.

Lorenz, E.N. (1963). The predictability of hydrodynamic flow. *Trans. New York Acad. Sci.* (Series 2) 25:409-432.

Loucks, O.L. (1979). New light on the changing forest. In: S.L. Flader, ed. *The Great Lakes Forest: An Environmental and Social History*, pp. 17-32. University of Minnesota Press, Minneapolis, in association with the Forest History Society, Santa Cruz, California.

Lovelock, J.E. (1972). Gaia as seen through the atmosphere. *Atmos. Environ.* 6:579-580.

Lovelock, J.E. (1979). *Gaia: A New Look at Life on Earth.* Oxford University Press, New York.

Lovelock, J.E. (1988). *The Ages of Gaia: A Biography of Our Living Earth.* Norton, New York.

Lovett, G.M. and J.D. Kinsman. (1990). Atmospheric pollutant deposition to high-elevation ecosystems. *Atmos. Environ.* 24A:2767-2786.

Lowe, J.W.G. (1985). *The Dynamics of Apocalypse: A Systems Simulation of the Mayan Collapse.* University of New Mexico Press, Albuquerque, New Mexico.

Lubchenco, J., A.M. Olson, L.B. Brubaker, S.R. Carpenter, M.M. Holland, S.P. Hubbell, S.A. Levin, J.A. MacMahon, P.A. Matson, J.M. Melillo, H. A. Mooney, C.H. Peterson, H.R. Pulliam, L.A. Real, P.J. Regal, and P.J. Risser. (1991). The sustainable biosphere initiative: an ecological research agenda. *Ecology* 72:317-412.

Ludwig, D.F. (1989). Anthropic ecosystems. *Bull. Ecol. Soc. Am.* 70:12-14.

Luecke, C., L.G. Rudstam, and Y. Allen. (1992). Inter-annual patterns of planktivory 1987-1989: an analysis of vertebrate and invertebrate planktivores. In: J.F. Kitchell, ed. *Food Web Management: A Case Study of Lake Mendota*, pp. 275-301. Springer-Verlag, New York.

Luken, J.O. (1990). *Directing Ecological Succession.* Chapman and Hall, New York

Luniak, M. (1980). Birds of allotment gardens in Warsaw, Poland. *Pol. Acta Ornithol.* (Warsaw) 17:297-309.

Lutz, H.J. (1928). Trends and silvicultural significance of upland forest succession in southern New England. *Yale Univ. School of Forest. Bull.* No. 22.

Lutz, R.J. (1938). The influence of forests on farm income and sub-marginality in the town of Petersham. *Land-use Planning Seminar*. Harvard Forest, Petersham, Massachusetts. (unpublished).

Lyell, C. (1830-33). *Principles of Geology*, 3 vols. John Murray, London.

MacConnell, W.P. (1975). Remote sensing 20 years of change in Massachusetts. *Mass. Agric. Exper. Sta. Bull.* No. 630.

MacConnell, W.P. and W. Niedzwiedz. (1974). Remote sensing 20 years of change in Worcester County, Massachusetts, 1951-1971. Mass. Agric. Exper. Sta., University of Massachusetts, Amherst, Massachusetts.

MacDonald, G.J. and L. Sertorio. (1989). *Global Climate and Ecosystem Change*. Plenum Press, New York.

Mack, R.N. (1990). Catalog of woes. *Nat. Hist.*, March, pp. 45-53.

Mackie, C. (1986). Disturbance and succession resulting from shifting cultivation in an upland rainforest in Indonesian Borneo. Ph.D. Thesis, Rutgers University, New Brunswick, New Jersey.

Mackie, C., T.C. Jessup, A.P. Vayda, and K. Kartawinata. (1987). Shifting cultivation and patch dynamics in an upland forest in East Kalimantan, Indonesia. In: Y. Hadi, K. Awang, N.M. Majid, and S. Mohamed, eds. *Proceedings, Regional Workshop on Impact of Man's Activities on Tropical Upland Forest Ecosystems*, pp. 465-518. Faculty of Forestry, Universiti Pertanian Malaysia, Serdang, Selangor.

Magnuson, J.J. (1991). Fish and fisheries ecology. *Ecol. Appl.* 1:13-26.

Magnuson, J.J. and R.C. Lathrop. (1992). Historical changes in the fish community. In: J.F. Kitchell, ed. *Food Web Management: A Case Study of Lake Mendota*, pp. 193-231. Springer-Verlag, New York.

Malthus, T.R. (1798). *A Essay on the Principle of Population, as It Affects the Future Improvement of Society*. J. Johnson, London.

Mandelbrot, V. (1983). *The Fractal Geometry of Nature*. W.H. Freeman and Co., New York.

Mann, C. (1991). Lynn Margulis: science's unruly earth mother. *Science* 252:378-381.

Mann, C.S. (1889). *Boyhood in the 1850s. Petersham and Other Items—A Journal*. Petersham Historical Society, Petersham, Massachusetts.

Mannion, A.M. (1991). *Global Environmental Change: A Natural and Cultural Environmental History*. Longman, Essex, United Kingdom.

Marco, G.J., R.M. Hollingworth, and W. Durham. (1987). *Silent Spring Revisited*. American Chemical Society, Washington, D.C.

Margulis, L. (1988). Jim Lovelock's Gaia. In: P. Bunyard and E. Goldsmith, eds. *Gaia: The Thesis, the Mechanisms, and the Implications*, pp. 50-65. Wadebridge Ecological Centre, Camelford, United Kingdom.

Marquardt, W.H. (1978). Advances in archaeological seriation. *Adv. Archaeol. Method Theory* 1:257-314.

Marrow, A.J., ed. (1969). *The Practical Theorist: The Life and Work of Kurt Lewin*. Basic Books, New York.

Marsh, G.P. (1848). Address delivered before the Agricultural Society of Rutland County, Sept. 30th, 1847. *Rutland Herald*, Rutland, Vermont.

Marsh, G.P. (1864). *Man and Nature; or, Physical Geography as Modified by Human Action*. Reprinted 1965, Belknap Press of Harvard University Press, Cambridge.

Marshall, A.J. (1966). *The Great Extermination*. William Heinemann, London.

Martin, P.S. and R.G. Klein. (1984). *Quaternary Extinctions: A Prehistoric Revolution*. University of Arizona Press, Tucson, Arizona.

Massachusetts General Court. (1846). Transcription of the Report of the General Court of Massachusetts, Boston.

Maxwell, H. (1910). The use and abuse of the forests by the Virginia Indians. *William Mary Coll. Quart.* 19:73-104.

May, R.M. (1974). Biological populations with nonoverlapping generations. *Science* 186:645-647.

Mayr, E. (1982). *The Growth of Biological Thought: Diversity, Evolution, and Inheritance.* Harvard University Press, Cambridge.

McAndrews, J.H. (1988). Human disturbance of North American forests and grasslands: the fossil pollen record. In: B. Huntley and T. Webb, III. *Vegetation History,* pp. 673–697. Kluwer Academic Publishers, Dordrecht, Netherlands.

McDonnell, M.J. and S.T.A. Pickett. (1990). The study of ecosystem structure and function along urban-rural gradients: an unexploited opportunity for ecology. *Ecology* 71:1231–1237.

McEvoy, A.F. (1986). *The Fisherman's Problem: Ecology and Law in the California Fisheries.* Cambridge University Press, New York.

McIntosh, R.P. (1967). The continuum concept of vegetation. *Bot. Rev.* 33:130-187.

McIntosh, R.P. (1975). H.A. Gleason—"Individualistic ecologist" 1882-1975: his contributions to ecological theory. *Bull. Torrey Bot. Club* 102:253-273.

McIntosh, R.P. (1985). *The Background of Ecology: Concept and Theory.* Cambridge University Press, New York.

McKinney, H.L. (1966). Alfred Russel Wallace and the discovery of natural selection. *J. Hist. Med. Allied Sci.* 21:333-357.

McLean, L.A. (1967). Pesticides and the environment. *BioScience* 17:613-617.

McNeely, J.A., K.R. Miller, W.V. Reid, R.A. Mittermeier, and T.B. Werner. (1990). *Conserving the World's Biological Diversity.* International Union for the Conservation of Nature and Natural Resources, World Resources Institute, Conservation International, World Wildlife Fund-US, World Bank, Washington, D.C.

McNeill, W.H. (1982). *The Pursuit of Power.* University of Chicago Press, Chicago.

McNulty, S.G., J.D. Aber, and R.D. Boone. (1991). Spatial changes in forest floor and foliar chemistry of spruce-fir forests across New England. *Biogeochemistry* 14:13-29.

McNulty, S.G., J.D. Aber, T.M. McLellan, and S.M. Katt. (1990). Nitrogen cycling in high elevation forests of the northeastern U.S. in relation to nitrogen deposition. *Ambio* 19:38-40.

Meadows, D.H., D.L. Meadows, and J. Randers, eds. (1992). *Beyond the Limits.* Earthscan Publications, London.

Medawar, P.R. (1967). *The Art of the Soluble.* Methuen and Co., Ltd., London.

Meentenmeyer, V. (1978). Macroclimate and lignin control of litter decomposition rates. *Ecology* 59:465-472.

Mehra, M.K. (1979). Kalman filters and their application to forecasting. *TIMS Stud. Manage. Sci.* 12:75-94.

Meier, R.L. (1976). A stable urban ecosystem. *Science* 192:962-968.

Meine, C. (1988). *Aldo Leopold: His Life and Work.* University of Wisconsin Press, Madison, Wisconsin.

Melillo, J.M., J.D. Aber, and J.M. Muratore. (1982). Nitrogen and lignin control of hardwood leaf litter decomposition dynamics. *Ecology* 63:621-626.

Merchant, C. (1980). *The Death of Nature: Women, Ecology, and The Scientific Revolution.* Harper & Row, New York.

Merchant, C. (1989). *Ecological Revolutions—Nature, Gender, and Science in New England.* University of North Carolina Press, Chapel Hill, North Carolina.

Metropolitan District Commission Division of Watershed Management (MDC). (1991). *Quabbin Reservation White-tailed Deer Impact Management Plan.* MDC, Boston.

Meybeck, M. (1982). Carbon, nitrogen, and phosphorus transport by world rivers. *Am. J. Sci.* 282:401-450.

Miess, M. (1979). The climate of cities. In: I.C. Laurie, ed. *Nature in Cities,* pp. 91-114. J. Wiley and Sons, New York.

Mikesell, M. (1968). Landscape. In: *International Encyclopedia of the Social Sciences,* pp. 575-580. The Macmillan Co. and the Free Press, New York.

Mills, E.L. (1989). *Biological Oceanography: An Early History, 1870-1960.* Cornell University Press, Ithaca, New York.

Mills, S. (1983). French farming: good for people, good for wildlife. *New Sci.* 100:568-571.

Milne, B.T. (1990). Lessons from applying fractal models to landscape patterns. In: M.G. Turner and R.H. Gardner, eds. *Quantitative Methods in Landscape Ecology. The Analysis and Interpretation of Landscape Heterogeneity*, pp. 199-238. Springer-Verlag, New York.

Minshall, W. G. (1984). Tributaries as modifiers of the river continuum concept: analysis by polar ordination and regression models. *Hydrobiologia* 99:208-220.

Mitsch, W.J. and S.E. Jorgensen, eds. (1989). *Ecological Engineering: An Introduction to Ecotechnology*. Wiley Interscience, New York.

Möbius, K.A. (1877). *Die Auster und die Austernwirtschaft*. Verlag von Wiegandt, Hempel, and Pary, Berlin. Trans. by H.J. Rice, in: *Report of the Commissioner for 1880*, Part VIII. U.S. Commission of Fish and Fisheries.

Moran, E.F. (1984). Limitations and advances in ecosystem research. In: E.F. Moran, ed. *The Ecosystem Concept in Anthropology*, pp. 3-32. Westview Press, Boulder, Colorado.

Moreno, C.A., J.P. Sutherland, and H.F. Jara. (1984). Man as a predator in the intertidal zone of southern Chile. *Oikos* 42:155-160.

Moreno, C.A., Lunecke, K.M., and M.I. Lepez. (1986). The response of an intertidal *Concholepas concholepas* population to the protection from man in southern Chile and the effects on benthic sessile assemblages. *Oikos* 46:359-364.

Mosley, M.E. and R.A. Feldman. (1988). Fishing, farming, and the foundation of Andean civilization. In: G. Bailey and J. Parkington, eds. *The Archeology of Prehistoric Coastlines*, pp. 125-134. Cambridge University Press, Cambridge.

Munyon, P.G. (1978). *A Reassessment of New England Agriculture in the Last Thirty Years of the Nineteenth Century: New Hampshire, a Case Study*. Arno Press, New York.

Murdoch, W.W. and E. McCauley. (1985). Three distinct types of dynamic behaviour shown by a single planktonic system. *Nature* 316:628-630.

Murphy, D.D. and S.B. Weiss. (1992). Predicting effects of climate change on biological diversity in western North America: species losses and mechanisms. In: R.L. Peters and T.E. Lovejoy, eds. *Consequences of Greenhouse Warming to Biodiversity*. Yale University Press, New Haven, Connecticut.

Myers, N. and R. Tucker. (1987). Deforestation in Central America: Spanish legacy and North American consumers. *Environ. Rev.* 11:55-71.

Naess, A. (1989). *Ecology, Community, and Lifestyle, An Outline of an Ecosophy*. Cambridge University Press, Cambridge.

Naiman, R.J., C.A. Johnston, and J.C. Kelly. (1988). Alteration of North American streams by beaver. *BioScience* 38:753-762.

National Research Council. (1989). *Alternative Agriculture*. Committee on the Role of Alternative Farming Methods in Modern Production Agriculture, Board on Agriculture, National Academy Press, Washington, D.C.

National Research Council. (1991). *Restoration of Aquatic Ecosystems: Science, Technology, and Public Policy*. National Academy Press, Washington, D.C.

National Oceanic and Atmospheric Administration. (1985). *Climates of the States*, Vol. 2: New York - Wyoming, 3rd edition. Gale Research Company, Detroit, Michigan.

Nelson, M. (1990). The biotechnology of space biospheres. In: M. Asahima and G. Malacinski, eds. *Fundamentals of Space Biology*. Japan Scientific Press, Springer-Verlag, Berlin.

Nelson, M. and G. Soffen, eds. (1990). *Biological Life Support Systems*. Synergetic Press, Oracle, Arizona. Reprinted from NASA, Office of Management, Scientific and Technical Information Divison, Proceedings of the Workshop on Biological Life Support Technologies: Commercial Opportunities, M. Nelson and G.A. Soffen, eds., 1989.

Nelson, M., L. Leigh, A. Alling, T. MacCallum, J. Allen, and A. Alvarez-Romo. (1991). *Biosphere 2 Test Module: A Ground-based Sunlight-driven Prototype of a Closed Ecological Life Support System*. COSPAR XXVIII Plenary Meeting, MF11.2.6, 1991. Advances in Space Research, Pergammon Press.

Nelson, M., T.L. Burgess, A. Alling, N. Alvarez-Romo, W.F. Dempster, R.L. Walford, and J.P. Allen. (1993). Initial results from Biosphere 2: a closed ecological system laboratory. *BioScience* (in press).

Nelson, R.K. (1983). *Make Prayers to the Raven: A Koyukon View of the Northern Forest.* University of Chicago Press, Chicago.

New York State Department of Environmental Conservation. (1989). Air Quality Report, Ambient Air Monitoring System, Annual 1988. Division of Air Resources, Albany, New York.

Nichols, G.E. (1913). The vegetation of Connecticut. II. Virgin forest. *Torreya* 13:199-215.

Nichols, S.A., R.C. Lathrop, and S.R. Carpenter. (1992). Long-term vegetation trends—a history. In: J.F. Kitchell, ed. *Food Web Management: A Case Study of Lake Mendota,* pp. 151-191. Springer-Verlag, New York.

Niering, W.A. (1987). Vegetation dynamics (succession and climax) in relation to plant community management. *Conserv. Biol.* 1:287-295.

Nihlgard, B. (1985). The ammonium hypothesis—an additional explanation to the forest decline in Europe. *Ambio* 14:2-8.

Nilsson, J. and P. Grennfelt. (1988). *Critical Loads for Sulphur and Nitrogen*—Report from a workshop held at Skokloster, Sweden, 19-24 March, 1988. Nordic Council of Ministers, Copenhagen, Denmark.

Nohl, W. and K.D. Neumann. (1986). Landschaftsbildbewertung im Alpenpark Berchtesgaden- Umweltpsychologische Untersuchungen zur Landschaftsästhetik. In Deutsches Nationalkomitee. MAB-Mitteilungen Nr. 23. MAB, Bonn.

Norgaard, R. (1987). Economics as mechanics and the demise of biological diversity. *Ecol. Econ.* 38:107-121.

Norgaard, R. (1989). Models and knowledge in ecology and economics. *Soc. Catalana D'Econ.* 7:188-199.

North American Regional Consultation of the IUCN, WRI, UNEP Biodiversity Strategy and Action Plan, July 14-17, 1991, not published.

Northcote, T.G., S.P. Morales, D.A. Levy, and M.S. Greaven. (1989). *Pollution in Lake Titicaca, Peru. Training, Research, and Management.* Westwater Research Centre, University of British Columbia, Vancouver.

Northeastern Timber Salvage Administration (NETSA). (1943). *Report of the U.S. Forest Service Programs resulting from the New England Hurricane of September 21, 1938.* NETSA, Boston.

Novick, N.J., T.M. Klein, and M. Alexander. (1984). Effect of simulated acid precipitation on nitrogen mineralization and nitrification in forest soils. *Water Air Soil Pollut.* 23:317-330.

Nuorteva, P. (1971). The synanthropy of birds as an expression of the ecological cycle disorder caused by urbanization. *Ann. Zool. Fen.* 8:547-553.

O'Neill, R.V., B.T. Milne, M.G. Turner, and R.H. Gardner. (1988a). Resource utilization scales and landscape pattern. *Landscape Ecol.* 2:63-69.

O'Neill, R.V., J.R. Krummel, R.H. Gardner, G. Sugihara, B. Jackson, D.L. DeAngelis, B.T. Milne, M.G. Turner, B. Zygmunt, S.W. Christensen, V.H. Dale, and R.L. Graham. (1988b). Indices of landscape pattern. *Landscape Ecol.* 1:153-162.

O'Neill, R.V., R.H. Gardner, B.T. Milne, M.G. Turner, and B. Jackson. (1991a). Heterogeneity and spatial hierarchies. In: J. Kolasa and S.T.A. Pickett, eds. *Ecological Heterogeneity,* pp. 85-96. Springer-Verlag, New York.

O'Neill, R.V., S.J. Turner, V.I. Cullinen, D.P. Coffin, T. Cook, W. Conley, J. Brunt, J.M. Thomas, M.R. Conley, and J. Gosz. (1991b). Multiple landscape scales: an intersite comparison. *Landscape Ecol.* 5:137-144.

Odum, E.P. and M.G. Turner. (1990). The Georgia landscape: a changing resource. In: I.S. Zonneveld and R.T.T. Forman, eds. *Changing Landscapes: An Ecological Perspective,* pp. 137-164. Springer-Verlag, New York.

Odum, H. (1982). *Systems Ecology.* Wiley, New York.

Oelschlaeger, M. (1991). *The Idea of Wilderness: From Prehistory to the Age of Ecology.* Yale University Press, New Haven, Connecticut.

Office of Technology Assessment. (1987). *Technologies to Maintain Biological Diversity 3.* Office of Technology Assessment, Washington, D.C.

Olivia, D. and J.C. Castilla. (1986). The effect of human exclusion on the population structure of keyhole limpets *Fissurella crassa* and *Fissurella limbata* on the coast of central Chile. *Mar. Ecol.* 7:201-217.

Olson, S.H. (1971). *The Depletion Myth: A History of Railroad Use of Timber.* Harvard University Press, Cambridge, Massachusetts.

Orlove, B.S. (1986). Barter and cash sale on Lake Titicaca: a test of competing approaches. *Curr. Anthropol.* 27:85-106.

Orlove, B.S., D.P. Levieil, H.P. Treviño. (1992). Social and economic aspects of the fisheries. In: C. Dejoux and A. Iltis, eds. *Lake Titicaca: A synthesis of Limnological Knowledge.* Kluwer Academic Publishers, Dordrecht, Netherlands.

Ormerod, S. J. and R. W. Edwards. (1987). The ordination and classification of macroinvertebrate assemblages in the catchment of the River Wye in relation to environmental factors. *Freshwater Biol.* 17:1.

Ortega, S. (1987). The effect of human predation on the size distribution of *Siphonaria gigas* (mollusca: Pulmonata) on the Pacific coast of Costa Rica. *Veliger* 29:251-255.

Overseas Development Administration. (1991). *Biological Diversity and Developing Countries: Issues and Options.* Overseas Development Administration, London.

Owen, D.F. (1971). Species diversity in butterflies in a tropical garden. *Biol. Conserv.* 3:191-198.

Pabst, M.R. (1941). Agricultural trends in the Connecticut Valley region of Massachusetts, 1800-1900. *Smith Coll. Stud. Hist.* 26:1-135.

Packard, S. (1988). Restoration and the rediscovery of the tallgrass savanna. *Restor. Manage. Notes* 6:13-22.

Paerl, H.W. (1985). Enhancement of marine primary production by nitrogen-enhanced acid rain. *Nature* 315:747-749.

Paillet, F. (1982). Ecological significance of American chestnut in the Holocene forests of Connecticut. *Bull. Torrey Bot. Club* 109:457-473.

Paine, R.T. (1963). Food web complexity and species diversity. *Am. Nat.* 100:65-75.

Paine, R.T., J.C. Castilla, and J. Cancino. (1985). Perturbation and recovery patterns of starfish-dominated intertidal assemblages in Chile, New Zealand, and Washington State. *Am. Nat.* 125:679-691.

Pantulu, V.R. (1986). The Mekong River system. In: B.R. Davies and K.F. Walker, eds. *The Ecology of River Systems*, pp. 695-720. Dr W. Junk, The Hague.

Parenti, L.R. (1984). A taxonomic revision of the Andean Killifish genus *Orestias* (Cyprinodontiformes, Cyprinodontidae). *Bull. Am. Mus. Nat. Hist.* 178:110-214.

Parsons, J.J. (1972). Spead of African pasture grasses to the American tropics. *J. Range Mange.* 25:12-17.

Pastor, J., R.J. Naiman, and B. Dewey. (1987). A hypothesis of the effects of moose and beaver foraging on soil carbon and nitrogen cycles, Isle Royale. *Alces* 23:107-124.

Pastor, J., B. Dewey, R.J. Naiman, P.F. McInnes, and Y. Cohen. (1993). Moose browsing and soil fertility in the boreal forests of Isle Royale National Park. *Ecology* 74:467-480.

Patrick, R., ed. (1983). Diversity. *Benchmark Papers in Ecology/13.* Hutchinson Ross, Stroudsbourg, Pennsylvania.

Patterson, W.A. and A.E. Backman. (1988). Fire and disease history of forests. In: B. Huntley and T. Webb III, eds. *Vegetation History*, pp. 603-622. Kluwer, The Hague.

Pearce, D. and R. Turner. (1990). *Economics of Natural Resources and the Environment.* Johns Hopkins Press, Baltimore.

Pearce, D., A. Markandya, and E. Barbier. (1989). *Blueprint for a Green Economy.* Earthscan Publications Ltd., London.

Pearson, C.J. (1976). Vegetation and environmental changes associated with intensification of agriculture near cities: a study of Sydney. *Agric. Environ.* 3:31-43.

Peierls, B.L., N.F. Caraco, M.L. Pace, and J.J. Cole. (1991). Human influence on river nitrogen. *Nature* 350:386-387.

Pepper, D. (1984). *The Roots of Modern Environmentalism.* Croom Helm, London.

Peters, A.H. (1890). The depreciation of farming land. *Quart. J. Econ.* 4:18-33.

Peters, R.H. (1991). *A Critique for Ecology.* Cambridge University Press, Cambridge, England.

Peters, R.L. and J.D.S. Darling. (1985). The greenhouse effect and nature reserves. *BioScience* 35:707-717.

Peters, R.L. and T.E. Lovejoy. (1990). Terrestrial fauna. In: B.L. Turner, II, W.C. Clark, R.W. Kates, J.F. Richards, J.T. Matthews, and W.B. Meyer, eds. *The Earth as Transformed by Human Action,* pp. 353-369. Cambridge University Press, Cambridge.

Peters, R.L. and T.E. Lovejoy, eds. (1992). *Consequences of Greenhouse Warming to Biodiversity.* Yale University Press, New Haven, Connecticut.

Petersen, J., A. Haberstock, T. Siccama, K. Vogt, D. Vogt, and B. Tusting. (1992). The making of Biosphere 2. *Rest. Manage. Notes* 10:158-168.

Peterson, R.O. (1988). The pit or the pendulum: issues in large carnivore management. In: J.K. Agee and D.R. Johnson, eds. *Ecosystem Management for Parks and Wilderness,* pp. 105-117. University of Washington Press, Seattle.

Pettine, M., T. La Noce, R. Pagnotta, and A. Puddu. (1985). Organic and trophic load of major Italian rivers. In: E.T. Degens, S. Kempe, and R. Herrera, eds. Transport of Carbon and Minerals in Major World Rivers, Part 3. *Mitt. Geol.-Paläont. Inst. Univ. Hamburg, SCOPE/UNEP Sonderbd.* 58:417-430.

Pickett, S.T.A. (1980). Non-equilibrium coexistence of plants. *Bull. Torrey Bot. Club* 107:238-248.

Pickett, S. (1989). Space-for-time substitution as an alternative to long-term studies. In: G.E. Likens, ed. *Long-term Studies in Ecology: Approaches and Alternatives,* pp. 110-135. Springer-Verlag, New York.

Pickett, S.T.A. and J. Kolasa. (1989). Structure and theory in vegetation science. *Vegetatio* 83:7-15.

Pickett, S.T.A. and M.J. McDonnell. (1989). Changing perspectives in community dynamics: A theory of successional forces. *Trends Ecol. Evol.* 4:241-245.

Pickett, S.T.A. and P.S. White. (1985). Patch dynamics: a synthesis. In S.T.A. Pickett and P.S. White, eds. *The Ecology of Natural Disturbance and Patch Dynamics,* pp. 371-384. Academic Press, New York.

Pickett, S.T.A. and P.S. White, eds. (1985). *The Ecology of Natural Disturbance and Patch Dynamics.* Academic Press, New York.

Pickett, S.T.A., V.T. Parker, and P.L. Fiedler. (1992). The new paradigm in ecology: implications for conservation biology above the species level. In: P.L. Fiedler and S.K. Jain, eds. *Conservation Biology,* pp. 65-88. Chapman and Hall, New York.

Pimm, S.L. (1982). *Food Webs.* Chapman and Hall, London.

Pisarski, B. and W. Czechowski. (1978). Influence de la pression urbaine sur la myrmecofaune. *Memorab. Zool.* (Polish Academy of Sciences) 29:109-128.

Polunin, N., ed. (1986). *Ecosystem Theory and Application.* John Wiley & Sons, New York.

Polunin, N. and J. Grinevald. (1988). Vernadsky and biospheral ecology. *Environ. Conserv.* 15:117-122.

Popper, K.R. (1959). *The Logic of Scientific Discovery.* Basic Books, New York.

Popper, K.R. (1962). *Conjectures and Refutations: The Growth of Scientific Knowledge.* Basic Books, New York.

Popper, K.R. (1965). *The Logic of Scientific Discovery.* Harper and Row, New York.

Population Reference Bureau (PRB). (1976). *World Population Growth and Response.* Population Reference Bureau, Inc., Washington, D.C.

Population Reference Bureau (PRB). (1989). *United States and World Population Data Sheets.* Population Reference Bureau, Inc., Washington, D.C.

Pouyat, R. V. (1992). Soil characteristics and litter dynamics in mixed deciduous forests along an urban-rural gradient. Ph.D. Thesis, Rutgers University, New Brunswick, New Jersey.

Pouyat, R.V. and M.J. McDonnell. (1991). Heavy metal accumulation in forest soils along an urban-rural gradient in southern New York, U.S.A. *Water Air Soil Pollut.* 57-58:797-807.

Prance, G. (1990). Flora. In: B.L. Turner, II, W.C. Clark, R.W. Kates, J.F. Richards, J.T. Matthews, and W.B. Meyer, eds. *The Earth as Transformed by Human Action*, pp. 387-391. Cambridge University Press, Cambridge.

Prescott, C.E. and D. Parkinson. (1985). Effects of sulphur pollution on rates of litter decomposition in a pine forest. *Can. J. Bot.* 63:1436-1443.

Price, B. (1982). Cultural materialism: a theoretical review. *Am. Ant.* 47:709-741.

Priestly, J. (1772). Observations on different kinds of air. *Roy. Soc. Lond. Philos. Trans.* 62:147-264.

Prodon, R. and J. D. Lebreton. (1981). Breeding avifauna of a Mediterranean succession: the holm oak and cork oak series in the eastern Pyrenees. 1. Analysis and modeling of the structure gradient. *Oikos* 37:21-38.

Pruitt, B.H. (1981). Agriculture and society in the towns of Massachusetts, 1771: a statistical analysis. Ph.D. Thesis, Boston University, Boston.

Pyne, S.J. (1982). *Fire in America: A Cultural History of Wild Land and Rural Fire*. Princeton University Press, Princeton, New Jersey.

Quinn, J.F. and A.E. Dunham. (1983). On hypothesis testing in ecology and evolution. *Am. Nat.* 122:602-617.

Rahel, F.J. (1990). The hierarchical nature of community persistence: a problem of scale. *Am. Nat.* 136:328-344.

Ramenofsky, A.F. (1987). *Vectors of Death: The Archaeology of European Contact*. University of New Mexico, Albuquerque, New Mexico.

Ramírez, J.M., N. Hermosilla, A. Jerardino, and J.C. Castilla. (1991). Análisis bio-arqueológico preliminar de un sitio de cazadores recolectores costeros: Punta Curaumilla-1, Valparaíso. In: H. Niemeyer, ed. *Actas del XI Congreso Nacional de Arqueología Chilena*, pp. 81-93. Imprenta Caballero, Santiago.

Randall, A. (1988). What mainstream economists have to say about the value of biodiversity. In: E.O. Wilson, ed. *Biodiversity*, pp. 217-233. National Academy Press, Washington, D.C.

Random House. (1966). *The Random House Dictionary of the English Language*. Unabridged edition. Random House, New York.

Rane, F.W. (1908). *Fourth Annual Report of the State Forester of Massachusetts for the Year 1907*. Potter Publishing, Boston.

Raper, D.W., J.W.S. Longhurst, and J. Gunn. (1989). Evidence of small scale variation in acid deposition: A study from the Derbyshire High Peak district. In: J.W.S. Longhurst, ed. *Acid Deposition: Sources, Effects, and Controls*, pp. 25-65. Air Science Co., Corning, New York.

Rapoport, E.H. (1977). Especies transportadas por el hombre: un tipo distinto de contaminación? *Fundación Bariloche & Centro Internacional de Formación en Ciencias Ambientales (Madrid), Bariloche, Argentina*. (unpublished).

Rapoport, E.H. (1979). Transporte y comercio de especies invasoras: un nuevo concepto de contaminación. *Ciencia y Desarrollo*, No. 27:24-29, CONACYT, Mexico.

Rapoport, E.H. (1989). Malezas exóticas y plantas escapadas de cultivo en el noroeste patagónico: primera aproximación. *Contribuciones del Laboratorio Ecotono No. 1, Universidad Nacional del Comahue, Bariloche, Argentina*.

Rapoport, E.H. (1991). Tropical versus temperate weeds: a glance into the present and future. In: P.S. Ramakrishnan, ed. *Ecology of Biological Invasions in the Tropics*, pp. 41-52. SCOPE Workshop, International Scientific Publications, New Delhi.

Rapoport, E.H. and E. Ezcurra. (1977). *Polución por especies en la zona de Alicura*.

Fundación Bariloche & Hidronor. Departamento de Recursos Naturales y Energía, Bariloche, Argentina.

Rapoport, E.H., M.E. Díaz Betancourt, and I.R. López Moreno. (1983). *Aspectos de la Ecología Urbana de la Ciudad de México. Parte I. Flora de las Calles y Baldíos.* Instituto de Ecología and MAB-UNESCO, Limusa, Mexico, D.F.

Rappaport, R.A. (1968). *Pigs for the Ancestors: Ritual in the Ecology of New Guinea People.* Yale University Press, New Haven, Connecticut.

Rappaport, R.A. (1984). *Pigs for the Ancestors: Ritual in the Ecology of New Guinea People,* enlarged edition. Yale University Press, New Haven.

Rappaport, R.A. (1990). Ecosystems, populations and people. In: E.F. Moran, ed. *The Ecosystem Approach in Anthropology,* pp. 41-72. University of Michigan Press, Ann Arbor, Michigan.

Rapport, D. and J. Turner. (1977). Economic models in ecology. *Science* 195:367-373.

Raup, H.M. and R.E. Carlson. (1941). The history of land use in the Harvard Forest. *Harvard Forest Bull.* No. 20.

Raup, H.M. (1966). The view from John Sanderson's farm: a perspective for the use of the land. *Forest Hist.* 10:2-11.

Ravetz, J. (1988). Gaia and the philosophy of science. In: P. Bunyard and E. Goldsmith, eds. (1988). *Gaia: The Thesis, the Mechanisms and the Implications,* pp. 133-144. Wadebridge Ecological Centre, Camelford, United Kingdom.

Ravines, R. (1982). *Panorama de la Arqueología Andina.* Instituto de Estudios Peruanos, Lima.

Ray, G.C. (1989). Sustainable use of the global ocean. In: D.B. Botkin, M.F. Caswell, J.E. Estes, and A.A. Orio, eds. *Changing the Global Environment,* pp. 69-87. Academic Press, Boston.

Ray, G.C. (1991). Coastal-zone biodiversity patterns. *BioScience* 41:490-498.

Ray, G.C. and J.F. Grassle et al. (1991). Marine biological diversity. *BioScience* 41:453-463

Ray, J. (1693). *Three Physico-Theological Discourses,* 2nd ed. Sam Smith, London.

Raymond, J.S. (1981). The maritime foundations of Andean civilizations: a reconsideration of the evidence. *Am. Ant.* 46:806-821.

Redford, K.H. (1990). The ecologically noble savage. *Orion Nat. Quart.* 9:25-29.

Regional Plan Association. (1987). *Open Space Preservation in the 31 County New York Urban Region from 1960 to the 21st Century. A Midpoint Review-1986.* Regional Planning Association, New York.

Reid, W.V. and K.R. Miller. (1989). *Keeping Options Alive: The Scientific Basis for Conserving Biodiversity.* World Resources Institute, Washington, D.C.

Renfrew, C. (1973). *The Explanation of Culture Change.* University of Pittsburgh Press, Pittsburgh.

Richerson, P.J. (1977). Ecology and human ecology: a comparison of theories in the biological and social sciences. *Am. Ethnol.* 4:1-26.

Richerson, P.J. and R. Boyd. (1987). Simple models of complex phenomena: the case of cultural evolution. In: J. Dupré, ed. *The Latest on the Best: Essays on Evolution and Optimality,* p. 27-52. MIT Press, Cambridge.

Richerson, P.J. and H.J. Carney. (1987). Patterns of temporal variation in Lake Titicaca, a high altitude tropical lake. II. Succession rate and diversity of the phytoplankton. *Verh. Int. Ver. Limnol.* 23:734-738.

Richerson, P.J., C. Widmer, and T. Kittel. (1977). *The Limnology of Lake Titicaca (Peru-Bolivia), A Large, High Altitude Tropical Lake.* Institute of Ecology Publ. #14, University of California, Davis, California.

Richerson, P.J., P.J. Neale, W. Wurtsbaugh, R. Alfaro, and W. Vincent. (1986). Patterns of temporal variation in Lake Titicaca. A high altitude tropical lake. I. Background, physical and chemical processes, and primary production. *Hydrobiologia* 138:205-220.

Richey, J.E., E. Salati, and U. Dos Santos. (1985). Biochemistry of the Amazon River: An update. In: E.T. Degens, S. Kempe, and R. Herrera, eds. Transport of Carbon and

Minerals in Major World Rivers, Part 3. *Mitt. Geol.-Paldont. Inst. Univ. Hamburg,* *SCOPE/UNEP Sonderbd.* 58:245-258.

Risser, P.G., J.R. Karr, and R.T.T. Forman. (1984). *Landscape Ecology: Directions and Approaches.* Spec. Pub. No. 2. Illinois Natural History Survey, Champaign, Illinois.

Roberts, L. (1991a). Learning from an acid rain program. *Science* 251:1302-1305.

Roberts, L. (1991b). How bad is acid rain? *Science* 251:1303.

Robles, C.D. (1983). Lobster predation determining mussel population structure on a warm temperate shore. *Bull. Ecol. Soc. Am.* 64:92.

Roosevelt, A. (1989). Resource management in Amazonia before the conquest: beyond ethnographic projection. *Adv. Econ. Bot.* 7:30-62.

Rosenberg, D.M. and D.R. Barton. (1986). The Mackenzie River system. In: B.R. Davies and K.F. Walker, eds. *The Ecology of River Systems,* pp. 425-434. Dr W. Junk, The Hague.

Rossiter, M.W. (1975). *The Emergence of Agricultural Science: Justus Liebig and the Americans, 1840-1880.* Yale University Press, New Haven, Connecticut.

Rostlund, E. (1957). The myth of a natural prairie belt in Alabama: an interpretation of historical records. *Ann. Assoc. Am. Geog.* 47:392-411.

Roszak, T. (1972). *Where the Wasteland Ends: Politics and Transcendence in Post Industrial Society.* Doubleday, Garden City, New York.

Rothenberg, W.B. (1981). The market and Massachusetts farms, 1750-1855. *J. Econ. Hist.* 41:283-314.

Rothschuh, K.E. (1973). *History of Physiology.* Risse, G.B. (transl. & bibliog). Krieger, Huntington, New York.

Roughgarden, J. (1986). A comparison of food-limited and space-limited animal competition communities. In: J. Diamond and T.J. Case, eds. *Community Ecology.* Harper and Row, New York.

Roughgarden, J., R.M. May, and S.A. Levin, eds. (1989). *Perspectives in Ecological Theory.* Princeton University Press, Princeton, New Jersey.

Rudel, T.K. (1989). Population, development, and tropical deforestation: a cross-national study. *Rural Sociol.* 54:327-338.

Rudnicky, J.L. and M.J. McDonnell. (1989). Forty-eight years of canopy change in a hardwood-hemlock forest in New York City. *Bull. Torrey Bot. Club* 116:52-64.

Rudstam, L.G., Y. Allen, B.M. Johnson, C. Luecke, J.R. Post, and M.J. Vanni. (1992). Food web structure of Lake Mendota. In: J.F. Kitchell, ed. *Food Web Management: A Case Study of Lake Mendota,* pp. 233-241. Springer-Verlag, New York.

Runte, A. (1990). Joseph Grinnel and Yosemite: rediscovering the legacy of a California conservationist. *Calif. Hist.* 69:170-181, 225.

Russ, G.R, and A.C. Alcalá. (1989). Effects of intense fishing pressure on an assemblage of coral reef fishes. *Mar. Ecol. Prog. Ser.* 56:13-27.

Russell, E.W.B. (1979). Vegetational change in northern New Jersey since 1500 A.D.: a palynological, vegetational, and historical synthesis. Ph.D. Dissertation, Rutgers University, New Brunswick, New Jersey.

Russell, E.W.B. (1980). Vegetational change in northern New Jersey from precolonization to the present: a palynological interpretation. *Bull. Torrey Bot. Club* 107:432-446.

Russell, E.W.B. (1983). Indian-set fires in the forests of the Northeastern United States. *Ecology* 64:78-88.

Ruttan, V. (1971). Technology and the environment. *Am. J. Agric. Econ.* 53:707-717.

Ruyle, E.E. (1973). Genetic and cultural pools: some suggestions for a unified theory of biocultural evolution. *Hum. Ecol.* 1:201-215.

Sack, R.D. (1990). The realm of meaning: the inadequacy of human-nature theory and the view of mass consumption. In: B.L. Turner, II, W.C. Clark, R.W. Kates, J.F. Richards, J.T. Matthews, and W.B. Meyer, eds. *The Earth as Transformed by Human Action,* pp. 659-671. Cambridge University Press, Cambridge.

Sagan, D. (1988). Gaia and biospheres. In: P. Bunyard and E. Goldsmith, eds. (1988).

Gaia: The Thesis, the Mechanisms and the Implications, pp. 237-242. Wadebridge Ecological Centre, Camelford, United Kingdom.

Sagan, D. (1990). *Biospheres: Metamorphosis of Planet Earth.* St. Martin's Press, New York.

Sauer, C.O. (1925). The Morphology of Landscape. *U. Calif. Publ. Geog.* 2 (No.2):19-54.

Sauer, C.O. (1956). The agency of man on the Earth. In: W.L. Thomas, Jr., ed. *Man's Role in Changing the Face of the Earth,* pp. 49-69. University of Chicago Press, Chicago.

Sauer, C.O. (1961). Fire and early man. *Paideuma: Mitteilungen zur Kulturkunde* 7:399-407.

Scavia, D., G.A. Lang, and J.F. Kitchell. (1988). Dynamics of Lake Michigan plankton: a model evaluation of nutrient loading, competition, and predation. *Can. J. Fish. Aquat. Sci.* 45:165-177.

Schaller, J. (1987). *The Geographical Information System ARC/INFO.* Proc. EUROCARTO No. VI, p. 170-179. Intern. Cartogr. Assoc., Brno.

Schiff, A.L. (1962). *Fire and Water: Scientific Heresy in the Forest Service.* Harvard University Press, Cambridge, Massachusetts.

Schindler, D.W. (1977). Evolution of phosphorus limitation in lakes. *Science* 195:260-262.

Schindler, D.W. (1981). Studies of eutrophication in lakes and their relevance to the estuarine environment. In: B.J. Neilson and L.E. Cronin, eds. *Estuaries and Nutrients.* Humana, New York.

Schindler, D.W. (1987). Detecting ecosystem response to anthropogenic stress. *Can. J. Fish. Aquat. Sci.* (suppl.) 44:6-25.

Schlesinger, W.H. (1991). *Biogeochemistry: An Analysis of Global Change.* Academic Press, San Diego, California.

Schneider, K. (1991). Ranges of animals and plants head north. *The New York Times,* August 13, 1991, p. C-1.

Schubel, J.R. and T.M. Bell. (1991). Population growth and the coastal ocean. *Mar. Sci. Res. Center Bull.* 1:1-6.

Schuberth, C.J. (1968). *The Geology of New York City and Environs.* Natural History Press, New York.

Schulze, E.D. (1989). Air pollution and forest decline in a spruce (*picea abies*) forest. *Science* 244:776-783.

Schwarz, M. and M. Thompson. (1990). *Divided We Stand: Redefining Politics, Technology, and Social Choice.* University of Pennsylvania Press, Philadelphia.

Science Advisory Board (SAB). (1990). *Reducing Risk: Setting Priorities and Strategies for Environment Protection.* United States Environmental Protection Agency, Washington, D.C.

Searle, J. (1984). *Minds, Brains, and Science.* Harvard University Press, Cambridge.

Searle, J. (1991). Intentionalistic explanations in the social sciences. *Phil. Social Sci.* 21:332-344.

Sears, P.B. (1964). Ecology—a subversive subject. *BioScience* 14:11-13.

Sears, P.B. (1988). *Deserts on the March.* Island Press, Washington, D.C. Reprint of 1935 edition.

Seastedt, T.R. (1984). The role of microarthropods in decomposition and mineralization processes. *Ann. Rev. Entomol.* 29:25-46.

Shapiro, J. (1990). Biomanipulation: the next phase—making it stable. In: R.D. Gulati, E.H.R.R. Lammens, M.-L. Meijer, and E. van Donk, eds. *Biomanipulation—Tool for Water Management,* pp. 13-27. Kluwer, Amsterdam.

Shapiro, J., V. Lamarra, and M. Lynch. (1975). Biomanipulation: an ecosystem approach to lake restoration. In: P.L. Brezonik and J.L. Fox, eds. *Water Quality Management Through Biological Control,* pp. 85-96. University of Florida, Gainesville, Florida.

Shehata, S.A. and S.A. Bader. (1985). Effect of Nile River water quality on algal distribution at Cairo, Egypt. *Environ. Int.* 11:465-474.

Shepard, P. and D. McKinley, eds. (1969). *The Subversive Science: Essays Toward an Ecology of Man.* Houghton Mifflin, Boston.

Shumway, R.H. (1988). *Applied Statistical Time Series Analysis.* Prentice Hall, Englewood Cliffs, New Jersey.

Siccama, T.G. (1974). Vegetation, soils, and microclimate on the Green Mountains of Vermont. *Ecol. Mong.* 44:325-349.

Siegfried, W.R., and P.A.R. Hockey (1985). Exploitation and conservation of brown mussel stocks by coastal people of Transkei. *Environ. Conserv.* 12:303-307.

Simberloff, D. (1980). A succession of paradigms in ecology: essentialism to materialism and probabilism. *Synthese* 43:3-39.

Simberloff, D. (1982). A succession of paradigms in ecology. Essentialism to materialism and probabilism. In: E. Saarinen, ed. *Conceptual Issues in Ecology*, pp. 63-99. Reidel (Kluwer), Boston.

Simmons, I.G. (1989). *Changing the Face of the Earth.* Basil Blackwell, Oxford.

Simms, S.R. (1992). Wilderness as a human landscape. In: S.I. Zeveloff, M.L. Vause, and W.H. McVaugh, eds. *A Wilderness Tapestry: An Eclectic Approach to Preservation*, pp. 183-202. University of Nevada Press, Reno, Nevada.

Simon, H.A. (1973). The organization of complex systems. In: H.H. Pattee, ed. *Hierarchy Theory: The Challenge of Complex Systems*, pp. 1-28. Braziller, New York.

Sipe, T.W. (1990). Gap partitioning among maples (Acer) in the forests of central New England. Ph.D. Thesis, Harvard University, Cambridge.

Skulberg, O.M. and A. Lillehammer. (1984). Glåma. In: B.A. Whitton, ed. *Ecology of European Rivers*, pp. 469-498. Blackwell, Oxford.

Smith, D.M. (1946). Storm damage in New England forests. M. Forest S. Thesis, Yale University, New Haven, Connecticut.

Smith, W.H. (1974). Air pollution—effects on the structure and function of temperate forest ecosystems. *Environ. Pollut.* 6:111-129.

Smith, W.H., F.H. Bormann, and G.E. Likens. (1968). Response of chemoautotrophic nitrifiers to forest cutting. *Soil Sci.* 106-47-473.

Sober, E. (1984). *The Nature of Selection: Evolutionary Theory in Philosophical Focus.* The Massachusetts Institute of Technology Press, Cambridge.

Soderstrom, B., E. Baath, and B. Lundgren. (1983). Decrease in soil microbial activity and biomasses owing to nitrogen amendments. *Can. J. Microbiol.* 29:1500-1506.

Soil Survey Staff. (1975). *Soil Taxonomy: A Basic System of Classification for Making and Interpreting Soil Surveys.* USDA- Soil Con. Serv. Agr. Handbook 436. U.S. Govt. Printing Office, Washington, D.C.

Solbrig, O.T. (1991). *Biodiversity: Scientific Issues and Collaborative Program of Research Proposals.* MAB Digest no. 9. UNESCO, Paris.

Soliman, H.A. (1982). The Nile River: study of carbon transport. In: E.T. Degens, ed. Transport of Carbon and Minerals in Major World Rivers, Part 1. *Mitt. Geol.-Paldont. Inst. Univ. Hamburg, SCOPE/UNEP Sonderbd.* 52:433-434.

Soulé, M.E. and K.A. Kohm. (1989). *Research Priorities for Conservation Biology.* Island Press, Washington, D.C.

Spencer, H. (1844). Remarks on the theory of reciprocal dependence in the animal and vegetable creations, as regards its bearing upon paleontology. *Philos. Mag.* 24:90-94 (reprinted 1904. *An Autobiography*, 2 v. D. Appleton, New York; see v. 1, pp. 624-630).

Sperber, D. (1985). Anthropology and psychology: towards an epidemiology of representations. *Man* 20:73-89.

Spurr, S.H. (1950). Stand composition in the Harvard Forest. Ph.D. Thesis, Yale University, New Haven, Connecticut.

Spurr, S.H. (1956). Plantation success in the Harvard Forest as related to planting site and clearing, 1907-1949. *J. Forest.* 54:577-579.

Stauffer, D. (1985). *Introduction to Percolation Theory.* Taylor and Francis, London.

Steadman, D.W. and S.L. Olson. (1985). Bird remains from an archaeological site on Henderson Island, South Pacific: Man-caused extinctions on an "uninhabited" island. *Proc. Nat. Acad. Sci. USA* 82:6191-6195.

Stearns, F. and T. Montag, eds. (1974). *The Urban Ecosystem: A Holistic Approach.* Dowden, Hutchinson and Ross, Inc., Stroudsburg.

Steele, J.H. (1991). Marine functional diversity. *BioScience* 41:470-474.

Steele, J.H. and E.W. Henderson. (1984). Modeling long-term fluctuations in fish stocks. *Science* 224:985-987.

Stephens, E.P. (1955). The historical-development method of determining forest trends. Ph.D. Thesis, Harvard University, Cambridge.

Stern, P.C., O.R. Young, and D. Druckman, eds. (1992). *Global Environmental Change: Understanding the Human Dimensions*. National Academy of Sciences Press, Washington, D.C.

Sternberg, H.O'R. (1987). Aggravation of floods in the Amazon River as a consequence of deforestation? *Geofrafiska Annaler* 69A:210-219.

Steudler, P.A., R.D. Bowden, J.M. Melillo, and J.D. Aber. (1989). Influence of nitrogen fertilization on methane uptake in temperate forest soils. *Nature* 341:314-316.

Stewart, D.J., and M. Ibarra. (1991). Predation and production by salmonine fishes in Lake Michigan, 1978-88. *Can. J. Fish. Aquat. Sci.* 48:909-922.

Stewart, O.C. (1956). Fire as the first great force employed by man. In: W.L. Thomas, ed. *Man's Role in Changing the Face of the Earth*, pp. 115-133. University of Chicago Press, Chicago.

Stewart, R. (1991). Biosphere 2 nerve system. *Commun. Assoc. Comput. Machin.* 34:69-71.

Stow, B.R. (1853). Journals and diary. Unpublished. Harvard Forest Archives.

Strayer, D., J.S. Glitzenstein, C.G. Jones, J. Kolasa, G.E. Likens, M.J. McDonnell, C.G. Parker, and S.T.A. Pickett. (1986). Long-term ecological studies: an illustrated account of their design, operation, and importance to ecology. *Occ. Publ. Inst. Ecosys. Stud.* 2:1-38.

Strong, D.R. (1982). Null hypothesis testing. In: E. Saarinen, ed. *Conceptual Issues in Ecology*, pp. 245-260. Reidel (Kluwer), Boston.

Sukopp, H. and S. Hejny, eds. (1990). *Urban Ecology. Plants and Plant Communities in Urban Environments*. SPB Academic Publishing bv, The Hague.

Summers, P.W., V.C. Bowersox, and G.J. Stensland. (1986). The geographical distribution and temporal variations of acidic deposition in eastern North America. *Water Air Soil Pollut.* 31:523-535.

Swanson, T. and Barbier, E.B. (1992). *Economics for the Wilds: Wildlife, Wild Lands, Diversity, and Development*. Earthscan, London.

Szekielda, K.-H. and D. McGinnis. (1985). Investigations with satellites on eutrophication of coastal regions. IV. The Changjiang River and the Huanghai Sea. In: E.T. Degens, S. Kempe, and R. Herrera, eds. Transport of Carbon and Minerals in Major World Rivers, Part 3. *Mitt. Geol.-Paläont. Inst. Univ. Hamburg, SCOPE/UNEP Sonderbd.* 58:49-84.

Tansley, A.G. (1935). The use and abuse of vegetational concepts and terms. *Ecology* 16:284-307.

Taylor, P.J. (1988). Technocratic optimism, H.T. Odum, and the partial transformation of ecological metaphor after World War II. *J. Hist. Biol.* 21:213-244.

Ter Braak, C.J.F. and I.C. Prentice. (1988). A theory of gradient analysis. *Adv. Ecol. Res.* 18:272-327.

Terborgh, J. (1971). Distribution on environmental gradients: theory and a preliminary interpretation of distributional patterns in the avifauna of the Cordillera Vilcabamba, Peru. *Ecology* 52:23-40.

The Times. (1973). *Atlas of the World*. Times Newspapers Limited, London.

The Nature Conservancy (TNC). (1991). *Biodiversity Network News* 4(2). The Nature Conservancy, Science Division, Arlington, Virginia.

Thom, R. (1975). *Structural Stability and Morphogenesis: An Outline of a General Theory of Models*. Benjamin-Cummings, Reading, Massachusetts.

Thomas, W.L., Jr., ed. (1956). *Man's Role in Changing the Face of the Earth*. University of Chicago Press, Chicago.

Thomason, R.H. (1987). The context sensitivity of belief and desire. In: M.P. Georgeff and A.L. Lansky, eds. *Reasoning about Actions and Plans*, pp. 341-360. Morgan Kaufmann, Los Altos, California.

Thompson, J.H. (1977). *Geography of New York State*. Syracuse University Press, Syracuse.

Thompson, L.G., M.E. Davis, E. Mosley-Thompson, and K-b Liu. (1988). Pre-Incan agricultural activity recorded in dust layers in two tropical ice cores. *Nature* 336:763-765.

Thompson, M. (1988). Socially viable ideas of nature: a cultural hypothesis. In: E. Baark and U. Svedin, eds. *Man, Nature, and Technology: Essays on the Role of Ideological Perceptions*, pp. 57-79. St. Martin's Press, New York.

Thompson, M., R. Ellis, and A. Wildavsky. (1990). *Cultural Theory*. Westview Press, Boulder, Colorado.

Thorbahn, P.F. (1984). Br'er elephant and the brier patch. *Nat. Hist.* 93:70-78.

Thoreau, H.D. (1962). *The Journals of Henry D. Thoreau*. Dover Publications, New York. Reprint of 1906 edition.

Tobey, R.C. (1981). *Saving the Prairies: The Life Cycle of the Founding School of American Plant Ecology, 1895-1955*. University of California Press, Berkeley, California.

Todd, D.K., ed. (1970). *The Water Encyclopedia*. Water Information Center, Port Washington, New York.

Tofflemire, T.J. and L. Hetling. (1969). Pollution sources and loads in the lower Hudson River. In: C. Howells and G. Laver, ed. *Hudson River Ecology*, pp. 78-146. 2nd Symp., N.Y.S. Dept. of Environmental Conservation, Albany, New York.

Torbert, E.N. (1935). The evolution of land utilization in Lebanon, New Hampshire. *Geog. Rev.* 25:209-230.

Trojan, P., ed. (1982). General problems of synanthropization. *Memorab. Zool.* (Warsaw) 37:1-147.

Tuan, Y. (1968). Discrepancies between environmental attitude and behaviour: examples from Europe and China. *Can. Geog.* 12:176-191.

Tuan, Y. (1970). Our treatment of the environment in ideal and actuality. *Am. Sci.* 58:246-249.

Turner, B.L., II. (1989). The human causes of global environmental change. In: R.S. DeFries and T.F. Malone, eds. *Global Change and Our Common Future: Papers from a Forum*, pp. 90-99. National Academy Press, Washington, D.C.

Turner, B.L., II. (1991). Thoughts on linking the physical and human sciences in the study of global environmental change. *Res. Explor.* 7:133-135.

Turner, B.L., II, and W.B. Meyer. (1991). Land use and land cover in global environmental change: considerations for study. *Int. Social Sci. J.* 130:669-679.

Turner, B.L., II and K.W. Butzer. (1992). The Columbian encounter and land-use change. *Envir.* 34:16-20, 37-44.

Turner, B.L., II, W.C. Clark, R.W. Kates, J.F. Richards, J.T. Mathews, and W.B. Meyer, eds. (1990a). *The Earth as Transformed by Human Action*. Cambridge University Press, Cambridge and New York.

Turner, B.L., II, R.E. Kasperson, W.B. Meyer, K.M. Dow, D. Golding, J.X. Kasperson, R.C. Mitchell, and S.J. Ratick. (1990b). Two types of global environmental change: definitional and spatial scale issues in their human dimensions. *Global Environ. Change* 1:14-22.

Turner, M.G. and R.H. Gardner, eds. (1990). *Quantitative Methods in Landscape Ecology: The Analysis and Interpretation of Landscape Heterogeneity*. Springer-Verlag, New York.

Turner, R.E. and N.N. Rabalais. (1991). Changes in Mississippi River water quality this century. *Bioscience* 41:140-147.

Tutin, T.G. (1940). Report no. XI. In: H.C. Gilson, ed. Reports of the Percy Sladen Trust Expedition. *Trans. Linn. Soc. Lond.* 1(ser. 3):191-202.

Tyler, G., A.M. Balsberg-Pahlsson, G. Bengtsson, E. Baath, and L. Tranvik. (1989). Heavy-metal ecology of terrestrial plants, microorganisms, and invertebrates. *Water Air Soil Pollut.* 47:189-215.

Ueno, M. (1967). Zooplankton of Lake Titicaca on the Bolivian side. *Hydrobiologia* 29:547-568.

Ulrich, B. (1984). Langzeitwirkungen von Luftverunreinigungen auf Waldökosysteme. *Düsseldorfer Geobot. Kolloq.* 1:11-23.

Ulrich, B. (1987). Einführung und Überblick über den Themenbereich F "Bodenchemie, Nährstoffhaushalt (incl. Meliorationsmaßnahmen). In: E. Stuttgen, ed. *Statusseminar zum BMFT-Förderschwerpunkt Ursachenforschung zu Waldschäden. Tagungsbericht, Projektleitung Biologie, Ökologie, Energie,* pp. 296-304. Kernforschungsanlage, Zentralbibliothek, Jülich, Germany.

United Nations (UN). (1973). *The Determinants and Consequences of Population Trends, Vol. 1. New Summary of Findings on Interaction of Demographic, Economic, and Social Factors.* Population Studies No. 50. United Nations Department of Economic and Social Affairs, New York.

UNEP/WHO/UNESCO/WMO. (1987). *GEMS/Water Data Summary 1982-1984.* WHO Collaborating Centre for Surface and Groundwater Quality, Burlington, Ontario.

United Nations Environment Programme (UNEP). (1987). *Environmental Data Report.* Blackwell, Oxford.

United Nations Environment Programme (UNEP). (1989). *Environmental Data Report.* Blackwell, Oxford.

United Nations Environment Programme (UNEP). (1990a). UNEP/Bio.Div./WG.2/1/3. UNEP, Nairobi.

United Nations Environment Programme (UNEP). (1990b). UNEP/Bio.Div./WG.2/1/4. UNEP, Nairobi.

United Nations Environment Programme (UNEP). (1991). *Draft Convention on Biological Diversity.* UNEP/Bio. Div./WG.2/3/3. UNEP, Nairobi.

United Nations General Assembly. (1989). A/C.2/44/L.86. UNGA, New York.

United States Agency for International Development (USAID). (1989). *Conserving Tropical Forests and Biological Diversity.* USAID, Washington, D.C.

United States Bureau of Census. (1980). Number of Individuals. New York. U.S. Department of Commerce. U.S. Government Press, Washington, D.C.

United States Department of Agriculture (USDA), Forest Service. (1990). *Conserving Our Heritage; America's Biodiversity.* USDA, Washington, D.C.

United States Environmental Protection Agency (EPA). (1990). *Threats to Biological Diversity In The United States.* United States Environmental Protection Agency, Washington, D.C.

United States Environmental Protection Agency (EPA). (1992). *Biodiversity Uncertainties and Research Needs.* 600/F/F-92/024. EPA, Washington, D.C.

United States House of Representatives (1991a) H.R. 585, "National Biological Diversity Conservation Act." U.S. Congress, Washington, D.C.

United States House of Representatives (1991b) H.R. 2082, "National Biological Diversity Conservation and Environmental Research Act." U.S. Congress, Washington, D.C.

United States Senate. (1991). S. 58, "National Biological Diversity Conservation and Environmental Research Act." U.S. Congress, Washington, D.C.

Urban, D.L., R.V. O'Neill, and H.H. Shugart, Jr. (1987). Landscape ecology. *BioScience* 37:119-127.

Urk, G. van. (1984). Lower Rhine-Meuse. In: B.A. Whitton, ed. *Ecology of European Rivers,* pp. 437-468. Blackwell, Oxford.

Vale, T. and G. Vale. (1976). Suburban bird populations in west-central California. *J. Biogeog.* 3:157-165.

van der Elst, R.P. (1979). A proliferation of small sharks in the shore-based Natal sport fishery. *Environ. Biol. Fish.* 4:349-362.

van Dorp, D. and P.F.M. Opdam. (1987). Effects of patch size, isolation, and regional abundance on forest bird communities. *Landscape Ecol.* 1:59-73.

van Melsen, A.G. (1952). *From Atomos to Atom: The History of the Concept Atom.* Duquesne University Press, Pittsburgh.

Van Blaricom, G. and J.A. Estes. (1987). *Community Ecology of Sea Otters*. Springer-Verlag, Berlin.

Van der Maarl, E. (1975). Man-made natural ecosystems in environmental management and planning, pp. 273-274. In: W.H. Van Dobbin and R.H. Lowe-McConnell, eds. *Unifying Concepts in Ecology*, pp. 273-274. Dr. W. Junk: The Hague.

Vanni, M.J., C. Luecke, J.F. Kitchell, Y. Rentmeester, J. Temte, and J.J. Magnuson. (1990). Cascading trophic interactions and phytoplankton abundance in a eutrophic lake: effects of massive fish mortality. *Nature* 344:333-335.

Varley, J.D. (1988). Managing Yellowstone Park into the twenty-first century: the park as an aquarium. In: J.K. Agee and D.R. Johnson, eds. *Ecosystem Management for Parks and Wilderness*, pp. 216-225. University of Washington Press, Seattle.

Vaux, P., W. Wurtsbaugh, H.P. Treviño, L. Mariño, E. Bustamante, J. Torres, P. Richerson, and R. Alfaro. (1988). Ecology of pelagic fishes of Lake Titicaca, Peru-Bolivia. *Biotropica* 20:220-229.

Vayda, A.P. (1983). Progressive contextualization: methods for research in human ecology. *Hum. Ecol.* 11:265-281.

Vayda, A.P. (1986). Holism and individualism in ecological anthropology. *Rev. Anthropol.* 13:295-313.

Vayda, A.P. (1988). Actions and consequences as objects of explanation in human ecology. *Environ. Technol. Society* 51:2-7. Also in: R.J. Borden et al., eds. *Human Ecology: Research and Applications*, pp. 9-18. Society for Human Ecology, College Park, Maryland.

Vayda, A.P. (1990). Actions, variations, and change: the emerging anti-essentialist view in anthropology. *Canberra Anthropol.* 13:29-45.

Vayda, A.P., and B.J. McCay. (1975). New directions in ecology and ecological anthropology. *Ann. Rev. Anthropol.* 4:293-306.

Vayda, A.P., C.J.P. Colfer, and M. Brotokusomo. (1980). Interactions between people and forests in East Kalimantan. *Impact Sci. Society* 30:179-190.

Vernadsky, V.I. (1926). *Biosphera*. Leningrad.

Vernadsky, V.I. (1929). *La biosphère*. Librarie Félix Alcan, Paris.

Vester, F. and A. von Hesler. (1980). *The Sensitivity Model*. RPU (Regionale Planungsgemeinschaft Untermain), Frankfurt.

Vincent, W.F., W. Wurtsbaugh, C.L. Vincent, and P.J. Richerson. (1984). Seasonal dynamics of nutrient limitation in a tropical high-altitude lake (Lake Titicaca, Peru-Bolivia): application of physiological bioassays. *Limnol. Oceanog.* 29:540-552.

Vincent, W.F., C.L. Vincent, M.T. Downes, and P.J. Richerson. (1985). Nitrate cycling in Lake Titicaca (Peru/Bolivia): the effects of high-altitude and tropicality. *Freshwater Biol.* 15:31-42.

Visauder, J. (1986). Philosophy and Human Ecology. In: R.J. Borden, J. Jacobs, and G.L. Young (eds). *Human Ecology: A Gathering of Perspectives*, pp. 117-127. Society for Human Ecology, College Park, Maryland.

Vitousek, P.M. and P.A. Matson. (1991). Gradient analysis of ecosystems. In J.J.Cole, G.M. Lovett, S.E.G. Findlay, eds. *Comparative Analysis of Ecosystems: Patterns, Mechanisms, and Theories*, pp. 287-298. Springer-Verlag, New York.

Vollenweider, R.A. (1968). Scientific fundamentals of the eutrophication of lakes and flowing waters with particular reference to nitrogen and phosphorus as factors in eutrophication. *Organization for Economic Co-operation and Development Tech. Report* DAS/CSI/68.27, Washington, D.C.

Wade, N. (1972). Theodore Roszak: visionary critic of science. *Science* 178:960-962.

Wagner, F.H. (1991). Concluding remarks. *Trans. N. Am. Wildlife Nat. Res. Conf.* 56:96-97.

Wagner, F.H. (1993). Changing institutional arrangements for setting natural-resources policy. In: W. Laycock, M. Vavra, and R. Pieper, eds. *Proceedings of Symposium on Ecological Implications of Livestock Herbivory in the West*. Springer-Verlag, New York (in press).

Waide, J.B. and W.T., Swank. (1977). Simulation of potential effects of forest utilization on the nitrogen cycle in different southeastern ecosystems. In: D.L. Correll, (ed.) *Watershed Research in North America*, pp. 767-789. Smithsonian Inst., Edgewater, Maryland.

Walker, K.F. (1986). The Murray-Darling system. In: B.R. Davies and K.F. Walker, eds. *The Ecology of River Systems*, pp. 631-660. Dr W. Junk, The Hague.

Wallace, A.R. (1859). On the tendency of varieties to depart indefinitely from the original type. *J. Linn. Soc. Zool.* 3:53-62.

Walsh, J.J., G.T. Rowe, R.L. Iverson, and C.P. McRoy. (1981). Biological export of shelf carbon is a sink of the global CO_2 cycle. *Nature* 291:196-201.

Walters, C. (1986). *Adaptive Management of Renewable Resources*. Macmillan Publishing Company, New York.

Walters, C.J., and C.S. Holling. (1991). Large-scale management experiments and learning by doing. *Ecology* 71:2060-2068.

Waring, R.H., A.J.S. McDonald, S. Larsson, T. Ericsson, A. Wiren, E. Arwidsson, A. Ericsson, and T. Lohammar. (1985). Differences in chemical composition of plants grown at constant relative growth rates with stable mineral nutrition. *Oecologia* 66:157-160.

Warren, R.J. (1991). Ecological justification for controlling deer populations in eastern national parks. *Trans. N. Am. Wildlife Nat. Res. Conf.* 56:56-66.

Weiner, D.R. (1988). *Models of Nature: Ecology, Conservation, and Cultural Revolution in Soviet Russia*. Indiana University Press, Bloomington, Indiana.

Welcomme, R.L. (1986). The Niger River system. In: B.R. Davies and K.F. Walker, eds. *The Ecology of River Systems*, pp. 9-24. Dr. W. Junk, The Hague.

Westveld, M. (1956). Natural forest vegetation zones of New England. *J. Forest.* 54:332-338.

Wheeler, R.G. and J.D. Black. (1954). Farm planning—a trial operation. *J. Farm Econ.* 36:198-209.

White, A.R. (1967). *The Philosophy of Mind*. Random House, New York.

White, C.S. and M.J. McDonnell. (1988). Nitrogen cycling processes and soil characteristics in an urban versus rural forest. *Biogeochemistry* 5:243-262.

White, E.H. (1974). Whole tree harvesting depletes soils nutrients. *Can. J. Forest Res.* 4:530-535.

White, L. (1967). The historical roots of our ecologic crisis. *Science* 155:1203-1207.

Whitehead, A.N. (1929). *The Aims of Education and Other Essays*. Macmillan, New York.

Whitehead, A.N. (1941). Explanatory note to introduce "The Philosopher's Summary." In: P.A. Schilpp, ed. *The Philosophy of Alfred North Whitehead*, pp. 663-665. Tudor, New York.

Whitmore, T.M. (1991). A simulation of the sixteenth-century population collapse in the Basin of Mexico. *Ann. Assoc. Am. Geog.* 81:463-487.

Whitmore, T.M., B.L. Turner, II, D.L. Johnson, R.W. Kates, and T.R. Gottschang. (1990). In: B.L. Turner, II, W.C. Clark, R.W. Kates, J.F. Richards, J.T. Matthews, and W.B. Meyer, eds. *The Earth as Transformed by Human Action*, pp. 25-39. Cambridge University Press, Cambridge.

Whitney, G.G. (1987). The ecological history of the Great Lakes forest of Michigan. *J. Ecol.* 75:667-684.

Whitney, G.G. (1990). The history and status of the hemlock-hardwood forests of the Allegheny Plateau. *J. Ecol.* 78:443-458.

Whitney, G.G. (1991). Relation of plant species to substrate, landscape position, and aspect in north central Massachusetts. *Can. J. Forest Res.* 21:1245-1252.

Whitney, G.G. and D.R. Foster. (1990). Overstorey composition and age as determinants of the understorey flora of woods in central New England. *J. Ecol.* 76:867-876.

Whitney, P. (1793). *Worcester County. America's First Frontier*. Isaiah Thomas, Worcester.

Whittaker, R.H. (1967). Gradient analysis of vegetation. *Bio. Rev.* 49:207-264.

Whittaker, R.H. (1973). *Ordination and Classification of Communities*. Dr. W. Junk, The Hague.

Whitton, B.A. and D.T. Crisp. (1984). Tees. In: B.A. Whitton, ed. *Ecology of European Rivers*, pp. 145-178. Blackwell, Oxford.

Whorton, J. (1974). *Before Silent Spring*. Princeton University Press, Princeton, New Jersey.

Wickendon, L. (1955). *Our Daily Poison: The Effects of DDT, Fluorides, Hormones, and Other Chemicals on Modern Man*. Devin-Adair, New York.

Wiens, J.A. (1984). On understanding a non-equilibrium world: myth and reality in community patterns and processes. In: D.R. Strong, D. Simberloff, L.G. Abele, and A.B. Thistle, eds. *Ecological Communities*, pp. 439-457. Princeton University Press, Princeton, New Jersey.

Wiens, J.A. (1985). Vertebrate responses to environmental patchiness in arid and semiarid ecosystems. In: S.T.A. Pickett and P.S. White, eds. *The Ecology of Natural Disturbance and Patch Dynamics*, pp. 169-193. Academic Press, New York.

Wiens, J.A., C.S. Crawford, and J.R. Gosz, Jr. (1985). Boundary dynamics: a conceptual framework for studying landscape ecology. *Oikos* 45:421-427.

Wiens, J.A. and J.T. Rotenberry. (1981). Habitat associations and community structure of birds in shrubsteppe environments. *Ecol. Mong.* 51:21-41.

Williams, G.C. (1966). *Adaptation and Natural Selection*. Princeton University Press, Princeton, New Jersey.

Williams, M. (1982). Clearing the United States forests: pivotal years 1810-1860. *J. Hist. Geog.* 8:12-28.

Williams, M. (1983). "The apple of my eye:" Carl Sauer and historical geography. *J. Hist. Geog.* 9:1-28.

Williams, M. (1989). *Americans and Their Forests: A Historical Geography*. Cambridge University Press, New York.

Williams, M. (1991). Agricultural impacts in temperate lands. In: M. Williams, ed. *Wetlands: A Threatened Landscape*, pp. 181-206. Basil Blackwell, Oxford.

Wilson, D.S. (1980). *The Natural Selection of Populations and Communities*. Benjamin/Cummings, Menlo Park, California.

Wilson, E.O. (1985). The biological diversity crisis. *BioScience* 35:700-706.

Wines, R.A. (1985). *Fertilizer in America: From Waste Recycling to Resource Exploitation*. Temple University Press, Philadelphia.

Winkler, M.G. (1985). A 1,200-year history of vegetation and climate for Cape Cod, Massachusetts. *Quat. Res.* 23:301-312.

Wirrman, D. and L.F. de Oliveira. (1987). Low Holocene level (7700-3650 years ago) of Lake Titicaca (Bolivia). *Paleogeog. Paleoclimatol. Paleoecol.* 59:315-323.

Wolman, A. (1965). The metabolism of cities. *Sci. Am.* 213:179-190.

Wolverton, B.C. (1986). Aquatic plants and wastewater treatment (an overview). In: K.R. Reddy and W.H. Smith, eds. *Aquatic Plants for Water Treatment and Resource Recovery*, pp. 3-15. Magnolia Publ., Orlando, Florida.

Wolverton, B.C. and R.C. McDonald. (1979). The water hyacinth: from prolific pest to potential provider. *Ambio* 8:1.

Woodburn, J. (1980). Hunters and gatherers today and reconstruction of the past. In: E. Gellner, ed. *Soviet and Western Anthropol.*, pp. 95-117. Duckworth, London.

Woodland, D.J. and J.N.A. Hooper. (1977). The effects of human trampling on coastal reefs. *Biol. Conserv.* 11:1-4.2

Woodwell, G.M., C.F. Wurster, Jr., and P.A. Isaacson. (1967). DDT residues in an east coast estuary: a case of biological concentration of a persistent pesticide. *Science* 156:-821-824.

World Commission on Environment and Development (WCED). (1987). *Our Common Future*. Oxford University Press, Oxford.

World Resources Institute, International Institute for Environment and Development, and UNEP. (1988). *World Resources 1988-89*. Basic Books, New York.

World Resources Institute. (1990). *World Resources Report 1990*. Oxford University Press, Oxford.

World Wildlife Fund-World Wide Fund for Nature (WWF). (1989). *The Importance of Biological Diversity*. WWF, Gland, Switzerland.

Worldmark Press. (1976). *Worldmark Encyclopedia of the Nations*. John Wiley and Sons, New York.

Worster, D. (1977). *Nature's Economy: A History of Ecological Ideas*. Cambridge University Press, New York.

Worster, D. (1990). The ecology of order and chaos. *Environ. Hist. Rev.* 14:1-18.

Worthington, E.B. (1983). *The Ecological Century: A Personal Appraisal*. Clarendon Press, Oxford.

Wurster, C.F., Jr. (1968). DDT reduces photosynthesis in marine phytoplankton. *Science* 159:1474-1475.

Wurtsbaugh, W.A., W.F. Vincent, T.R. Alfaro, and C.L. Vincent. (1985). Nutrient limitation of algal growth and nitrogen fixation in a tropical alpine lake, Lake Titicaca (Peru/Bolivia). Freshwater Biol. 15:185-195.

Ybert, J.P. (1987). Spectres palynologiques de tourbières et de sédiments lacustres de la fin du Pléistocène et de l'Holocène des Andes de Bolivie. *Géodynamique* 2:108-109.

Yoffee, N. and G.L. Cowgill. (1988). *The Collapse of Ancient States and Civilizations*. University of Arizona Press, Tucson, Arizona.

Young, J.M.R. (1990). *Sustaining the Earth*. Harvard University Press, Cambridge, Massachusetts.

Young, S., P. Benjamin, B. Jokisch, Y. Ogneva, and A. Garren. (1990). *Global Land Use/cover: Assessment of Data And Some General Relationships*. Report to the land-use working group, Committee for Research on Global Environmental Change. Social Science Research Council, New York.

Young, R.M. (1985). *Darwin's Metaphor: Nature's Place in Victorian Culture*. Cambridge University Press, New York.

Zebryk, T. (1991). Holocene paleoecology of a forested peatland in central New England, USA. M.F.S. Thesis, Harvard University, Cambridge.

Zeigler, B.P. (1979). Structuring principles for multifaceted systems modeling. In: B. Zeigler and T. Oren, eds. *Methodology in Systems Modeling and Simulation*, pp. 93-135. Elsevier, Amsterdam.

Zemba, S.G., D. Golomb, and J.A. Fay. (1988). Wet sulfate and nitrate deposition patterns in eastern North America. *Atmos. Environ.* 22:2751-2761.

Index